Figs
The Genus *Ficus*

Traditional Herbal Medicines for Modern Times

Each volume in this series provides academia, health sciences, and the herbal medicines industry with in-depth coverage of the herbal remedies for infectious diseases, certain medical conditions, or the plant medicines of a particular country.

Series Editor: Dr. Roland Hardman

Volume 1
Shengmai San, edited by Kam-Ming Ko

Volume 2
Rasayana: Ayurvedic Herbs for Rejuvenation and Longevity, by H.S. Puri

Volume 3
Sho-Saiko-To: (Xiao-Chai-Hu-Tang) Scientific Evaluation and Clinical Applications, by Yukio Ogihara and Masaki Aburada

Volume 4
Traditional Medicinal Plants and Malaria, edited by Merlin Willcox, Gerard Bodeker, and Philippe Rasoanaivo

Volume 5
Juzen-taiho-to (Shi-Quan-Da-Bu-Tang): Scientific Evaluation and Clinical Applications, edited by Haruki Yamada and Ikuo Saiki

Volume 6
Traditional Medicines for Modern Times: Antidiabetic Plants, edited by Amala Soumyanath

Volume 7
Bupleurum *Species: Scientific Evaluation and Clinical Applications,* edited by Sheng-Li Pan

Volume 8
Herbal Principles in Cosmetics: Properties and Mechanisms of Action, by Bruno Burlando, Luisella Verotta, Laura Cornara, and Elisa Bottini-Massa

Traditional Herbal Medicines for Modern Times

Figs
The Genus *Ficus*

Ephraim Philip Lansky
Helena Maaria Paavilainen

CRC Press
Taylor & Francis Group
Boca Raton   London   New York

CRC Press is an imprint of the
Taylor & Francis Group, an informa business

CRC Press
Taylor & Francis Group
6000 Broken Sound Parkway NW, Suite 300
Boca Raton, FL 33487-2742

First issued in paperback 2017

© 2011 by Taylor and Francis Group, LLC
CRC Press is an imprint of Taylor & Francis Group, an Informa business

No claim to original U.S. Government works

ISBN 13: 978-1-138-11520-0 (pbk)
ISBN 13: 978-1-4200-8966-0 (hbk)

Library of Congress Cataloging-in-Publication Data

Lansky, Ephraim P.
　　Figs : the genus Ficus / Ephraim Philip Lansky, Helena Maaria Paavilainen.
　　　　p. cm. -- (Traditional herbal medicines for modern times ; v. 9)
　　Includes bibliographical references and index.
　　ISBN 978-1-4200-8966-0 (hardcover : alk. paper)
　　1. Fig--Therapeutic use. I. Paavilainen, Helena M. II. Title. III. Series: Traditional herbal medicines for modern times ; v. 9.

RM666.F47L36 2010
615'.321--dc22　　　　　　　　　　　　　　　　　　　　　　　　　　　　2010017142

Visit the Taylor & Francis Web site at
http://www.taylorandfrancis.com

and the CRC Press Web site at
http://www.crcpress.com

For my father and my mother, for Rainbow, and for Rob

E.P.L.

In memory of Raquel Garcia, a true scholar and a true friend

H.M.P.

Contents

List of Figures

List of Tables

Fig Phylogeny

Ficus constitutes one of the largest genera of flowering plants with about 750 species. Although the extraordinary mutualism between figs and their pollinating wasps has received much attention, the phylogeny of both partners is only beginning to be reconstructed. As part of a Marie Curie Fellowship awarded to Nina Ronsted (Kew) and Vincent Savolainen (Kew/Imperial), and in collaboration with James Cook (University of Reading) and George Weiblen (University of Minnesota), a large-scale analysis of figs has been undertaken using DNA data. Of the six subgenera traditionally recognized as based on morphology and distribution, only subgenus *Scidium* is supported as monophyletic. Section *Malvanthera* was recovered as monophyletic if *Ficus elastica* (rubber fig) is excluded. The results do not conform to any previously proposed taxonomic subdivision of this section, and characters previously used for previous classification are homoplasious. Geographic division, however, is highly informative.* (Ronsted et al., 2008a,b)

REFERENCES

Rønsted, N., G.D. Weiblen, W.L. Clement, N.J.C. Zerega, and V. Savolainen. 2008a. Reconstructing the phylogeny of figs (Ficus, Moraceae) to reveal the history of the fig pollination mutualism. *Symbiosis* 45: 45–56.

Rønsted, N., G.D. Weiblen, V. Savolainen, and J.M. Cook. 2008b. Phylogeny, biogeography, and ecology of Ficus section *Malvanthera* (Moraceae). *Mol Phyl Evol.* 48: 12-22.

---

* This is an excerpt from a letter, dated December 17, 2008, kindly provided by Roland Hardman, Series Editor, Map. 10, Book Series, *Genus Ficus*; contact rhardman@bath.ac.uk or jcook@reading.ac.uk or v.savolainen@kew.org,)

Foreword

Figs, the fruits and trees of the *Ficus* genus, are among the oldest and most successful species of higher plants on earth. Humans and our relatives have eaten of the fruits of these trees from the earliest times, and utilized them and other parts of the tree—its leaves, its latex, its bark, and its roots—for medicinal purposes. Ease of accessibility for benefiting humans is from the fruits downward and outward on the tree, since the further down, fewer really edible products are found and the more substances there are that are suitable only for medicines, drugs, and pharmaceutical products.

The fig tree is imbued with a peculiar character that delineates its medicinal province. The mechanisms of the genomic–transcriptomic–proteomic–metabolomic cascades that allow a fig to be a fig—to adapt to threats, gain advantage, and evolve—are also of value to analogous circumstances in human medicine. These include control of the fig's own growth and dominance in order to protect the greater good of the species, a requirement that is mirrored in our own interest to control overgrowth (i.e., neoplasia, in ourselves). The fig pulls off this extraordinary feat in part by providing a safe harbor for complex communities of fig wasps, at least one species of which is selected by the peculiar species of fig as the dedicated pollinator. The others are "professional imposters" who, it could be said, through counterfeiting, tricks, and chicanery, establish their positions in relation to the fig and the chosen pollinator, and in so doing, provide a counterpoint that helps control the great reproductive exuberance of the fig–pollinator dyad, and in so doing, to contain its naturally aggressive expansiveness and its expansion.

Our hope is that, by recognizing patterns in the fig's metabolic output for achieving specific evolutionary and short-term adaptive advantages in its own various existential scenarios, a drug designer will be better equipped to create new types of complex drugs and medicinal nutrients from the fruits and other parts of these trees that are of ongoing and enduring benefit to the members of his own species. *Ficus* trees should then enjoy and further fortify their own deeply honored position of seniority in the ancient hierarchy of trees through the emergence of a new type of pharmaceutics, based upon and employing, simultaneously, the biochemical responsiveness and internal logic of the adaptive pharmacology of these trees and their fruits.

Acknowledgments

I wish to first thank my co-author, Dr. Helena Paavilainen, for all of her excellent skills in achieving a managerial, investigative, editorial, and aesthetic balance to help ensure that this book could become a reality. My wife, Dr. Shen Yu, aided in many ways through infinite kindnesses and the auguries of traditional Chinese medicine, conveying a perspective for the evolution of *Ficus* in the Far East, and translations of Chinese scientific and traditional medical writings into English. Zipora Lansky contributed the original watercolor paintings and drawings in pencil, ink, and charcoal, as well as hundreds of color photos of *Ficus* species in Israel. Krina Brandt-Doekes and Kaarina Paavilainen kindly trekked through botanical gardens in their native Jerusalem and Finland, expertly tracking living species of *Ficus* with digital cameras. Shifra Lansky provided strategic discourse to help set priorities, organize material, and achieve objectives. Professor Eviatar Nevo, my distinguished colleague and the founder of the Institute of Evolution at the University of Haifa, helped me to better understand the complexity of stems and to identify some very large *Ficus* trees in Israel, while Professor Abraham Korol, director of the Institute of Evolution, and Professor Solomon Wasser, medical mycologist *par excellence*, exhibited extra patience with me during this period of frequent absence from the bustle of laboratory research due to writing. Dr. S. W. Kaplan read the manuscript in various stages and contributed valuable suggestions, encouragement, and, as ever, his own inestimable point of view. Professor Michael Heinrich, our reviews editor at the *Journal of Ethnopharmacology*, helped us to both focus and expand the work at an early stage. Eli Merom, Meir Lipschitz, and Kibbutz Sde Eliahu provided the gift of fresh, early season, perfect, multi-colored organically grown figs of *Ficus carica* in time to be photographed prior to being consumed by family and patients in the summer of 2009.

Ephraim Lansky

My gratefulness is due to all the above-mentioned and, in particular, to Dr. Lansky whose initiative and enthusiasm were the starting point for the book. Thank you for letting me participate in this! We both thank Barbara Norwitz and Jill Jurgensen from the Taylor & Francis Group for accepting the idea of the book and helping, with patience and kindness, to bring it to fruition. I am also deeply indebted to Krina Brandt-Doekes for her efficient help in a time of need and to Margo (Malka) Karlin for her assistance in a crucial stage of the work. My mother and sister, Maija and Kaarina Paavilainen, were extremely supportive during the whole project, helping in all possible ways, and also provided part of the photographic material. And finally, my deepest gratitude goes to Prof. Samuel Kottek (Hebrew University of Jerusalem) for his continuous support, encouragement, and advice during all these years.

Helena Paavilainen

1 Introduction

Figs, along with pomegranates, grapes, dates, olives, barley, and wheat, comprise the seven indigenous species of ancient Israel. Each of these groups (and not strictly species, because all of them possess numerous and biodiverse subgroups, especially figs) possess medical properties. Pomegranates, a distant cousin that is covered in another volume in this series (*Medicinal and Aromatic Plants—Industrial Profiles*), are related to figs by a complementary and inverse mechanism of formation, but otherwise are quite parallel in their morphology. Figs, which almost certainly predate pomegranates for human use, are likely the earliest cultivated plants, predating even the grains and the (date) honey, pomegranate, wine, and olives previously mentioned (Kislev et al. 2006).

Figs have evolved into over 800 different species, making the *Ficus* genus one of the most populous in number of species of all plant genera. Edible figs figure prominently in both human and animal nutrition around the globe in parts warm and moist enough to sustain them. The diversity in the number of species concerning their morphology is exceptional, a range extending from small bushes to massive and undoubtedly the widest single trees, the curtain and banyan trees (*Ficus benghalensis*); and also diversity within a single species, for example, *Ficus carica*, the common fig of commerce, with thousands upon thousands of cultivars spread throughout

FIGURE 1.1   Three *Ficus carica* fruits (watercolor by Zipora Lansky).

FIGURE 1.2   *Ficus carica* fruit (watercolor by Zipora Lansky).

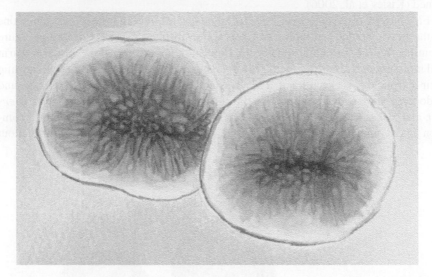

FIGURE 1.3   **(See color insert after page 128.)** *Ficus carica* fruit, sagittal section (water-color by Zipora Lansky).

the Levant, Mediterranean, Indian subcontinent, Far East, Latin America, Texas, and Southern California, especially.

Coevolving with humans, as well as with insects, specifically with the fig wasps that pollinate them, figs have produced a deep figurative impression on humankind, and, we can only suppose, conversely, also humankind on figs. Noteworthy are the role of the fig leaf in the Old Testament as a symbol of modesty, the English slang "giving a fig" (to care about something), and the uses of fig-ures and the process of fig-uring that have become buried deep in the earth of our linguistic unconscious. Even ficolins, upon which we shall expound at some length in Chapter 3, are a case in point. The word *ficolin*, which appears superficially similar to *Ficus* and refers to a lectinlike compound that helps define and point the way to the real depth of the fig's physiological potency, is actually a neologism combining the first parts of the words for *fibrinogen* and *collagen*. Formally ficolins have nothing to do with figs, yet figs provide a living botanical context for translational pharmacognostic research involving ficolins. Of course, this is just a coincidence, similar to figs and figuring, etc. Nevertheless, we shall make the point in this treatise, only partially tongue in cheek, that just as fig trees have the uncanny ability to send their powerful, sensitive, and highly adaptive roots through concrete jungles and virtually any natural impediments, so too have they sent their roots through us, to colonize us too—to powerfully enlist us through mysterious effects on our beings, our language, our mythology, as they have enlisted their dedicated pollinating wasps, to serve them, to use our feet and our emerging technologies to help them as one of the world's most successful genera to colonize favorable or possible habitats throughout the globe. For reasons and chemistry still not entirely known, *Ficus* latex has been used as a shamanic inebriant by Peruvian shamans, the *Ficus* trees dishing up the potion (*renaca*) to serve as a powerful botanic "teacher of medicine" for the shamans themselves (Luna 1984). Figs and fig trees throughout the world and the *Ficus* genus were also very likely one of the earliest and best sources of cultivated medicine as well as of food for people, and for their domesticated animals (Flaishman et al. 2008).

It is pretty much generally agreed that figs originated in Asia Minor and have been successful in spreading around the world. With spread came specification, so morphs of the original came into being in order to exploit diverse natural habitats, and as such, they are of many shapes and sizes. Many *Ficus* produce aerial roots that descend to the ground. In some species originating in the rainforests, the small *Ficus* plant takes up residence in a tree top, and dropping aerial roots, gradually overcomes and strangles its host (Bolay 1979). All *Ficus* have in common a peculiar reproductive characteristic, elaborated in the following text.

We are deeply indebted to Flaishman et al. (2008) for their penetrating and all-encompassing review of *F. carica* (the common fig, the edible fig, the fig of commerce) and how cultivars of this fig have evolved from wild figs, the exact species of which is not specified. Probably this "wild fig" is also *F. carica*, with the edible fig, *per se*, being a particular mutant of the wild type.

The mutation can be understood best in terms of specific changes in the reproductive structures of the fruit. They concern the length of the pistils, which are the tubular structures within the fruit that bear seeds and also accommodate the dedicated pollinating wasps for the bearing of their young. In short, there is a dynamic within the syconium, the name for the fruit sac of the fig, in which pistils can accommodate either a wasp egg until it hatches or seed of the fig, but not both (Wang et al. 2008). The symbionts (the wasps) must "restrain themselves" from occupying selected pistils so that the host (the fig) can also have some space for its own seeds. In at least some scenarios, if the wasps take too many sites, the survival of the host could be compromised. The endangerment of the host would then reciprocally endanger the wasps, and should not the wasps "know" this?

The mechanism that has evolved, so as not to depend entirely on the self-restraint of the wasps, is to produce the shortest pistils for quick wasp ingress and egress in the center of the syconium. The longer ones, much more difficult for a wasp to get in and out of, are left to the periphery, less accessible for receiving wasps, and kept for the sole internal needs of the fig. In dry periods, the competition for the pistils is keener, and so the mechanism of allotment becomes relatively more important (Wang and Sun 2009).

The mutant of wild fig that became the edible fig has longer pistils, and correspondingly less or even no requirement for a pollinator in order to go and keep producing fruit. This gets to the heart of the matter because, in the parthenogenic state, the fruits are more edible (though there are also apparently advantages to pollinated figs, which may be bigger and stronger) and the trees more productive from the human's point of view. The long pistils make for pulpier fig jam and bolder fig wine. By further suppression of the androecium (i.e., the full complement of the male parts, the stamens, of the flowers), all edible figs have essentially become females.

Wild, or nearly wild, figs are reported throughout much of the Middle East and the Mediterranean region and are distinguished from edible figs by two important features: first, a mutation in the wild fig gave rise to the long-styled pistils and succulent fruitlets of the edible fig and, second, as a consequence of either an a pleiotropic effect or a mutation in a tightly linked gene, the edible fig also displays a suppression of the androecium (Storey 1975). Due to suppression of the androecium, all "edible" figs are functionally female. Chromosome number and morphology in the genus *Ficus* have been studied mainly by Condit (1928, 1934, 1964), who states that the chromosomes of the various fig species are similar to each other in appearance, and 2n = 26 is the basic chromosome number in all figs. The genome size of fig is small, less than three times that of *Arabidopsis* (Ohri and Khoshoo 1987).

The historical as well as the potential future use of figs in medicine is the subject of this book, which is an attempt to present some intuitive order not just to a single fig species, such as *F. carica*, but rather to the entire *Ficus* genus, taking examples from different *Ficus* species worldwide. Since the historiographic sources used, ancient Levantine, medieval Europe, and New World, do not classify anything close to the full biodiversity within the genus, we rely less on these and more on modern research conducted during a 20-year period, from the end of the last millennium into the first decade of the present one, to guide us in selecting a few species for further study.

Each chapter, which is devoted to a different part of fig trees (Chapters 3–7) within the *Ficus* genus as a whole, discusses the botany of the most important species that have been related to that particular part pharmacologically. It goes on to consider the chemistry and pharmacology of that part in selected *Ficus* species; also, modern, medieval, and ancient methods for obtaining and preparing the beneficial components from that plant part for medicinal use are discussed, when known or supposed.

Figs, the *Ficus* trees, are an understudied genus in modern pharmacognosy. For many centuries, however, figs have been used in medicine, and this use was recorded in classical Middle Eastern and European medical writings. One of their uses, well known because it occurs in the Old Testament, is the placement of poultices of figs on tumors as treatment for abnormal swellings. Such swellings, according to reports of experts, could have been due to infection or, alternatively, cancer (Ben-Noun 2003). In either case, figs might be effective medical treatments. In short, there exists a great deal of information on classical medical usage of figs, recorded by the ancient

Greek and Middle Eastern physicians and later translated by German and English writers, involving what was probably cancer but also swellings due to infections. This work traverses these ancient and medieval discussions to review former specific means of employing the fig in medicine, especially for cancer and other medical conditions of inflammatory origin, diabetes, infectious disease, parasites, and gastrointestinal and pulmonic disorders. Its pharmacological actions include antibacterial, antioxidant, anti-inflammatory, gastroprotective, antidiarrheal, vulnerary, antitumor, anticancer, antispasmodic, immunobalancing/immunoharmonizing, and nutritive *par excellence*.

Our task will be to integrate the two major streams of knowledge and to bring them into a common flow to support and highlight the ongoing complex pharmaceutical research and development. One stream is the great tradition of herbal use, both folk medicine and the medicine prescribed by physicians of antiquity, and the other is the modern stream of scientific research, most of which, as noted, occurred only in the last 20 years. These twin elements, the medicine of the past and our fledgling modern state of scientific research, will almost certainly guide the emerging complex botanic medicine of the near future.

We use the term *complex medicine* to differentiate it from a medicament consisting of a more or less purified single compound. Almost all present-day medicines are of this latter type; that is, they are pure chemical compounds. In fact, for the most part, pharmaceutical medicines and pure chemicals are nearly synonymous. However, from time immemorial, before the pharmaceutical era of the last 150 years, only complex medicines had been employed. These were sometimes "real complex medicines" consisting of a mixture of parts or extracts from different plant species, or sometimes "simple complex medicines" consisting of just one single plant. But even if only one single plant were employed as a botanical "simple," the resultant medicine was still "complex" because the comminuted plant part or extract, even if filtered and stabilized, contained hundreds if not thousands of different chemical compounds, not just a single one. Still, though, to the medical herbalist, the simple medicine would perform in a singular and unified way.

Nowadays we know that even single pure chemicals can exert pleiotropic effects on genes. This means that the compound can induce or suppress a gene to transcribe proteins and that this effect can reverberate downward to affect multiple human organs and multiple physiological systems. If such a multiple or pleiotropic effect is possible from a pure chemical, how much more so, then, is the potential for multiple physiological targeting from a mixture of different compounds (Newman and Lansky 2007, Lansky 2006). For the most part, the pharmacological explorations of the past century have focused on the purifying of single compounds from plants, and using these individual compounds as leads for further pharmaceutical development, such as by synthetic improvements in the organic chemistry laboratory. Perhaps because it has taken so long to begin to get some handle on the pure chemicals in plants, it is only now that some attention is being given back to the idea of recombining these compounds in new ways to produce new complex drugs.

We previously commented on one example of a complex drug taken from modern medicine and, in the end, in spite of our progress, are left more or less with a mystery about its special efficacy and advantage. The example (Lansky and Von Hoff 2005) is of a pharmaceutical product that is over 60 years old, and is actually one of the most financially successful and longest-lived modern drugs, which, in spite of periodic waves through the years of cancer scare and bad press, continues to be sold at a rate of about 1.5 billion dollars per year. This sales volume represents about half the annual market for hormone replacement drugs used by perimenopausal women to hold back aging, fight hot flashes, prevent osteoporosis, and in general to compensate the body as well as the sensorium to the effects of declining endogenous estrogen levels. The drug is called Premarin in most parts of the world, the name being an acronym for Pre-Mar-In, standing for "pregnant mare urine." No one knows exactly why, but women generally prefer this product to the pure synthetic hormones made in the laboratory, although the potency of each product, the complex and the simple, may be, formally speaking, of equivalent estrogenic potency. Women preferring Premarin, in spite of the animal suffering inflicted, in spite of the warnings about cancer, do so because, they say, the complex one "feels better" or "works better."

In pomegranate, we were able to show that a combination of complex chemical fractions from anatomically discrete sections of the fruit, for example, pomegranate juice with pomegranate seed oil, resulted in an anticancer or anti-inflammatory product that was synergistically more potent than one deriving from the juice or oil phase alone. When the oil fraction was combined with not only juice but also with a peel-derived fraction, the effect was even better. Synergy and therapeutic potency were enhanced when the mixture's complexity was enhanced; the complex was consistently more effective than the single part, even if the added material was less potent on its own than the "active component" (Lansky et al. 2005a). When additional experiments were performed to compare the effects of combining chemicals that are normally sourced in the different pomegranate compartments, such as combining conjugated linolenic acids from pomegranate seed oil with flavonoids, phenolic acids, or tannins from pomegranate juice or peel, the same synergistic phenomenon was observed (Lansky et al. 2005b), indicating that pharmacologically weak compounds may enhance the activity of pharmacologically strong compounds, even if some of the amount of the strong compound is substituted with some of the weak compound. The whole extracts, such as pomegranate juice concentrate, were frequently more active than a single type of chemical fraction from the extract (such as total tannins) or a single (most) active compound from the active fraction (such as ellagic acid). The full complex milieu, rather than detracting from the effect of the single chemical within, actually contributes either to its activity or tolerance, as well as likely also to its safety. This is because other components in complex agents may produce warning signs of toxicity long before the active ingredients actually begin to cause observable toxic signs. Further, these supplementary, perhaps "inactive," accessory or "entourage" compounds (Ben-Shabat et al. 1998) may help to reduce the toxic effects, to take the "edge" off the active compound. These elements provide the basis for compounding different herbs within formulas in traditional herbal medicine, such as that from China, India, and Latin America, for example.

In the medieval and ancient worlds touched by Arabic medicine the figs of commerce, from the species *F. carica* L, or alternatively from its relative *F. sycomorus*, were chiefly employed. However, other fig trees such as *F. racemosa* and *F. religiosa* were employed as sources for medicines in India and throughout the Far East, though

FIGURE 1.4 (See color insert after page 128.) *Ficus carica* fruit (27.7.2009, Hebrew University Botanical Garden, Mount Scopus, Jerusalem, Israel: by Krina Brandt-Doekes).

FIGURE 1.5 *Ficus carica* fruit (27.7.2009, Hebrew University Botanical Garden, Mount Scopus, Jerusalem, Israel: by Krina Brandt-Doekes).

the fruits, the "figs," may have been less a source of medicines than the leaves or bark. For this reason, we have chosen the entire *Ficus* genus as the subject of our study. The focus is guided, as discussed earlier, by both citations of traditional use, in which the specific species are identified, and by chemical or pharmacological explorations of the extracts of specific plant parts of specific species.

The comparative chemistry and pharmacology of plant parts from different *Ficus* species have only been explored in very few cases. Usually, a single species is studied, and almost always only a single part from the tree. Occasionally, there are comparisons, say, between the activity of bark from the roots versus activity of bark from the stem, but this is unfortunately the exception and not the rule.

We must assume, though, from what we do know about the chemistry of different species and from the genetic commonalities between different *Ficus* species, that considerable overlap and similarities between the corresponding chemistry, say, of a leaf, from different *Ficus* species do occur. However, it also may not. If we find a particular class of alkaloids in the leaf of one species of *Ficus*, we can realistically explore the leaves of other (perhaps more easily obtainable) *Ficus* species as alternative sources of the same type of alkaloid. However, only careful exploration will reveal the truth. We cannot assume that because the alkaloid appeared in the leaf of one species of *Ficus*, it will appear in the leaves of all *Ficus* species. We must check, but the experience with any species of *Ficus* will help and guide subsequent study of other *Ficus* species.

Having the whole genus available does provide insight into the commonalities among species, where divergence occurred, and the specific advantages from any single type, pharmacologically speaking. If there is, hypothetically speaking, some advantage from combining, for example, an aqueous decoction of the leaves with a wine prepared from the fruits and all seasoned with ashes from the heartwood in a potential anticancer complex medicament, it may work even better by combining the special parts from one species with those of another. One species may have the best grouping of cytotoxic alkaloids in its aerial roots, and another a jelly from the achenes of its fruits that provides an ideal pharmaceutical matrix for natural controlled delivery, while the latex from a third may have the richest accumulation of antiangiogenic flavonoids. Thus, in the end, a single medicine could actually be constructed, perhaps advantageously, from extracts from parts of different *Ficus* species. The diversity within the genus would highlight the points of specialization to exploit, while the common origin of the different species might provide biochemical bases for the maximizing of complementarity and redundancy in modulating physiological systems, and provoking and maintaining therapeutic responsiveness in the patient.

Traditional uses of figs are considered, including how figs or fig tree parts were processed as

- Poultices from fresh or dried figs
- Poultices from fig leaves
- Fig wines
- Lye from fig tree bark
- Latex from stems and leaves

The pharmacological bases underlying the possible efficacy of these preparations are related to their chemical compositions. The presence of alkaloids in *Ficus* spp. has been recognized for several decades (Saxton 1971). Recently, the pharmacological aspects of *F. carica* have been specifically reviewed and shown to include antitumor effects; the ability to mediate body metabolism, hyperglycemia, hyperlipidemia, and cholesterol levels; to enhance oxidative resistance; to act against bacteria and viruses; to mediate immunity; to activate blood coagulation; and to support tumor treatment by reducing toxicity and side effects in actinotherapy and chemotherapy (Zhang and Jiang 2006). Some of the key compounds underlying these effects, to be developed further in the chapters that follow, include

- Alkaloids in the stems and leaves of *F. septica* and *F. hispida*
- Coumarins in the leaves of *F. carica, F. ruficaulis,* and *F. septica*
- Flavonoids in the leaves of *F. carica, F. ruficaulis,* and *F. septica*
- Sterols in *F. carica* leaves and latex
- Anthocyanins in *F. carica* fruits
- Triterpenoids in *F. microcarpa* aerial roots

Chapters 3–7 are each devoted to a different part of figs or fig trees and traverse the botany of the most important trees related to that particular part, that is, pharmacologically speaking. They also include the chemistry and the pharmacology of that part in specific *Ficus* species and, in most cases, modern, medieval, or ancient methods for obtaining and preparing the beneficial components from the plant part for medicinal uses.

In Chapters 8 and 9, finally, the issue of coevolution with the animal kingdom is considered: between figs and their wasps in Chapter 8, and between figs and us in Chapter 9. These last two chapters specifically consider how the evolutionary needs of one species (us) as it evolves affect the evolution of a second species (or genus!): *Ficus* spp., especially *F. carica*. This phenomenon is nicely treated in Michael Pollan's popular book *The Botany of Desire*, which considers how human desires result in a kind of opportunity for plants able and responsive enough to these desires in order to develop biochemical strategies for fulfilling them. His chapters follow the model for fulfillment of the desire for sweetness with apple, for beauty by tulip, for intoxication with cannabis, and for control with potato (Pollan 2001). According to Pollan's persuasive logic, the inherent immobility of plants forces them to develop strategies for defense and reproductive success that depend on the production of a functional biochemistry for its own purposes. When this biochemistry coincides with the "desires" of a second species, probably from a different kingdom, it results in evolutionary changes in the original species that enable it to better meet the desires of its target. In the case of humans and other higher animals, the purpose of the fig might be to use our legs and wings and technological tricks to move its seeds around and to colonize new realms

The book includes

1. Preparation and utilization of specific remedies
2. Pharmacology related to cancer medicine and inflammation
3. The coevolution of fig trees with insects and humans

4. Chemical details and physiological requirements of the trees and how their chemistry provides a parallel and complementary benefit to mammals, particularly man
5. Drug leads derivative from the phytochemistry of these trees

Some leaders in the field of pharmacological fig research include:

Alison Pawlus, Ph.D., M.D. Anderson Cancer Center
Robert A. Newman, Ph.D., M.D. Anderson Cancer Center, Punisyn Pharmaceuticals, Ltd.
M.A. Flaishman, Ph.D., University of Haifa
S. Pashapoor, Ph.D., Ardabil University of Medical Sciences, Iran
S.J. Lee, Ph.D., National Health Research Institutes, Taipei
H.M. Chang, Ph.D., National Taiwan University
Raphael Mechoulam, Ph.D., Hebrew University
Jacob Vaya, Ph.D., Migal Institute, Israel
Rachel Li, Ph.D., University of Queensland, Australia
M. Giday, Ph.D., Addis Ababa University, Ethiopia
M. Deepak, Ph.D., Al-Ameen College of Pharmacy, Bangalore
W.A. Marussich, University of Arizona, Tucson
Y.H. Kuo, Academia Sinica, Taipei
S.A. Adesanya, Obafemi Awolowo University, Nigeria
S. Akin, Medical Faculty of Uludag University, Turkey

REFERENCES

Ben-Noun, L.L. 2003. Figs—the earliest known ancient drug for cutaneous anthrax. *Ann Pharmacother.* 37: 297–300.
Ben-Shabat, S., E. Fride, T. Sheskin et al. 1998. An entourage effect: Inactive endogenous fatty acid glycerol esters enhance 2-arachidonoyl-glycerol cannabinoid activity. *Eur J Pharmacol.* 353: 23–31.
Bolay, E. 1979. Figs and strangler figs. *Pharm Unserer Zeit* 8: 97–112.
Condit, I.J. 1928. Cytological and morphological studies in the genus *Ficus.* I. Chromosome number and morphology in seven species. *Univ Calif Publ Bot.* 11: 233–44.
Idem. 1934. Cytological and morphological studies in the genus *Ficus.* II. Chromosome number and morphology in thirty-one species. *Univ Calif Publ Bot.* 17: 61–74.
Idem. 1964. Cytological studies in the genus *Ficus.* III. Chromosome numbers in sixty-two species. *Madrono* 17: 153–4.
Flaishman, M.A., V. Rodov, and E. Stover. 2008. The fig: Botany, horticulture, and breeding. *Hortic Rev.* 34: 113–96.
Kislev, M.E., A. Hartmann, and O. Bar-Yosef. 2006. Early domesticated fig in the Jordan Valley. *Science* 312: 1372–5.
Lansky, E.P. 2006. Beware of pomegranates bearing 40% ellagic acid. *J Med Food* 9: 119–22.
Lansky, E.P., G. Harrison, P. Froom, and W.G. Jiang. 2005b. Pomegranate (*Punica granatum*) pure chemicals show possible synergistic inhibition of human PC-3 prostate cancer cell invasion across Matrigel. *Invest New Drugs* 23: 121–2, Erratum 23: 379.

Lansky, E.P., W. Jiang, H. Mo et al. 2005a. Possible synergistic prostate cancer suppression by anatomically discrete pomegranate fractions. *Invest New Drugs* 23: 11–20.

Lansky, E.P., and D.D. Von Hoff. 2005. Complex and simple. *Leuk Res.* 29: 601–2.

Luna, L.E. 1984. The concept of plants as teachers among four Mestizo shamans of Iquitos, northeastern Peru. *J Ethnopharmacol.* 11: 135–56.

Newman, R.A., and E.P. Lansky. 2007. *Pomegranate: The Most Medicinal Fruit.* Laguna Beach, CA: Basic Health Publications.

Ohri, D., and T.N. Khoshoo. 1987. Nuclear DNA contents in the genus *Ficus* (Moraceae). *Plant Syst Evol.* 156: 1–4.

Pollan, M. 2001. *The Botany of Desire.* New York: Random House.

Saxton, J.E. 1971. The indolizidine group of alkaloids. *Alkaloids (London)* 1: 76–85.

Storey, W.B. 1975. Figs. In *Advances in Fruit Breeding*, ed. J. Janick, and J.N. Moore, 568–588. West Lafayette: Purdue University Press.

Wang, R.W., L. Shi, S.M. Ai, and Q. Zheng. 2008. Trade-off between reciprocal mutualists: Local resource availability-oriented interaction in fig/fig wasp mutualism. *J Anim Ecol.* 77: 616–23.

Wang, R.W., and B.F. Sun. 2009. Seasonal change in the structure of fig-wasp community and its implication for conservation. *Symbiosis* 47: 77–83.

Zhang, K., and R. Jiang. 2006. Pharmacological study of *Ficus carica*. *Zhongguo Linchuang Kangfu* 10: 226–8.

2 Overview of *Ficus* genus

The genus *Ficus* is of the Mulberry family (Moraceae) and possibly the most illustrious member of this family's genera. It is an extremely aggressive genus both by its penetration throughout the globe, establishing new species where numerous and varied niches present themselves, and in the behavior of its trees in claiming their individual territories, most dramatically in the various "strangler" species that begin as an aerial epiphyte atop some other well-established host tree. The *Ficus* then sends down its long aerial roots using the stem of the host as a supporting structure, eventually redoubling its rooting behavior until it has completely encircled its host. Eventually, the encirclement results in the obliteration of the host tree within its grasp, but for the *Ficus* it has become a moot point. The *Ficus* then enjoys the total height of the jungle canopy, for example, and has established itself upon the back of another tree species of a different genus. Such a feat and other evolutionary sleights of hand have enabled *Ficus* to maintain itself with ongoing adaptability to increasingly varied habitats for over 60 million years, following the shifting structures of the earth's crust, the breaking-off of continents from the Southern base of an original supercontinent. When the structure broke sending parts of itself northward, *Ficus* was on board. And it was not alone!

Along with *Ficus* for the whole 60-million-year ride were its loyal wasps, both the true pollinators and the many imposters who chose to join the great *Ficus* genus,

FIGURE 2.1   Dried Turkish figs in a plastic basket; pencil, by Zipora Lansky.

TABLE 2.1
Overview of *Ficus* Species Covered in This Book

Ficus Species	Region	Traditional and/or Economic Uses	Research	References
asperifolia Miq.	Africa	Wound healing, treatment of infertility	Increased uterine growth in rats, anti-HIV (antiviral), stimulate growth of dermal fibroblasts	Abdullahi 2006, Ajose 2007, Annan and Houghton 2008, Watcho et al. 2009
awkeotsang Makino (*F. pumila* var. *awkeotsang* Corner)	China, Singapore, Taiwan, Thailand	Fig produces a natural gel with outstanding value for food and pharmaceutical preparation	Nutritive proteins in the gel, pectinerase inhibition, antifungal	Chua et al. 2007, 2008, Hsiao et al. 2009, Li et al. 2005, Miyazaki et al. 2004, Wu et al. 2005
benghalensis L.	India	Hypoglycemic, phytoremediation	Magnetic properties, hypoglycemic effects	Kar et al. 2003, Prajapati and Tripathi 2008
benjamina L.	India, worldwide	Houseplant, decorative	Antinociceptive	Parveen et al. 2009
capensis Thunb.	Southern Africa, California	For threatened abortion, antisickling	Tocolytic, antisickling	Mpiana et al. 2008, Owolabi et al. 2009
carica L.	Middle East, Mediterranean, and worldwide	Nutritive, antitumor, digestive	Digestive, immunological adjuvant, trace elements; antispasmodic, antiplatelet	Devaraj et al. 2009, Gilani et al. 2008, Waheed and Siddique 2009, Yang et al. 2009
chlamydocarpa Mildbr. & Burret	Cameroon, Africa	—	Antimicrobial	Kuete et al. 2008
citrifolia Mill.	Florida, Caribbean, South America	Decorative; for infection, vulnerary	Modulation of multidrug resistance in cancer therapy	Duke 1975, *GRIN* 1994, Simon et al. 2001
cordata Thunb.	Africa	—	Antibacterial	Kuete et al. 2008, Poumale et al. 2008
cunia Buch.-Ham. e Rokb.	India, Myanmar, Thailand	Bast fibers	Bacterial recognition agglutinin	Adhya et al. 2006, Hanelt et al. 2001

TABLE 2.1 (continued)
Overview of *Ficus* Species Covered in This Book

Ficus Species	Region	Traditional and/or Economic Uses	Research	References
deltoidea Jack	Thailand, Indonesia, Malaysia	Decorative	Safety, photosensitizers antinociceptive	Fazliana et al. 2008, *GRIN* 1994, Ong et al. 2009, Sulaiman et al. 2008
elastica Roxb ex Hornem.	India, Nepal, Indonesia	Rubber; shellac; houseplant	Anti-inflammatory, polyprenols	*GRIN* 1994, Hanelt et al. 2001, Sackeyfio and Lugeleka 1986, Stone et al. 1967
erecta Thunb.	China, Japan and the Far East	Fiber; decorative	Antiosteoporotic	*GRIN* 1994, Yoon et al. 2007
exasperata Vahl (leaf)	Tropical Africa, Yemen, India	Analgesic; antiarthritic; diuretic; animal fodder	Safety	Bafor and Igbinuwen 2009, Burkill 1985, *GRIN* 1994
fistulosa Reinw. ex Blume (leaves and stem bark)	China, India, Indochina, Malesia	—	Antimalarial	Wu et al. 2003, Zhang et al. 2002
glabrata Kunth (latex)	Amazon, Mexico	Parasiticide	Anticoagulant, anthelminthic	Azevedo 1949, *GRIN* 1994, Hansson et al. 1986, Martinez 1969
glomerata Roxb. aka *racemosa* L., (bark, leaves, unripe fruit, fruit)	India, Indochina, South America	Antioxidant dysentery, hypoglycemic, menorrhagia, hemoptysis	Hepato-, gastro-, radioprotective, mosquito larvicidal	Channabasavaraj et al. 2008, Hanelt et al. 2001, Jahan et al. 2009, Rao et al. 2008, Rahuman et al. 2007, Subhaktha et al. 2007, Veerapur et al. 2009
gnaphalocarpa (Miq) Steud.	Southern Africa	Food; guttapercha; stimulant	—	Hanelt et al. 2001

(continued on next page)

TABLE 2.1 (continued)
Overview of *Ficus* Species Covered in This Book

Ficus Species	Region	Traditional and/or Economic Uses	Research	References
hirta Vahl (roots)	China, India, Indochina	—	Hepatoprotective, pharmacokinetics	Cai et al. 2007, *GRIN* 1994, Li et al. 2009, Lv et al. 2008
hispida L.d.	Southeast Asia, Malesi, Australia, United States	Tonic, antidiarrheal, febrifuge, lactagogue	Hepatoprotective, anticancer, antidiarrheal	Burkill 1966, Ghafoor 1985, Mandal et al. 2000, Mandal and Ashok Kumar 2002, Peraza-Sánchez et al. 2002, *Anon.* 1948–1976
infectoria Roxb.	India, China, Indonesia	Phytoremediation, food	Phytoremediation	Hanelt et al. 2001, Pandey et al. 2005, Yateem et al. 2008
lyrata Warb.	Tropical West Africa	Decorative, magic	Chemical characterization	Basudan et al. 2005, Burkill 1985, Hanelt et al. 2001
microcarpa L.d. (aerial roots, latex)	South Asia, Australia, North Africa	Medicinal use	Antifungal, phenylpropanoids	Chiang et al. 2005, Ghafoor 1985, Hanelt et al. 2001, Ouyang et al. 2007, Taira et al. 2005
nymphaeifolia Mill. (stem bark)	Central and Southern America	Anodyne	Isoflavones	Darbour et al. 2007, Wong 1976
ovata Vahl (stem bark)	Sub-Saharan Africa	Fiber; gum	Antimicrobial	Berg 1991, Burkill 1985, Kuete et al. 2009
palmata Forssk.	Middle East, South-West Asia	Food, fodder; laxative, demulcent	—	*GRIN* 1994, Hanelt et al. 2001, *Anon.* 1948-1976
platyphylla Del. (stem bark)	Nigeria, Tropical Africa	Rubber, tannins, fiber, astringent	Anticonvulsant, smooth muscle relaxant	Amos et al. 2001, Burkill 1985, Chindo et al. 2003, 2009, *GRIN* 1994

TABLE 2.1 (continued)
Overview of *Ficus* Species Covered in This Book

Ficus Species	Region	Traditional and/or Economic Uses	Research	References
pumila L. (fruits, leaves)	China, Japan, Southeast Asia	Food; anticancerous, antidysenteric, diuretic	Terpenoids	Burkill 1985, *GRIN* 1994, Kitajima et al. 2000, Hanelt et al. 2001, Ragasa et al. 1999
religiosa L. (fruits)	India, Southeast Asia, Middle East	Consumption, vomiting, oral ulcers, burns, gynecological	Antineuroinflam-matory, anticonvulsant	Ghafoor 1985, *GRIN* 1994, Jung et al. 2008, Prasad et al. 2006, Singh and Goel 2009
ruficaulis Merr (leaves)	Malesia, Taiwan	—	Anticancer furocoumarins	Chang et al. 2005, Wu et al. 2003
semicordata Buch.-Ham ex Sm.	Pakistan, India, Indochina	Food; fiber; for bladder diseases	Wasp pollination	Ghafoor 1985, Song et al. 2008
septica B.u.r.m. f. (stems, roots)	Indonesia, Australia	Antidote, purgative	Anticancer alkaloids	Burkill 1966, Damu et al. 2005, 2009, *GRIN* 1994, Uphof 1968
stenocarpa F. Muell. ex Beuth. (latex)	Australia, South-Western Pacific	—	Strongest proteolytic activity of latex from 46 species	Chew 1989, Williams et al. 1968
sur L. (leaf, stem bark, root bark)	Tropical Africa, Yemen	Food, fiber, tannin, lactagogue, febrifuge, diuretic	Antimalarial	Burkill 1985, *GRIN* 1994, Muregi et al. 2003, 2007
sycomorus L. (stem bark)	Tanzania, Nigeria, Tropical and South Africa, Middle East	Wood, decorative, shade tree	Antiviral, antimicrobial, antimalarial, inhibit smooth muscle contraction	*GRIN* 1994, Hanelt et al. 2001, Maregesi et al. 2008, Sandabe et al. 2006, Sanon et al. 2003

(continued on next page)

TABLE 2.1 (continued)
Overview of *Ficus* Species Covered in This Book

Ficus Species	Region	Traditional and/or Economic Uses	Research	References
thonningii Blume (stem bark, root)	Ethiopia, Sub-Saharan Africa	Latex, bast fiber, lactagogue, antidiarrheal, against malnutrition and debility	Antibacterial, antidiabetic, antiparasitic, gastro-vulnerary	Burkill 1985, *GRIN* 1994, Koné et al. 2007, Hanelt et al. 2001, Musabayane et al. 2007, Teklehaymanot and Giday 2007
thunbergii Maxim.	India, Pakistan, Far East	Food	—	Wu et al. 2003, Ghafoor 1985
virgata Reinw. ex Blume	Malesia, Australia, South-Western Pacific	—	Self-defense of tree	*GRIN* 1994, Konno et al. 2004, Tambunan et al. 2006

FIGURE 2.2   (See color insert after page 128.) Figs from *Ficus carica*, including a coronal section (by Kaarina Paavilainen).

FIGURE 2.3 Dried figs from *Ficus carica* (by Kaarina Paavilainen).

to mimic and benefit from the evolutionary synergy it established with its obligate pollinators. Yet the counterfeit wasps also play a role in the overall regulation of the complex wasp communities that travel with and are part of the overall *Ficus* poly-wasp ecosystem. These issues are discussed in depth in Chapter 8.

Ficus possesses a number of species estimated conservatively at 700 and some-times over 1000; an about average estimation is 850 species. In any event, this is an extremely large number of species for a genus to possess, and *Ficus* is probably the leader of all plant genera in the number of species it has. As this book was written and the scientific studies on the pharmacology of fig tree parts delineated, several *Ficus* species stood out. Although our selection represents less than 10% of the total number of *Ficus* species, we felt that these species, selected for their ethno-pharmacological interest, constituted a reasonable sampling of the species that com-prise *Ficus*. Accordingly, we decided to use this sample as the basis for discussing the range of species in the genus. The following discussion is therefore a species-by-species description encompassing the ethnography, distribution, and local uses, both common and medical, of each of the species alluded to in the pages that follow. As a neutral and random means of arranging the species, alphabetical order is used. The rest of the chapter focuses on modern ethnomedicinal uses of the plants, whereas the morphology and chemical composition of the various species of *Ficus* will be discussed in detail in Chapters 3–7.

A PERSONAL ACCOUNT OF AN ENCOUNTER WITH A FIG TREE

Now I understand how you can write an entire book on the fig. I have this fig tree in my tiny backyard. It became huge and went out over the fence to the neighbors. It produced a massive amount of fruit and attracted insects and bats in profusion. The neighbors have this brand new awning that collects guano in quantity. Hundreds of bats come all night and circle endlessly, catching the bugs that are eating the figs. They started to call and complain and threaten— the neighbors, not the bats. I've never even seen their faces, but they call on the phone all the time. So last fall I took out my saw and decapitated the tree, cut off its branches one by one. All the neighbors were happy, so was my wife. I stripped it down to a stump. Then it started to grow back. Dozens of small branches, like writhing snakes, in a great frenzied ball, stretching up and out in all directions. Now its like a miniature huge fig tree, and I think it'll produce 100 kilos of fruit this summer, but this time they'll all be low enough to pick without trouble; the neighbors can go to ...., I'm not going to cut it down again. It makes these fat delicious figs, and I feed them to my kids every morning all through August and well into September. I get up early and go outside in an ecstatic state to pick figs. It gives me strength and courage and keeps the kids healthy, and my youngest boy loves them. It's amazing how much strength they have; they grow so fast, it's unbelievable. Even if I stay until I'm 70 to live out my last days, I'll still have time to plant a new fig and enjoy its magical fruit, they grow that fast. And they need absolutely no irrigation. They have tremendously powerful roots that bore way down through the earth, they get into the smallest cracks and force the rocks apart, and they split the earth asunder and find sweet live groundwater.

TABLE 2.2
Ethnographic Data on Various Species of *Ficus*

Ficus asperifolia Miq.

		References
Genus	*Ficus* section: *Sycidium*	*GRIN* 1994
Synonyms	*F. acutifolia* Hutch.	*GRIN* 1994, *ALUKA*
	F. asperifolia Hook. ex Miq.	2003, Burkill 1985,
	F. cnestrophylla Warb.	Berg 1991
	F. colpophylla Warb.	
	F. irumuensis De Wild.	
	F. niokoloensis Berhaut	
	F. paludicola Warb.	
	F. pendula Hiern	
	F. pendula Welw. ex Hiern	
	F. scolopophora Warb.	
	F. spirocaulis Mildbr.	
	F. storthophylla Warb.	
	F. storthophylla Warb. var. *cuneata* De Wild.	
	F. urceolaris Hiern	
	F. urceolaris Welw. ex Hiern	
	F. urceolaris Welw. ex Hiern var. *bumbana* Hiern	
	F. warburgii H.Winkl.	
	F. xiphophora Warb.	
Common names	English: Sandpaper tree	Burkill 1985
	Fula-Pulaar (Guinea): Niénié	
	Manding-Mandinka (Guinea): Sutro	
	Manding-Mandinka (Sierra Leone): Kamakor	
	Mende (Sierra Leone): Kagami, Kamaami, Kame	
	'Safen' (Senegal): Tiăgtiăgad	
	Susu (Guinea): Nioyeniye	
	Susu (Sierra Leone): Nyoin	
Habitat and distribution	Evergreen rainforest and forest margins, riverine vegetation along streams, wooded savanna grassland. Altitude range: 0–1800 m.	Van Noort and Rasplus 2004, *GRIN* 1994, Berg 1991, Burkill 1985, Hutchinson and Dalziel 1958, *ALUKA* 2003
	Africa: From Senegal to southern Sudan, Kenya, Uganda and Tanzania, westwards to Angola and Zambia	
	Sudan, Kenya, Tanzania, Uganda, Cameroon, Zaire; Benin, Cote D'Ivoire, Ghana, Guinea, Guinea-Bissau, Nigeria, Senegal, Sierra Leone, Togo, Angola, Zambia	
Economic importance	Fruit: food; dyes, stains, inks, tattoos, and mordants	Burkill 1985
	Seed: agrihorticulture; fodder for birds; social: sayings, aphorisms	

(continued on next page)

TABLE 2.2 (continued)
Ethnographic Data on Various Species of *Ficus*

Ficus asperifolia Miq.   References

Economic importance (continued)	Leaf: arrow poisons, miscellaneously poisonous or repellent; carpentry and related applications; household, domestic, and personal items Twig: chew-sticks, etc. Latex: dyes, stains, inks, tattoos, and mordants Bark: exudations—gums, resins, etc.; fiber Bark-fiber: household, domestic, and personal items Plant: social: sayings, aphorisms Wood ash: dyes, stains, inks, tattoos, and mordants	
Local medicinal uses	Parts used: fruit, seed, leaf, latex, bark, root, root bark Local uses: diuretic, painkiller, tumors, cancers, sores, headache, eye problems (conjunctivitis), nasopharyngeal afflictions, for kidneys, liver (megalospleny), stomach, colic, ringworm, parasitic infections (cutaneous, subcutaneous), for conditions involving the anus, hemorrhoids, venereal diseases (gonorrhea), dropsy, swellings, edema, gout, febrifuges, abortifacients, ecbolics, regulating menstrual cycle, arrow poisons, miscellaneously poisonous or repellent, antidote, against malnutrition, debility, tonic, generally healing	Burkill 1985, Duke 1972, Ayensu 1978, Hartwell 1967–1971, *EthnobotDB* 1996
Toxicity	Leaf: used in the preparation of arrow poison Roots: contain alkaloids	Burkill 1985

Ficus awkeotsang Makino

Genus	*Ficus* section: *Erythrogyne*	*GRIN* 1994
Synonyms	*F. pumila* L. var. *awkeotsang* (Makino) Corner *F. nagayamae* Yamamoto	*GRIN* 1994, Wu et al. 2003
Common names	Chinese: Ai yu zi	Wu et al. 2003
Habitat and distribution	China, Taiwan	*GRIN* 1994, Wu et al. 2003
Economic importance	Fruit: food	Wu et al. 2003
Medicinal uses	Parts used: root Local uses: liver disease	Yang et al. 1987

Ficus benghalensis L.

Genus	*Ficus* section: *Conosycea*	*GRIN* 1994
Synonyms	*F. cotonaeifolia* Vahl *F. crassinervia* hort. ex Kunth *F. indica* L.	*MMPND* 1995, Hanelt et al. 2001

TABLE 2.2 (continued)
Ethnographic Data on Various Species of *Ficus*

Ficus benghalensis L.		References
Synonyms (continued)	*F. lasiophylla* Link	
	F. procera Salisb.	
	Urostigma benghalense (L.) Gasp.	
Common names	English: Banyan, Banyan fig, Banyan tree, Bengal banyan, Bengal fig, East Indian fig tree, Horn fig, Indian banyan, Weeping Chinese banyan	*MMPND* 1995, Hanelt et al. 2001, *GRIN* 1994
	Danish: Indisk figen	
	German: Banyanbaum, Banyan-Feige, Bengalischer Feigenbaum	
	French: Arbre banian, Banian, Figuier des pagodes, Figuier d'Inde	
	Spanish: Baniano, Higuera de Bengala	
	Portuguese: Bargá, Figueira-bargá, Figuera-Banyan	
	Serbian: Indijska smokva	
	Arabic: Fikus banghali, Tin banghali	
	Sanskrit: Akshaya vruksham, Avaroha, Bahupada, Bat, Bhringi, Jatalo, Nyagrodha, Nyagrodhah, Vat, Vata, Vatah	
	Punjabi: Bar	
	Hindi: Bad, Bar, Bargad, Bargat, Barh, Bat, Bot	
	Kannada: Aalada mara, Ala, Alada, Alada mara, Goli, Nyagrodha, Vata	
	Bengali: Bar, Bat, Bath, Bot	
	Marathi: Marri, Peddamarri, Vad, Wad	
	Gujarati: Vad, Vadlo	
	Oriya: Baragachha	
	Telugu: Mari peddamari, Marri, Peddamarri, Vata vrikshamu	
	Malayalam: Ala, Peraal: Peral (Kerala), Vatam, Vatavarksam	
	Sinhalese: Maha nuga	
	Tamil: Aal, Aalam vizhudhu, Al, Alam, Alamaram, Peral	
	Burmese: Pyi nyaung, Pyin vaung	
	Thai: Krang, Ni khrot	
	Vietnamese: Cây đa, Cây dong, Cây sanh, Đa lá tròn	
	Malay: Ara (Indonesia), Pokok ara	
	Nepalese: Bar	
	Chinese: Mong jia la rong	
	Japanese: Bengaru bodaiju	

(continued on next page)

TABLE 2.2 (continued)
Ethnographic Data on Various Species of *Ficus*

Ficus benghalensis L.		References
Habitat and distribution	Disturbed thickets.	*GRIN* 1994, Berg 1991,
	Tropical Asia: Pakistan, India, Bangladesh. Cultivated elsewhere, for example, Zambia, Jamaica, Florida.	Wunderlin 1997, *ALUKA* 2003
Economic importance	Fruit: famine food	Hanelt et al. 2001, *GRIN* 1994
	Leaf: tanning ; wax	
	Latex: birdlime	
	Twigs: food for lac insects	
	Stem: wood; fiber	
	Bark: tanning	
	Tree: ornamental; shade/shelter; social: religious	
Local medicinal uses	Parts used: aerial parts, fruit, leaf, latex (unspecified part), branches, bark, stem bark, root, roots (aerial), root bark	*EthnobotDB* 1996, Duke 1972, Hanelt et al. 2001
	Local uses: anodyne, astringent, hemostat, alternative, diuretic, bruise, wounds, ulcers, sores, abscess, boils, tumor, warts, scabies, hair, stomatitis, toothache, toothbrush, epistaxis, eye problems, ophthalmia, sore throat, bronchitis, cough, expectorant, asthma, mental problems, nausea, muscular pains, rheumatism, lumbago, feet, diabetes, cholera, diarrhea, dysentery, purgative, vermicide, menorrhagia, postpartum/ postabortion healing, abortifacient, venereal ulcers, aphrodisiac (male), spermatorrhea, atrophy, cachexia, fever, rinderpest, tonic	

FIGURE 2.4   Weeping leaves of *Ficus benjamina* (18.7.2009, Loimaa, Finland, by Kaarina Paavilainen).

FIGURE 2.5   *Ficus benjamina* as ornamental plant in an office room (18.7.2009, Loimaa, Finland, by Kaarina Paavilainen).

TABLE 2.2 (continued)
Ethnographic Data on Various Species of *Ficus*

Ficus benjamina L.

		References
Genus	*Ficus* section: *Conosycea*	*GRIN* 1994
Synonyms	*F. benjamina* var. *benjamina*	*GRIN* 1994, *MMPND*
	F. benjamina var. *comosa* (Roxb.) Kurz	1995, Hanelt et al.
	F. benjamina var. *nuda*	2001, Ghafoor 1985,
	F. benyamina L.	Wunderlin 1997
	F. comosa Roxb.	
	F. cotonaeifolia Vahl	
	F. neglecta Decne.	
	F. nitida auct.	
	F. nitida Thunb.	
	F. nuda (Miq.) Miq.	
	F. pendula Link	
	F. pyrifolia Salisb.	
	F. retusa auct.	
	F. retusa var. *nitida* (Thunb.) Miq.	
	F. schlechteri auct.	
	F. waringiana auct.	
	Urestigma benjamina (L.) Miq.	
	U. haemotocarpum Miq.	
	U. nudum Miq.	
Common Names	English: Benjamin tree, Golden fig, Java fig, Java fig tree, Java tree, Malayan banyan, Oval-leaf fig tree, Tropic laurel, Weeping Chinese banyan, Weeping fig, Weeping laurel	*MMPND* 1995, *GRIN* 1994
	Danish: Birkefigen	
	German: Benjamin-Feige, Benjamin-Gummibaum, Birkenfeige	
	Dutch: Wariengien (Dutch Indies), Waringin	
	French: Figuier de Indes	
	Spanish: Árbol benjamín, Benjamina, Ficus benyamina, Matapalo	
	Portuguese (Brazil): Beringan, Ficus-banjamina, Figueira-banjamina	
	Serbian: Bendžamin, Fikus benjamina	
	Arabic: Fikus banjamina, Tin banyamin	
	Sanskrit: Mandara	
	Hindi: Sunonija	
	Tamil: Vellal	
	Burmese: Kyet kadut, Nyaung lun, Nyaung thabye	
	Thai: Sai yoi, Sai yoi bai laem	
	Vietnamese: Cây sanh, Sanh	

(continued on next page)

TABLE 2.2 (continued)
Ethnographic Data on Various Species of *Ficus*

Ficus benjamina L.		References
Common names (continued)	Malay: Beringin (Indonesia), Mendera, Waringin (Java, Sumatra) Sundanese: Caringin Nepalese: Conkar, Kabra, Samii, Svaamii Chinese: Bai rong, Chui ye rong Japanese: Shidare gajumaru	
Habitat and distribution	Moist mixed forests, disturbed thickets and hammocks, altitude range 400–800 m. Nepal, N. India, Bangladesh, Burma, S. China, Malaysia to the Solomon Islands and N. tropical Australia. China, Taiwan, Bhutan, India, Bangladesh, Nepal, Cambodia, Laos, Myanmar, Thailand, Vietnam, Indonesia, Malaysia, Papua New Guinea, Philippines, Australia, Solomon Islands. Cultivated also in Pakistan, Sri Lanka, the United States (Florida), West Indies (Lesser Antilles), Malacca, Zambia	*GRIN* 1994, Berg 1991, Hanelt et al. 2001, Ghafoor 1985, Wunderlin 1997, Wu et al. 2003, Burkill 1985
Economic importance	Twigs: food for lac insects Latex: rubber Bark: tanning; gums; resins; fiber Tree: shade trees; ornamental	Hanelt et al. 2001, Burkill 1985, *ALUKA* 2003
Local medicinal uses	Parts used: leaf, bark Local uses: anodyne, arthritis, rheumatism, liver medicine Nonspecific use in India mentioned.	*EthnobotDB* 1996, Duke 1972, Burkill 1985

Ficus capensis Thunb.		
Genus	*Ficus* section: *Sycomorus*	*GRIN* 1994
Synonyms	*F. brachypus* Warb. *F. brassii* R. Br. *F. capensis* var. *guineensis* Miq. *F. capensis* var. *pubescens* Warb. *F. capensis* var. *trichoneura* Warb. *F. caulocarpa* Warb. *F. erubescens* Warb. *F. grandicarpa* Warb. *F. guineensis* Stapf *F. kiboschensis* Warb. *F. kwaiensis* Warb. *F. lichtensteinii* Link *F. mallotocarpa* Warb. *F. matabelæ* Warb.	*GRIN* 1994, *MMPND* 1995, Burkill 1985, Hutchinson 1925, *ALUKA* 2003

TABLE 2.2 (continued)
Ethnographic Data on Various Species of *Ficus*

Ficus capensis Thunb.		References
Synonyms (continued)	*F. munsæ* Warb.	
	F. oblongicarpa Warb.	
	F. plateiocarpa Warb.	
	F. sarcipes Warb.	
	F. sericeogemma Warb.	
	F. simbilensis Warb.	
	F. stellulata var. *glabrescens* Warb.	
	F. sur Forssk.	
	F. sycomorus var. *alnea* Hiern	
	F. sycomorus var. *polybotrya* Hiern	
	F. sycomorus var. *prodigiosa* Welw. ex Hiern	
	F. thonningiana Miq.	
	F. umbonigera Warb.	
	F. villosipes Warb.	
	S. capensis Miq.	
	S. guineensis Miq.	
	Sycomorus thonningiana Miq.	
Common names	English: Broom cluster fig, Bush fig, Cape fig, Fig, Fig of heaven, Fire sticks, Heritage tree, Kooman, Wild fig	*MMPND* 1995, *GRIN* 1994, Burkill 1985
	German: Kap-Feige	
	French: Petit sycomore	
	Badyara [Senegal]: Ibé	
	Balanta [Senegal]: Brozi, Bruzi	
	Banyun [Senegal]: Dièntèd, Ki bundal, Si bundèl	
	Basari [Senegal]: A-nak, A-nak o gva	
	Bedik [Senegal]. Ba-nák, Ga-nák	
	Crioulo [Senegal]: Défay	
	Diola [Senegal]: Bu rirun, Ku bun	
	Swahili: Mkuyu	
	Kikuyu: Mu-kuyu	
	Hausa: Uwar yara	
Habitat and distribution	Widespread at most altitudes in wooded grassland, by water courses and evergreen forest (secondary jungle) 0–2300 m, especially in the more open country. Bioko, Príncipe, São Tomé, widely distributed in Sub-Saharan Africa, from Cape Verde to the Yemen and south to Angola and S. Africa (Cape Province). Cape Verde, Eritrea, Ethiopia, Cameroon, Zaire, Côte D'Ivoire, Gambia, Ghana, Guinea, Guinea-Bissau, Liberia, Nigeria, Senegal, Sierra Leone, Togo, Angola, Malawi, Mozambique, Zambia, Zimbabwe, South Africa, Swaziland, Yemen, California	*GRIN* 1994, Berg 1991, Burkill 1985, Hutchinson 1925, Hutchinson and Dalziel 1958, *ALUKA* 2003

(continued on next page)

TABLE 2.2 (continued)

Ethnographic Data on Various Species of *Ficus*

Ficus capensis Thunb.		References
Economic importance	Fruit: food; masticatory; agrihorticulture: fodder for wild animals, birds; pastimes: carving; musical instruments; games; toys; etc.	Burkill 1985
	Leaf: food; food for special diets	
	Bark: tannins, astringents; dyes, stains, inks, tattoos, and mordants; fiber; food; masticatory; food for special diets	
	Root, aerial: food	
	Wood, poles: building materials	
	Wood, sticks: fuel and lighting	
	General: agrihorticulture: indicators (soil, water); ornamental: cultivated or partially tended; carpentry and related applications; farming, forestry, hunting and fishing apparatus; household, domestic, and personal items; social: religion, superstitions, magic	
Local medicinal uses	Parts used: fruit, seed, leaf, leaf (young), bark, bark sap, root	*EthnobotDB* 1996, Duke 1972, Ayensu 1978, Burkill 1985
	Local uses: generally healing, painkiller, debility, malnutrition, nasopharyngeal afflictions, convulsions, spasm, epilepsy, paralysis, dropsy, edema, swellings, leprosy, parasitic infection (cutaneous, subcutaneous), gout, lumbago, yaws, stiffness, eye problems, headache, bronchitis, pulmonary troubles, lactation stimulants (incl. veterinary), food (special diets), rickets, kidney problems, diarrhea, dysentery, laxative, diuretic, abortifacients, antiabortifacients, aphrodisiac, sterility, pregnancy, menstrual cycle, genital stimulants/depressants, ecbolics, gonorrhea, venereal diseases, febrifuges, food poisoning, antidotes	

FIGURE 2.6 *Ficus carica* growing out of a cliff (29.7.2009, Hebrew University Botanical Garden, Mount Scopus, Jerusalem, Israel, by Krina Brandt-Doekes).

FIGURE 2.7 A well-grown *Ficus carica* (27.7.2009, Seven Species Garden, Binyamina, Israel, by Helena Paavilainen).

TABLE 2.2 (continued)
Ethnographic Data on Various Species of *Ficus*

Ficus carica L.

		References
Genus	*Ficus* section: *Ficus*	*GRIN* 1994
Synonyms	*Caprificus insectifera* Gasp.	*GRIN* 1994, *MMPND*
	Ficus caprificus Risso	1995, Ghafoor 1985
	F. carica subsp. *carica*	
	F. carica subsp. *rupestris* (Hausskn. ex Boiss.) Browicz	
	F. carica var. *caprificus* (Risso) Tschirch & Ravasini	
	F. colchica Grossh.	
	F. communis Lam.	
	F. hyrcana Grossh.	
	F. kopetdagensis Pachom.	
	F. latifolia Salisb.	
	F. leucocarpa Gasp.	
	F. macrocarpa Gasp.	
	F. pachycarpa Gasp.	
	F. praecox Gasp. ex Guss.	
	F. sativa Poit. & Torpin	
Common names	English: Caprifig, Common fig, Cultivated fig, Edible fig, Fig, Fig tree, Wild fig, Wild Mediterranean fig	*MMPND* 1995, *GRIN* 1994, Hanelt et al. 2001, Wu et al. 2003, *EthnobotDB* 1996, Burkill 1985
	Dutch: Gewone vijgeboom, Vijg, Vijgeboom	
	German: Echte Feige, Echter Feigenbaum, Essfeige, Feige, Feigenbaum	
	Danish: Almindelig figen	
	Swedish: Fikon, Fikonträd	
	French: Carique, Figue, Figue commune, Figuier, Figuier commun	
	Spanish: Higo, Higuera, Higuera común	
	Italian: Fico, Fico comune	
	Portuguese: Behereira, Caprifigos (Brazil), Figo, Figueira, Figueira-comum, Figueira-da-europa, Figueira-do-reino	
	Russian: Figovoje derevo, Inžir	
	Serbian: Glušica, Smoka, Smokovnica, Smokva, Smokva-gluhača	
	Croatian: Divja smokva, Divja smokva glušica, Gluha smokva, Glušica, Smokva, Smokva-gluhača, Smokvenica, Smokvina	
	Slovenian: Figa, Figovec, Figovina, Smokvovec	
	Bosnian: Glušac	
	Macedonian: Smoka	
	Finnish: Viikuna	

TABLE 2.2 (continued)
Ethnographic Data on Various Species of *Ficus*

Ficus carica L.		References
Common names (continued)	Arabic: Bakur (Mali), Barshumi fikus, Tin barshomi Persian: Anjeer, Anjir Urdu: Anjeer Sanskrit: Anjeer, Anjir, Anjira, Kakodomar, Phalgu Hindi: Anjeer, Anjir Kannada: Anjura, Simeyam Marathi: Anjir Gujarati: Anjir Tamil: Cimaiyatti, Simaiyatti Burmese: Thaphan, Thinbaw thapan Vietnamese: Quả và, Vô hoa quả (transliteration of Chinese) Nepalese: Anjiir Chinese: Wu hua guo Japanese: Ichijiku Korean: Mu hwa gwa, Mu hwa gwa na mu	
Habitat and distribution	Disturbed sites, 0–300 m. Among rocks, in woods and scrub on hot dry soils. Cultivated and subspontaneous in India, Pakistan, Afghanistan, Russia, Iran, Middle East, N. Africa, and Europe; introduced in cultivation elsewhere. Algeria, Morocco, Tunisia, Afghanistan, Cyprus, Egypt—Sinai, Iran, Iraq, Israel, Jordan, Lebanon, Syria, Turkey, Azerbaijan, Tajikistan, Turkmenistan, Pakistan, Russia, Greece, Crete, Italy, Sardinia, Sicily, France, Corsica, Spain—Baleares, Portugal, India, China, South Asia, Somalia, South Africa, Australia, New Zealand, United States, West Indies, Mexico, Argentina, Brazil, Macaronesia, Galapagos	*GRIN* 1994, Berg 1991, Hanelt et al. 2001, Ghafoor 1985, Wunderlin 1997, Wu et al. 2003, *Plants for A Future* 1996, Friis 1999
Economic importance	Fruit: food; dry figs; fresh figs for local consumption; fig preparations, preserves, etc.; coffee substitute; for flavoring liqueurs and tobacco; alcohol (vine and brandy); beverage base; for fodder; folk medicine Leaf: for fodder Latex: curdling agent; coagulates plant milks Wood: for hoops, garlands, ornaments, etc.; as a hone (saturated with oil and covered with emery) Tree: agrihorticulture: ornamental, cultivated or partially tended; social: religion, superstitions, magic All parts: folk medicine	Hanelt et al. 2001, *GRIN* 1994, Wunderlin 1997, Wu et al. 2003, *Plants for A Future* 1996, Burkill 1985

(continued on next page)

TABLE 2.2 (continued)
Ethnographic Data on Various Species of *Ficus*

Ficus carica L.

		References
Local medicinal uses	Parts used: fruit, unripe green fruit, syrup of figs, leaf, young branches, latex from stems	*Plants for A Future* 1996, *EthnobotDB*
	Local uses: generally healing, analgesic, tonic, restorative, demulcent, emollient, disinfectant, fumigant, aperient, cancer, tumor, tumor (abdomen), tumor (uterus), scrofula, pimples, abscesses, corns, warts, dental abscesses, gumboils, stomatitis, gingivitis, galactogogue, sore throat, asthma, cough, pertussis, pectoral, expectorant, flu, digestive, stomachic, laxative, diuretic, piles, measles, insect stings, insect bites, vermifuge, rennet.	1996, Duke 1993, Burkill 1985, Brutus and Pierce-Noel 1960, Liogier 1974, *Anon.* 1978, Hartwell 1967-1971, Pittier 1926, Steinmetz 1957, Al-Rawi 1964, Font Query 1979, Latorre and Latorre 1977
Toxicity	Sap and the half-ripe fruits said to be poisonous. Sap can be a serious eye irritant.	*Plants for A Future* 1996, Riotte 1978, Polunin and Huxley 1987, Bown 1995

Ficus chlamydocarpa Mildbr. & Burret

Genus	*Ficus* section: *Galoglychia*	Van Noort and Rasplus 2004
Synonyms	*Ficus clarencensis*	Hutchinson and Dalziel 1958
Common names	—	
Habitat and distribution	In montane forest. Cameroon, Bioko (= Fernando Po), Annobon, São Tomé, Príncipe	Van Noort and Rasplus 2004, Hutchinson and Dalziel 1958, Hutchinson 1917, *ALUKA* 2003
Economic importance	—	
Local medicinal uses	—	

Ficus citrifolia Mill.

Genus	*Ficus* section: *Americana*	*GRIN* 1994
Synonyms	*Ficus brevifolia* Nutt.	*GRIN* 1994, *MMPND*
	F. citrifolia Mill. var. *brevifolia* (Nutt.) D'Arcy	1995, Wunderlin 1997
	F. gentlei Lundell	
	F. gigantea Kunth	
	F. hemsleyana Standl.	
	F. laevigata Vahl	
	F. populnea Willd.	
	F. laevigata Vahl. var. *brevifolia* (Nutt.) D'Arcy	

TABLE 2.2 (continued)
Ethnographic Data on Various Species of *Ficus*

Ficus citrifolia Mill.		References
Synonyms (continued)	*F. laevigata* Vahl. var. *brevifolia* (Nutt.) Warb. ex Rossberg	
	F. laevigata Vahl. var. *lentiginosa* (Vahl) Urb.	
	F. lentiginosa Griseb.	
	F. lentiginosa Vahl	
	F. pedunculata Willd.	
	F. populifolia Desf.	
	F. populnea var. *brevifolia* (Nutt.) Warb.	
	F. populnea Willd.	
	F. populnea Willd. var. *bahamensis* Warb.	
	F. populnea Willd. var. *brevifolia* (Nutt.) Warb.	
	F. populnea Willd. subvar. *floridana* Warb.	
	F. populoides Warb.	
Common names	English: Barbados strangler fig, Barrenfig, Bearded fig (Barbados), Bearded fig tree, Florida strangler fig, Giant bearded fig, Guadalupe ficus, Short leaf fig (Bahamas), Shortleaf wild fig, Short-leaved wild fig (Bahamas), White fig, Wild banyan, Wild banyan tree, Wild fig	*MMPND* 1995, Wunderlin 1997
	French: Aralie-cerise, Arali siriz, Figuier (Guadeloupe), Figuier barbu, Figuier blanc, Figuier maudit (Guadeloupe), Figyé blan, Figyé modi (Martinique)	
	Spanish: Higuillo (Cuba, Dominican Republic), Jagueillo (Puerto Rico), Jagüerillo (Cuba), Jagüey (Cuba, Puerto Rico), Jagüey blanco, Jagüey macho (Puerto Rico), Jigüeillo, Jigüerillo, Jugüeillo (Puerto Rico)	
Habitat and distribution	In tropical hammocks.	*GRIN* 1994, Wunderlin 1997
	Florida, Mexico, West Indies, Central America, South America.	
	United States (Florida), Mexico, Belize, Costa Rica, El Salvador, Guatemala, Honduras, Nicaragua, Panama, Anguilla, Antigua and Barbuda, Bahamas, Barbados, Cuba, Dominica, Dominican Republic, Grenada, Guadeloupe, Haiti, Jamaica, Martinique, Montserrat, Netherlands Antilles, Puerto Rico, St. Kitts and Nevis, St. Lucia, St. Vincent and Grenadines, Trinidad and Tobago, Virgin Islands (British), Virgin Islands (U.S.), French Guiana, Guyana, Suriname, Venezuela, Brazil, Bolivia, Colombia, Ecuador, Peru, Argentina, Paraguay	

(continued on next page)

TABLE 2.2 (continued)
Ethnographic Data on Various Species of *Ficus*

Ficus citrifolia Mill.		References
Economic importance	Tree: agrihorticulture: ornamental, shade tree	*GRIN* 1994, Wunderlin 1997
Local medicinal uses	Local uses: infection, wounds, masticatory, vermifuge	*EthnobotDB* 1996, Duke 1993, 1975, 1972

Ficus cordata Thunb.		
Genus	*Ficus* section: *Urostigma*	*GRIN* 1994
Synonyms	*Ficus atrovirens* hort. Berol. ex Mildbr. & Burret	*GRIN* 1994, *MMPND*
	F. cordata subsp. *lecardii*	1995, Berg 1991,
	F. cordata subsp. *salicifolia*	Hutchinson 1925,
	F. cordata Thunb. subsp. *salicifolia* (Vahl) C. C. Berg	*ALUKA* 2003
	F. cordata var. *marlothii* Warb.	
	F. glaucophylla Desf.	
	F. lecardii Warb.	
	F. pretoriae Burtt Davy	
	F. salicifolia Vahl	
	F. salicifolia Vahl var. *australis* Warb.	
	F. teloukat Batt. & Trab.	
	F. tristis Kunth et Bouché	
	Urostigma cordatum (Thunb.) Gasp.	
	Urostigma thunbergii Miq.	
Common names	*Ficus cordata* Thunb.:	*MMPND* 1995
	English: Namaqua fig, Namaqua fig tree	
	German: Wunderbaum-Feige	
	Ficus cordata Thunb. subsp. *Salicifolia* (Vahl)	
	C. C. Berg:	
	English: African fig, Wonderboom fig (South Africa)	
	German: Wunderbaum-Feige	
	Afrikaans: Wonderboomvy	
Habitat and distribution	Arid habitats. Almost always growing on rock outcrops or cliff faces, at altitudes up to 1500 m.	*GRIN* 1994, Van Noort and Rasplus 2004, Hutchinson 1925
	Southwest South Africa, through Namibia to south western Angola.	
	Algeria, Egypt, Libya, Chad, Eritrea, Ethiopia, Somalia, Sudan, Kenya, Tanzania, Uganda, Cameroon, Central African Republic, Zaire, Cote D'Ivoire, Guinea, Guinea-Bissau, Mali, Nigeria, Senegal, Angola, Malawi, Mozambique, Zambia, Zimbabwe, Botswana, Namibia, Oman, Saudi Arabia, Yemen	

TABLE 2.2 (continued)
Ethnographic Data on Various Species of *Ficus*

Ficus cordata Thunb.		References
Economic importance	—	
Local medicinal uses	—	

Ficus cunia Buch.-Ham. ex Roxb.		
Genus	*Ficus* section: *Hemicardia*	Van Noort and Rasplus 2004
Synonyms	*Covellia conglomerata* Miq. *C. cunia* (Buch.-Ham. ex Roxb) Miq. *C. inaequiloba* Miq. *Ficus conglomerata* Roxb. *F. cunia* Ham. ex Roxb. *F. semicordata* Buch.-Ham. ex Sm. *Tremotis cordata* Raf.	*MMPND* 1995, Hanelt et al. 2001, Ghafoor 1985
Common names	Arabic: tin barri Sanskrit: kharapatra Hindi: khewnau Thailand: manod nam	Hanelt et al. 2001, *MMPND* 1995, *EthnobotDB* 1996
Habitat and distribution	India, Myanmar, Thailand, Malacca	Hanelt et al. 2001
Economic importance	Leaf: fodder Bark: bast fibers; medicinal use Tree: host tree for lac insects (*Tachardia lacca*)	Hanelt et al. 2001
Local medicinal uses	Parts used: bark Local uses: aphtha, bladder problems	Hanelt et al. 2001, *EthnobotDB* 1996, Duke 1993, Al-Rawi 1964, *Anon.* 1948–1976

FIGURE 2.8   (See color insert after page 128.) *Ficus deltoidea* branchlet (19.7.2009, University of Turku, Botanical Garden, Ruissalo, Turku, Finland, by Kaarina Paavilainen).

FIGURE 2.9   (See color insert after page 128.) A small *Ficus deltoidea* (19.7.2009, University of Turku, Botanical Garden, Ruissalo, Turku, Finland, by Kaarina Paavilainen).

TABLE 2.2 (continued)
Ethnographic Data on Various Species of *Ficus*

Ficus deltoidea Jack		References
Genus	*Ficus* section: *Erythrogyne*	*GRIN* 1994
Synonyms	*F. deltoidea* subsp. *deltoidea*	*GRIN* 1994,
	F. deltoidea subsp. *motleyana*	Ghafoor 1985
	F. deltoidea var. *lutescens* (Desf.) Corner	
	F. diversifolia Blume	
	F. lutescens Desf.	
	F. motleyana Miq.	
	F. ovoidea Jack.	
Common names	English: Delta fig, Fig shrub, Mistletoe fig, Rusty leaved bush fig	*GRIN* 1994, Ghafoor 1985,
	Dutch: Mistelvijg	*ZipcodeZoo* 2004
Habitat and distribution	A native of Malayan Archipelago, introduced elsewhere. Thailand, Indonesia, Malaysia	*GRIN* 1994, Ghafoor 1985
Economic importance	Tree: agrihorticulture: ornamental	*GRIN* 1994
Local medicinal uses	—	

FIGURE 2.10 *Ficus elastica* up close (27.7.2009, Binyamina, Israel, by Helena Paavilainen).

FIGURE 2.11 *Ficus elastica* on the street. (27.7.2009, Binyamina, Israel, by Helena Paavilainen).

TABLE 2.2 (continued)
Ethnographic Data on Various Species of *Ficus*

Ficus elastica Roxb. ex Hornem.

		References
Genus	*Ficus* section: *Conosycea*	*GRIN* 1994
Synonyms	*Ficus cordata* Kunth & Bouch.	*MMPND* 1995, *GRIN*
	F. decora hort.	1994, Hanelt et al.
	F. skytinodermis Summerh.	2001, Wunderlin 1997,
	F. taeda Kunth & Bouch.	Wu et al. 2003
	Macrophthalma elastica (Roxb. ex Hornem.) Gasp.	
	Urostigma circumscissum Miq.	
	U. elasticum (Roxb. ex Hornem.) Miq.	
	U. karet Miq.	
	U. odoratum Miq.	
	Visiania elastica (Roxb. ex Hornem.) Gasp.	
Common names	English: India rubber fig, India rubber plant, Indian rubber tree, Indian rubberplant, Indian rubbertree, Karet tree, Ornamental rubber tree, Rubber plant, Rubberplant	*MMPND* 1995, *GRIN* 1994, Hanelt et al. 2001, Ghafoor 1985, Wunderlin 1997, Wu
	German: Gummibaum	et al. 2003, Burkill
	Danish: Gummifigen	1985

(continued on next page)

TABLE 2.2 (continued)
Ethnographic Data on Various Species of *Ficus*

Ficus elastica Roxb. ex Hornem.   References

**Common
names
(continued)**
	French: Arbre á caoutchouc, Caoutchouc,
	  Caoutchoutier, Ficus, Figuier élastique, Gommier
	Spanish: Amate (Salvador), Árbol del caucho, Cauchú
	  comun, Goma de la India (Cuba), Goma
	  elastica (Cuba), Gomero, Gomero (Argentina),
	  Higuera cauchera, Higuera de la India, (Cuba,
	  Mexico), Higuiera cauchera, Planta del caucho
	Portuguese: Árvore da borracha (Brazil), Figueira-da-
	  borracha, Figueira-indiana, Seringueira, Seringueira-
	  de-jardim
	Serbian: Fikus, Fikus zmajevac, Gumijevac, Tropska
	  smokva
	Croatian: Fikus, Gumijevac
	Slovenian: Gumovec
	Sanskrit: Vat
	Hindi: Attaabor, Atta bor, Bargad, Bor
	Kannada: Rabra chovad
	Bengali: Bor
	Assamese: Athabor, Attah
	Tamil: Cimaiyal
	Burmese: Ganoi, Kanoi, Nyaung kyetpaung
	Thai: Yang india
	Vietnamese: Đa búp đỏ
	Chinese: Jin du jiao shu, Yin du rong
	Japanese: Indo gomu no ki
	Korean: In do go mu na mu	

**Habitat and
distribution**
	In disturbed sites.
	India, Nepal, Myanmar, Malacca to Sumatra, Java.
	  Introduced in Sri Lanka.
	    Bhutan, India, Sri Lanka, Nepal, Myanmar, Burma,
	    Indonesia, Java, Malaysia, China, Western Africa,
	    Florida, West Indies (Antilles). Widely cultivated.
		GRIN 1994, Berg 1991,
		Hanelt et al. 2001,
		Ghafoor 1985,
		Wunderlin 1997, Wu
		et al. 2003

**Economic
importance**
	Leaf-bud: vegetable
	Latex: Assam caoutchouc (rubber)
	Bark: exudations—gums, resins, etc.
	Tree: agrihorticulture: ornamental, cultivated, or
	  partially tended in all tropical regions; shade trees;
	  decorative house plant; host plant for lac insects for
	  the shellac production; shellac
		Hanelt et al. 2001, *GRIN*
		1994, Ghafoor 1985,
		Wu et al. 2003, Burkill
		1985

Local medicinal uses	—	

Ficus erecta Thunb.

Genus	*Ficus* section: *Erythrogyne*	*GRIN* 1994

TABLE 2.2 (continued)
Ethnographic Data on Various Species of *Ficus*

Ficus erecta Thunb.

		References
Synonyms	*Ficus beecheyana* Hook. & Arn.	*GRIN* 1994, *MMPND*
	F. beecheyana f. *koshunensis* (Hayata) Sata	1995, Wu et al. 2003
	F. beecheyana f. *tenuifolia* Sata	
	F. beecheyana var. *koshunensis* (Hayata) Sata	
	F. erecta f. *sieboldii* (Miq.) Corner	
	F. erecta var. *beecheyana* (Hook. & Arn.) King	
	F. erecta var. *beecheyana* f. *koshunensis* (Hayata) Corner	
	F. erecta var. *sieboldii* (Miq.) King	
	F. japonica Blume	
	F. koshunensis Hayata	
	F. maruyamensis Hayata	
	F. sieboldii Miq.	
	F. tenax Blume.	
Common names	Chinese: Ai xiao tian xian guo	*MMPND* 1995, *GRIN*
	Korean: Cheon seon gwa, Cheon seon gwa na mu, Gin kkok ji cheon seon gwa, Kkok ji cheon seon gwa	1994, Wu et al. 2003
Habitat and distribution	Forests, along streams.	*GRIN* 1994, Wu et al.
	China, Japan, South Korea, Taiwan, Vietnam	2003
Economic importance	Bark: fiber; paper	*GRIN* 1994, Wu et al.
	Tree: agrihorticulture: ornamental	2003
Local medicinal uses	—	

Ficus exasperata Vahl

Genus	*Ficus* section: *Sycidium*	*GRIN* 1994
Synonyms	*Ficus asperrima* Roxb.	*GRIN* 1994, *MMPND*
	F. punctifera Warb.	1995, Berg 1991
	F. scabra Willd.	
	F. silicea Sim	
Common names	English: Sandpaper (leaf) tree, white fig tree (The Gambia)	*MMPND* 1995, Burkill 1985
	French: Papier de verre	
	Swahili: Msasa	
	Banyun (Senegal): Ki ink, Si ès, Si vès	
	Bedik (Senegal): Ga-nyè	
	Crioulo (Senegal): Karda	
	Diola (Senegal): Bu pundun, Bu sas, Bu sasa, Bu ves, Bu yaya, Hu sas, Ñiña	
	Sanskrit: Kharapatra	
	Malayalam: Teregam, Therakam	
	Tamil: Irambarathan, Irambarattam, Maramthinni atthi	

(continued on next page)

TABLE 2.2 (continued)
Ethnographic Data on Various Species of *Ficus*

Ficus exasperata Vahl		References
Habitat and distribution	In evergreen forests and forest margins, also in secondary forest, and riverine vegetation, often as a strangler, sometimes persisting in cleared places. Altitude range: 0–2000 m.	*GRIN* 1994, Berg 1991, Van Noort and Rasplus 2004, Burkill 1985, Hutchinson and Dalziel 1958, *ALUKA* 2003
	Widespread in tropical Africa, from Mozambique, Zambia, and northern Angola to Senegal and Ethiopia. Also in the southern part of the Arabian Peninsula and India.	
	Djibouti, Ethiopia, Sudan, Kenya, Tanzania, Uganda, Cameroon, Equatorial Guinea, Bioko, São Tomé and Príncipe, Cote D'Ivoire, Gambia, Ghana, Guinea, Guinea-Bissau, Liberia, Mali, Nigeria, Senegal, Sierra Leone, Togo, Angola, Malawi, Mozambique, Zambia, Zimbabwe, Comoros, Madagascar, Yemen, India, Sri Lanka	
Economic importance	Fruit: food; agrihorticulture: fodder for pigeons	*GRIN* 1994, Burkill 1985
	Leaf: agrihorticulture: fodder for elephant; abrasives, cleaners, etc.; carpentry and related applications; household, domestic and personal items; arrow poisons; medicinal use	
	Sap: medicinal use; poisonous/repellent	
	Bark: exudations—gums, resins, etc.; medicinal use	
	Root: medicinal use	
	Tree: agrihorticulture: ornamental, cultivated or partially tended; shade trees	
	Wood ash, charcoal: medicinal use	
Local medicinal uses	Parts used: fruit, leaf, sap, bark, root	*GRIN* 1994, Burkill 1985
	Local uses: generally healing, painkillers; malnutrition, debility, skin, mucosae, parasitic infection (cutaneous, subcutaneous), leprosy, eye treatments, oral treatments, nasopharyngeal afflictions, arteries, veins, arthritis, rheumatism; gout, dropsy, swellings, edema, kidneys, diuretics, vermifuges, stomach troubles, diarrhea, dysentery, anus, hemorrhoids, abortifacients, ecbolics; venereal diseases, febrifuges, arrow poisons, miscellaneously poisonous or repellent	

Ficus fistulosa Reinw. ex Blume		
Genus	*Ficus* section: *Sycocarpus*	Van Noort and Rasplus 2004
Synonyms	*Ficus harlandii* Bentham.	Wu et al. 2003

TABLE 2.2 (continued)
Ethnographic Data on Various Species of *Ficus*

Ficus fistulosa Reinw. ex Blume		References
Common names	Chinese: Shui tong mu	Wu et al. 2003
Habitat and distribution	Forests, along streams, on rocks; altitude range 200–600 m. China, Taiwan, Bangladesh, NE India, Indonesia, Malaysia, Myanmar, Philippines, Thailand, Vietnam.	Wu et al. 2003
Economic importance	Medicinal uses	*EthnobotDB* 1996, Duke 1993
Local medicinal uses	Nonspecific use in India mentioned.	*EthnobotDB* 1996, Duke 1993

Ficus glabrata V. Kunth [= *Ficus insipida* Willd.]		
Genus	*Ficus* section: *Pharmacosycea*	*GRIN* 1994
Synonyms	*Ficus anthelminthica* Mart. non Hassler	*MMPND* 1995, *GRIN* 1994
	F. crassa Klotzsch & H. Karst. ex Dugand	
	F. insipida Willd.	
	F. insipida subsp. *insipida*	
	F. insipida subsp. *radulina*	
	F. krugiana Warb.	
	F. laurifolia Duss non Lam.	
	F. mexicana (Miq.) Miq.	
	F. radulina S. Watson	
	F. segoviae Miq.	
Common names	English: Gouti fig, Red fig (Belize)	*MMPND* 1995, *GRIN* 1994
	French: Figuier agouti, Figuier blanc, Figuier maudit, Figuier médicinal, Figyé agouti (Martinique), Figyé modi (Martinique)	
	Spanish: Amate (Mexico), Amate de hijo grande (Colombia), Amete de montaña (Honduras), Bibosi (Bolivia), Chilamate (Costa Rica, Nicaragua), Cocoba (Bolivia), Doctor ojé (Peru), Gomelero (Bolivia), Higo (Honduras, Nicaragua), Higuero (Honduras), Higuerón (Colombia, Nicaragua, Venezuela), Higuerón de rio (Ecuador), Leche de oje, Ojé (Bolivia, Peru), Siranda (Mexico)	
Habitat and distribution	Mexico, Belize, Costa Rica, El Salvador, Guatemala, Honduras, Nicaragua, Panama, French Guiana, Guyana, Suriname, Venezuela, Brazil, Bolivia, Colombia, Ecuador, Peru	*GRIN* 1994
Economic importance	Medicinal use	*EthnobotDB* 1996, Duke 1993

(continued on next page)

TABLE 2.2 (continued)
Ethnographic Data on Various Species of *Ficus*

Ficus glabrata V. Kunth [= *Ficus insipida* Willd.]		**References**
Local medicinal uses	Local uses: parasiticide, vermifuge	*GRIN* 1994, *EthnobotDB* 1996, Duke 1993, Martinez 1969, Pittier 1926, Standley and Steyermark 1952

Ficus glomerata Roxb. [= *Ficus racemosa* L.]		
Genus	*Ficus* section: *Sycomorus*	Van Noort and Rasplus 2004
Synonyms	*Covellia glomerata* (Roxb.) Miq.	Ghafoor 1985, *GRIN* 1994, *MMPND* 1995, Hanelt et al. 2001
	Ficus goolereea Roxb.	
	F. mollis Miq.	
	F. racemosa L.	
	F. vesca F. Muell. ex Miq.	
Common names	English: Cluster fig, Cluster tree, Country fig, Gular fig, Redwood fig	*MMPND* 1995, Hanelt et al. 2001
	Urdu: Dimiri	
	Sanskrit: Gular, Hemadugdhaka, Jantuphala, Sadaphalah, Udumbar, Udumbara, Udumbarah, Yajnanga	
	Hindi: Ambar (Bombay), Domoor, Doomar, Gular, Udumbara, Umar, Umbar	
	Kannada: Alhi, Atthimara, Atti	
	Bengali: Dumur, Hpak-lu, Jagyadumbar, Mayen, Taung tha phan, Thapan, Ye thapan	
	Marathi: Audumbar, Umbar	
	Gujarati: Gular, Umardo	
	Oriya: Dimri	
	Telugu: Arri, Athi, Bodda, Maydi, Paidi, Udumbaramu	
	Malayalam: Athi (Kerala), Athiathial, Atthi	
	Sinhalese: Attikka	
	Tamil: Anai, Athi, Attee marum, Atthi, Atti, Malaiyin munivan, Utumparam, Vellaiatthi	
	Burmese: Hpak-lu, Jagyadumbar, Mayen, Taung tha phan, Thapan, Ye thapan	
	Thai: Duea kliang, Duea nam, Ma duea, Ma duea chumphon, Ma duea utum phon (Central Thailand)	
	Vietnamese: Cây sung, Sung	
	Laotian: Kok dua kieng	
	Malay: Ara, Elo (Indonesia)	
	Nepalese: Dumrii	
	Chinese: Ju guo rong	

TABLE 2.2 (continued)
Ethnographic Data on Various Species of *Ficus*

Ficus glomerata Roxb. [= *Ficus racemosa* L.]		References
Habitat and distribution	India, Thailand, Southern Myanmar, Indochina, Southern China, Southern Himalaya, Northern Australia, South America	Hanelt et al. 2001
Economic importance	Fruit: food Leaf: vegetable; fodder; soil cover Latex: waterproof paper; plasticizer for *Hevea*-rubber; rubber Tree: agrihorticulture: shade tree in coffee plantations; host plant for lac insects; lac; as rootstock for *Ficus carica*; social: religious, one of the holy trees of the Hindus	Hanelt et al. 2001
Local medicinal uses	Local uses: astringent, bilious, fumitory, wounds, cancer, toothache, backache, expectorant, anorectic, carminative, stomachic, diarrhea, dysentery, cholera, hemorrhoids, antidote (*Datura*)	*EthnobotDB* 1996, Duke 1993, Hirschhorn 1983, *Anon*. 1948-1976, Hartwell 1967-1971

Ficus gnaphalocarpa (Miq.) Steud. ex A. Rich.		
Genus	*Ficus* section: *Sycomorus*	Van Noort and Rasplus 2004
Synonyms	*Ficus gnaphalocarpa* (Miq.) A. Rich. *F. gnaphalocarpa* (Miq.) Steud. ex A. Rich. *F. sycomorus* subsp. *gnaphalocarpa* (Miq.) C.C. Berg *F. trachyphylla* (Miq.) Miq. *Sycomorus gnaphalocarpa* Miq. *Sycomorus trachyphylla* Miq.	Hanelt et al. 2001, *GRIN* 1994, *MMPND* 1995, Hutchinson and Dalziel 1958, Friis 1999, *ALUKA* 2003
Common names	Portuguese: figueira brava Nigeria: baure Senegal: iwi Ghana: kilumpi, lombang Sudan: turu gaul	Hanelt et al. 2001
Habitat and distribution	In woodland, by streams in the savannah regions, but more often in hilly country or sometimes open grassland. Altitude range: c. 1100 m. Widespread in the drier parts of tropical Africa, from South Africa and Namibia to Ethiopia and west to Senegal. South Africa (Cape Verde), Namibia, Angola, Nigeria, Senegal, Gambia, Benin, Sudan, Somalia, Ethiopia	Hanelt et al. 2001, Van Noort and Rasplus 2004, Hutchinson and Dalziel 1958, Friis 1999, *ALUKA* 2003
Economic importance	Fruit: food; alcoholic drink Young leaves: food Latex: rubber (guttapercha)	Hanelt et al. 2001

(continued on next page)

TABLE 2.2 (continued)

Ethnographic Data on Various Species of *Ficus*

***Ficus gnaphalocarpa* (Miq.) Steud. ex A. Rich.**   **References**

Economic importance (continued)	Bark: social: chewed with cola nuts (stimulant) Tree: agrihorticulture: rootstock for *F. carica*	
Local medicinal uses	—	

***Ficus hirta* Vahl**

Genus	*Ficus* section: *Eriosycea*	*GRIN* 1994
Synonyms	*Ficus dumosa* King	*GRIN* 1994, Wu et al.
	F. hibiscifolia Champ. ex Benth.	2003
	F. hirsuta Roxb.	
	F. hirta subsp. *dumosa*	
	F. hirta subsp. *roxburghii*	
	F. hirta var. *brevipila* Corner	
	F. hirta var. *dumosa* (King) Corner	
	F. hirta var. *hibiscifolia* (Champ. ex Benth.) Chun	
	F. hirta var. *imberbis* Gagnep.	
	F. hirta var. *palmatiloba* (Merr.) Chun	
	F. hirta var. *roxburghii* (Miq.) King	
	F. katsumadae Hayata	
	F. palmatiloba Merr.	
	F. porteri H. Lév. & Vaniot	
	F. quangtriensis Gagnep.	
	F. roxburghii Miq.	
	F. simplicissima Lour. var. *hirta* (Vahl) Migo	
	F. tridactylites Gagnep.	
	F. triloba Buch.-Ham. ex J. O. Voigt	
Common names	Chinese: Cu ye rong	*GRIN* 1994, Wu et al. 2003
Habitat and distribution	Forests, forest margins, low elevations. China, Bhutan, India, Nepal, Myanmar, Thailand, Vietnam, Indonesia	*GRIN* 1994, Wu et al. 2003
Economic importance	—	
Local medicinal uses	—	

***Ficus hispida* L. f.**

Genus	*Ficus* section: *Sycocarpus*	*GRIN* 1994
Synonyms	*Covellia hispida* (L. f.) Miq.	*MMPND* 1995, Ghafoor
	Ficus compressa S. S. Chang	1985, Wu et al. 2003
	F. daemonum Koenig ex Vahl	
	F. heterostyla Merr.	
	F. hispida var. *badiostrigosa* Corner	

TABLE 2.2 (continued)
Ethnographic Data on Various Species of *Ficus*

Ficus hispida L. f.		References
Synonyms (continued)	*F. hispida* var. *rubra* Corner	
	F. letaqui H. Lév. & Vaniot	
	F. oppositifolia Roxb.	
	F. oppositifolia Willd.	
	F. sambucixylon H. Lév.	
Common names	English: River fig, Rough-leaved fig, Soft fig	*MMPND* 1995, *GRIN*
	German: Fluss-Feige	1994, Ghafoor 1985,
	Sanskrit: Kaakodumbara, Kakodumbarika	Wu et al. 2003
	Hindi: Kala umbar	
	Sinhalese: Kota dimbula	
	Tamil: Peyatti	
	Burmese: Hpauwu, Ka-aung, Kadut, Kha-aung, Mai-nawt-hpu	
	Thai: Ma duea plong	
	Laotian: Kok dua pong	
	Vietnamese: Ngái	
	Chinese: Dui ye rong	
Habitat and distribution	Along streams, plains; altitude range 700–1500 m.	*GRIN* 1994, Ghafoor
	Southeast Asia, Malesia, Papua New Guinea, and Australia	1985, Wu et al. 2003, Van Noort and Rasplus 2004
	China, Bhutan, India, Nepal, Sri Lanka, Laos, Cambodia, Myanmar, Thailand, Vietnam, Indonesia, Malaysia, Papua New Guinea, Australia, Pakistan, Bangladesh, United States	
Economic importance	Fruit: food; medicinal use	*GRIN* 1994, Ghafoor 1985
Local medicinal uses	Local uses: tonic, boils, warts, lactagogue, emetic, stomachache, diarrhea, dysuria, parturition, fever	*GRIN* 1994, Ghafoor 1985, *EthnobotDB* 1996, Duke 1993, Burkill 1966, *Anon.* 1948-1976, Hartwell 1967-1971

Ficus infectoria Roxb. [= *Ficus virens* Aiton]

Genus	*Ficus* section: *Urostigma*	Van Noort and Rasplus 2004
Synonyms	*Ficus ampla* Kunth	Hanelt et al. 2001,
	F. caulobotrya (Miq.) Miq. var. *fraseri* (Miq.) Miq.	*MMPND* 1995,
	F. caulocarpa (Miq.) Miq.	Ghafoor 1985, Wu
	F. glabella Blume	et al. 2003
	F. infectoria auct.	
	F. infectoria auct. non Willd.	
	F. infectoria var. *lambertiana* (Miq.) King	

(continued on next page)

TABLE 2.2 (continued)
Ethnographic Data on Various Species of *Ficus*

Ficus infectoria Roxb. [= *Ficus virens* Aiton]   References

Synonyms (continued)	*F. infectoria* Willd. *F. infectoria* Willd. var. *caulocarpa* (Miq.) King *F. lacor* Buch.-Ham. *F. lucens* Blume *F. saxophila* Blume var. *sublanceolata* Miq. *F. venosa* Wall. *F. virens* Aiton *F. virens* Dryand. *F. virens* var. *sublanceolata* (Miq.) Corner *F. wightiana* Wall. ex Benth. *Urostigma fraseri* Miq. *U. infectorium* Miq. *U. lambartiana* Miq. *U. wightianum* Miq.	
Common names	English: Cuvi white fig, White fig, White-fruited wavy leaf fig tree Urdu: Pakodo Sanskrit: Phagu, Phalgu, Plaksa Hindi: Kahimal, Kahimmal, Kaim, Keol, Pakar, Pakri, Pakur, Pilkahan, Pilkhan, Ramanjir Kannada: Basari, Basarigoli, Juvvimara, Karibasari, Karibasuri, Matai ichchi Bengali: Pakur Marathi: Bassari, Pakari, Pipli Telugu: Badijuvvi, Banda juvvi, Jati, Jatijuvi Malayalam: Bakri, Chakkila, Chela, Cherala, Cherla, Chuvannal, Itti, Jati Tamil: Cuvalai pipal, Itti, Jovi, Kallal, Kurugatti, Kurugu, Matai ichchi, Suvi	*MMPND* 1995, Hanelt et al. 2001, Ghafoor 1985, Wu et al. 2003
Habitat and distribution	By stream sides in subtropical China; 300–2700 m. Tropical South and Southeast Asia: from India to China and Indonesia. China, Bhutan, Cambodia, India, Indonesia, Japan, Laos, Malaysia, Myanmar, New Guinea, Philippines, Sri Lanka, Thailand, Vietnam, Northern Australia	Hanelt et al. 2001, Ghafoor 1985, Wu et al. 2003
Economic importance	Fruits: food Leaf: fodder Young sprouts: food Tree: agrihorticulture: shade tree in coffee plantations; ornamental tree along avenues; host tree for lac insects	Hanelt et al. 2001, Ghafoor 1985

TABLE 2.2 (continued)
Ethnographic Data on Various Species of *Ficus*

Ficus infectoria Roxb. [= *Ficus virens* Aiton]		References
Local medicinal uses	Local uses: cancer	*EthnobotDB* 1996, Duke 1993, Hartwell 1967-1971

Ficus lyrata Warb.

Genus	*Ficus* section: *Galoglychia*	*GRIN* 1994, Van Noort and Rasplus 2004
Synonyms	*Ficus pandurata* Sander cx Watson *F. pandurata* Sander non Hance	*MMPND* 1995, Ghafoor 1985
Common names	English: Banjo fig, Fiddleleaf fig, Fiddle-leaf fig, Fiddle-leaved fig tree, Lyre-leaf figtree, Lyre-leaved fig tree German: Geigenblättriger Feigenbaum, Geigenfeige French: Figuier à feuilles lyrées Japanese: Kashiwa bagomu	*MMPND* 1995, *GRIN* 1994, Hanelt et al. 2001, Burkill 1985
Habitat and distribution	Lowland rainforest Tropical West Africa: from Sierra Leone to Gabon. Cameroon, Gabon, Benin, Cote D'Ivoire, Liberia, Nigeria, Sierra Leone, Togo	*GRIN* 1994, Berg 1991, Hanelt et al. 2001, Ghafoor 1985, Van Noort and Rasplus 2004, Burkill 1985, Hutchinson and Dalziel 1958, *ALUKA* 2003
Economic importance	Bark: exudations—gums, resins, etc. Tree: social: religion, superstitions, magic (fetish tree); agrihorticulture: ornamental, cultivated or partially tended; shade trees; indoor houseplant	Hanelt et al. 2001, Ghafoor 1985, Burkill 1985
Local medicinal uses	—	

FIGURE 2.12   *Ficus microcarpa*, trunk (14.5.2004, Kipahulu, Maui, by Forest & Kim Starr; Image 040514-0204; *Plants of Hawaii*).

FIGURE 2.13   *Ficus microcarpa*, leaves and fruit (8.6.2008, Citrus grove Sand Island, Midway Atoll, by Forest & Kim Starr; Image 080608-7463; *Plants of Hawaii*).

TABLE 2.2 (continued)
Ethnographic Data on Various Species of *Ficus*

Ficus microcarpa L. f. [not to be confused with *F. microcarpa* Vahl = *F. thonningii* Blume]

		References
Genus	*Ficus* section: *Conosycea*	*GRIN* 1994
Synonyms	*Ficus amblyphylla* (Miq.) Miq.	*GRIN* 1994, *MMPND*
	F. benjamina Willd.	1995, Hanelt et al.
	F. cairnsii Warburg	2001, Ghafoor 1985,
	F. condaravia Buch.-Ham.	Wunderlin 1997, Wu
	F. littoralis Blume	et al. 2003

Genus   *Ficus* section: *Conosycea*

Synonyms   *Ficus amblyphylla* (Miq.) Miq.
F. benjamina Willd.
F. cairnsii Warburg
F. condaravia Buch.-Ham.
F. littoralis Blume
F. magnoliaefolia Blume
F. microcarpa L. f. var. *crassifolia* (W. C. Shieh) Liao
F. microcarpa var. *fuyuensis* J. C. Liao
F. microcarpa var. *latifolia* (Miq.) Corner
F. microcarpa var. *oluangpiensis* J. C. Liao
F. microcarpa var. *pusillifolia* J. C. Liao
F. nitida auct.
F. reticulata (Miq.) Miq., not Thunb.
F. retusa auct., non L.
F. retusa Heyne ex Roth
F. retusa hort., non L.
F. retusa King.
F. retusa L.
F. retusa L. var. *crassifolia* W. C. Shieh
F. retusiformis H. Lév.
F. rubra Roth
F. sarmentosa Buch.-Ham. ex Sm. var. *sarmentosa*
Urostigma accedens var. *latifolia* Miq.
U. amblyphyllum Miq.
U. microcarpa (L. f.) Miq.

Common names   English: Chinese banyan, Coffee, Curtain fig, Indian laurel, Indian laurel fig, Laurel fig, Malay banyan
German: Chinesische Feige, Indischer Lorbeer, Lorbeerfeige, Vorhang-Feige
Danish: Laurbærfigen
Spanish: Laurel de Indias
Portuguese: Figueira-lacerdinha, Laurel-da-Índia
Sanskrit: Gajapadapa, Kantalaka, Ksavataru, Kuberaka, Kuni, Kunjarapadapa, Nandivriksha, Plaksah, Sthalivrksa, Tunna
Hindi: Chilkan, Kamarup
Kannada: Hillala, Hinala, Kirgoli, Kirigoli, Kirugoli, Nankipipri, Peelaalada mara, Peeladamara, Pilal
Marathi: Nandruk

MMPND 1995, *GRIN* 1994, Hanelt et al. 2001, Wunderlin 1997, Wu et al. 2003

(continued on next page)

TABLE 2.2 (continued)
Ethnographic Data on Various Species of *Ficus*

Ficus microcarpa L. f. [not to be confused with *F. microcarpa* Vahl = *F. thonningii* Blume]

		References
Common names (continued)	Telugu: Billa juvvi, Hema vudaga, Konda pillara, Nandireka, Plaksa, Yerra juvvi	
	Malayalam: Itti, Itti arealou, Ittiyal, Itty alu, Kallithi	
	Tamil: Icci, Ichi, Kallicchi, Kallicci, Kalluichi, Malaiyichi	
	Burmese: Nyaung ok	
	Thai: Sai yoi, Sai yoi bai laem (Central Thailand)	
	Vietnamese: Cây gùa	
	Nepalese: Jaamu	
	Chinese: Rong shu	
	Japanese: Gajumaru	
Habitat and distribution	Disturbed sites, mountains, plains, below 1900 m. South India, Sri Lanka, Southern China, Australia, New Caledonia and Japan (Ryukyu Islands); introduced and cultivated in Pakistan, Iraq, and N. Africa.	*GRIN* 1994, Berg 1991, Hanelt et al. 2001, Ghafoor 1985, Wunderlin 1997, Wu et al. 2003
	China, Japan, Taiwan, Bhutan, India, Nepal, Sri Lanka, Cocos (Keeling) Islands, Cambodia, Laos, Myanmar, Thailand, Vietnam, Christmas Island, Indonesia, Malaysia, Papua New Guinea, Philippines, Australia, Micronesia, Palau, New Caledonia, Solomon Islands, South-East United States (Florida), West Indies, tropical Southern America, Hawaii, Pakistan, Iraq, North Africa	
Economic importance	Leaf: medicinal use	Hanelt et al. 2001, *GRIN* 1994, Wu et al. 2003
	Root: medicinal use	
	Tree: agrihorticulture: shade/shelter tree in coffee plantations; ornamental tree	
Local medicinal uses	Parts used: leaf, root	Hanelt et al. 2001, *GRIN* 1994
	Nonspecific use in India, Malaysia, and Southern China mentioned.	

Ficus nymphaeifolia Mill.

Genus	*Ficus* section: *Americana*	*GRIN* 1994
Synonyms	*Ficus anguina* Benoist	*GRIN* 1994, *MMPND* 1995, Jørgensen 2008
	F. cabusana Standley & Steyerm.	
	F. duquei Dugand	
	F. duquei var. *obtusiloba* Dugand	
	F. ierensis Britton	
	F. involuta var. *urbaniana* (Warb.) Dugand	
	F. nymphoides Thunb.	
	F. urbaniana Warb.	
	Urostigma nymphaeifolium (Mill.) Miq.	

TABLE 2.2 (continued)
Ethnographic Data on Various Species of *Ficus*

Ficus nymphaeifolia Mill.

		References
Common names	English: Nymphaea leaf fig	*MMPND* 1995
	French: Figuier à feuilles de nymphéa, Figuier grandes feuilles (Martinique), Figyé gran fèy (Martinique)	
Habitat and distribution	Lowlands, altitude range 0–500 m.	*GRIN* 1994, Jørgensen 2008
	Belize, Costa Rica, Nicaragua, Panama, Antigua and Barbuda, Dominica, Guadeloupe, Martinique, Montserrat, Netherlands Antilles, St. Kitts and Nevis, St. Lucia, St. Vincent and Grenadines, Trinidad and Tobago, French Guiana, Guyana, Suriname, Venezuela, Brazil, Bolivia, Colombia, Ecuador, Peru	
Economic importance	Medicinal uses	*EthnobotDB* 1996, Duke 1993, Wong 1976
Local medicinal uses	Local uses: anodyne, erysipelas	*EthnobotDB* 1996, Duke 1993, Wong 1976

Ficus ovata Vahl

Genus	*Ficus* section: *Galoglychia*	*GRIN* 1994
Synonyms	*Ficus asymmetrica* Hutch.	Hutchinson and Dalziel
	F. baoulensis A. Chev.	1958, Berg 1991,
	F. brachypoda Hutch.	*ALUKA* 2003
	F. buchneri Warb.	
	F. megaphylla Warb.	
	F. octomelifolia Warb.	
	F. pseudo-elastica Hiern	
	F. sapinii De Wild.	
	F. tuberculosa var. *elliptica* Hiern	
	Urostigma ovatum (Vahl) Miq.	
Common names	Badyara [Senegal]: Kobo	Burkill 1985
	Balanta [Senegal]: Bio ibi, Diogé	
	Basari [Senegal]: A-mbombelor, A-tagur	
	Crioulo [Senegal]: Figuiéra	
	Diola [Senegal]: Bu kholèn, Bu rirum	
	Diola-flup [Senegal]: Bo ñañ dièk	
	Fula-pulaar [Senegal]: Sakkaréhi	
	Konyagi [Senegal]: Dira ñas	
Habitat and distribution	Swamp forest edges, riverine forest, secondary forest, lake shores, savanna woodlands. Altitude range: 0–2100 m.	*GRIN* 1994, Berg 1991, Van Noort and Rasplus 2004, Burkill 1985, Hutchinson and Dalziel 1958, *ALUKA* 2003
	Sub-Saharan Africa from Senegal to Ethiopia and Kenya, and southwards to Northern Angola and Mozambique.	

(continued on next page)

TABLE 2.2 (continued)
Ethnographic Data on Various Species of *Ficus*

Ficus ovata Vahl		References
Habitat and distribution (continued)	Eritrea, Ethiopia, Sudan, Kenya, Tanzania, Uganda, Cameroon, Zaire, Benin, Cote D'Ivoire, Fernando Po, Ghana, Guinea, Guinea-Bissau, Liberia, Nigeria, Senegal, Sierra Leone, Togo, Angola, Malawi, Mozambique, Zambia	
Economic importance	Fruit: agrihorticulture: fodder; medicinal use; increases milk production of cows	Burkill 1985
	Latex: farming, forestry, hunting and fishing apparatus	
	Bark: masticatory; exudations—gums, resins, etc.; fiber	
	Tree: agrihorticulture: hedges, markers; ornamental, cultivated or partially tended; shade trees; social: religion, superstitions, magic	
Local medicinal uses	Parts used: fruit	Burkill 1985
	Local uses: lactation stimulant (incl. veterinary)	

Ficus palmata Forssk.		
Genus	*Ficus* section: *Ficus*	*GRIN* 1994
Synonyms	*Ficus caricoides* Roxb.	*GRIN* 1994, *MMPND*
	F. forsskalii Vahl	1995, Hanelt et al.
	F. morifolia Forssk.	2001, Ghafoor 1985,
	F. palmata var. *petitiana* (A. Rich.) Fiori	Friis 1999, Hutchinson
	F. petitiana A. Rich.	1917, *ALUKA* 2003
	F. pseudocarica Miq.	
	F. pseudocarica var. *tomentosa* A. Rich.	
	F. pseudosycomorus Decne.	
	F. robecchii Warb. ex Mildbr. & Burret	
	F. virgata Roxb.	
Common names	English: Punjab fig, Wild fig	*MMPND* 1995, Hanelt
	Spanish: Higuieron de Kabul (Costa Rica)	et al. 2001, Ghafoor
	Hindi: Anjir, Anjiri, Anzori, Bedu, Jangli anjir, Kak, Khamri, Khemri, Kimri	1985
	Gujarati: Pepri	
	Telugu: Manjimedi	
Habitat and distribution	Thickets along watercourses, between altitudes of 600 and 2700 m. Occasionally found in forests, more commonly around villages.	*GRIN* 1994, Hanelt et al. 2001, Ghafoor 1985, *Plants for A Future*
	From East and North-East Africa (Ethiopia, Somalia, Sudan, Egypt) to South-West Asia (Arabia, Afghanistan, Pakistan, North India).	1996, Van Noort and Rasplus 2004, Friis 1999, Hutchinson
	Algeria, Egypt, Eritrea, Ethiopia, Somalia, Sudan, Oman, Saudi Arabia, Yemen, Afghanistan, Egypt, Iran, Israel, Jordan, India, Nepal, Pakistan, United States (California), Central America	1917, *ALUKA* 2003

TABLE 2.2 (continued)
Ethnographic Data on Various Species of *Ficus*

Ficus palmata Forssk.   **References**

Economic importance	Fruit: food; dried figs; medicinal use Leaf: food; agrihorticulture: fodder Tree: rootstock for *Ficus carica* in breeding *Ficus* Wood: hoops, garlands, ornaments, etc.	Hanelt et al. 2001, *GRIN* 1994, Ghafoor 1985, *Plants for A Future* 1996
Local medicinal uses	Parts used: fruit, sap, latex Local uses: emollient, demulcent, warts, lung diseases, laxative, constipation, dysentery, bladder diseases, splinters lodged in the flesh	Ghafoor 1985, *Plants for A Future* 1996, *EthnobotDB* 1996, Duke 1993, *Anon.* 1948–1976, Shah and Joshi 1972
Toxicity	The sap and the half-ripe fruits are said to be poisonous.	*Plants for A Future* 1996

Ficus platyphylla Del.

Genus	*Ficus* section: *Galoglychia*	*GRIN* 1994
Synonyms	*Ficus kotschyana* (Miq.) Miq. *F. lateralis* Warb. *F. somalensis* auct. *F. umbrosa* Warb. ex Mildbr. & Burret *Urostigma kotschyanum* Miq.	*MMPND* 1995, Friis 1999, Hutchinson 1917, *ALUKA* 2003
Common names	English: Broadleaf fig, Flake rubber tree, Red Kano rubber tree, West African fig tree Russian: Fikus ploskolistnyi Hausa: Gamji Bambara (Senegal): Gaba, Kobo, Nkaba, Onan bolo Manding-bambara (Senegal): Gaba, Kobo, Nkaba, Onan bolo, Tamba nualé Fula-pulaar (Senegal): Bappéhi, Dèndévi, Surèy Maninka (Senegal): Nkobo Ndut (Senegal): Dob, Ngébèl Non (Senegal): Mbadat, Mbart Serer (Senegal): Mbada	*MMPND* 1995, Burkill 1985
Habitat and distribution	Savanna woodland, often in rocky places, up to an altitude of 750m. Senegal to Ethiopia and Somalia. Ethiopia, Somalia, Sudan, Nubia, Uganda, Cote D'Ivoire, Cameroon, Gold Coast, Gambia, Ghana, Guinea-Bissau, Mali, Niger, Nigeria, Senegal, Togo	*GRIN* 1994, Van Noort and Rasplus 2004, Friis 1999, Hutchinson and Dalziel 1954, Hutchinson 1917, *ALUKA* 2003
Economic importance	Fruit: food Leaf: agri-horticulture: fodder; farming, forestry, hunting and fishing apparatus; dyes, stains, inks, tattoos and mordants; medicinal use	*GRIN* 1994, Burkill 1985, Hutchinson 1917

(continued on next page)

TABLE 2.2 (continued)
Ethnographic Data on Various Species of *Ficus*

Ficus platyphylla Del.   References

Economic **importance** (continued)	Gum: masticatory; household, domestic and personal items; farming, forestry, hunting and fishing apparatus	
	Latex: rubber	
	Bark: astringent; tannins; exudations—gums, resins, etc.; fibre; medicinal use	
	Wood: fuel and lighting	
	Tree: agri-horticulture: shade-trees; fence-posts, poles, sticks; social: ceremonial; religion, superstitions, magic	
Local medicinal uses	Parts used: leaf, bark	Burkill 1985
	Local uses: generally healing, astringent, leprosy, antidote (venomous stings, bites, etc.)	

Ficus pumila L.

Genus	*Ficus* section: *Erythrogyne*	*GRIN* 1994
Synonyms	*Ficus awkeotsang* Makino	*GRIN* 1994, Hanelt et al.
	F. hanceana Maxim.	2001, *MMPND* 1995,
	F. minima hort. ex Gard.	Ghafoor 1985
	F. pumila var. *awkeotsang*	
	F. pumila var. *pumila*	
	F. repens hort.	
	F. repens Rottler	
	F. scandens Lam.	
	F. stipulata hort.	
	F. stipulata Thunb.	
	Plagiostigma pumila Zucc.	
	P. stipulata Zucc.	
	Urostigma scandens Liebm.	
Common **names**	English: Climbing fig, Creeping fig, Creeping rubber plant, Figvinel Tropical ivy (Ceylon)	*MMPND* 1995, *GRIN* 1994, Hanelt et al.
	German: Kletterfeige	2001, Wunderlin 1997,
	Danish: Hængefigen	Wu et al. 2003, Burkill
	French: Figuier rampant	1985
	Italian: Fico rampicante	
	Spanish: Higuera trepadora, Paja de colchón (Puerto Rico), Paz y justicia (Puerto Rico), Uña (Colombia, Guatemala)	
	Portuguese (Brazil): Falsa-hera, Hera de China, Hera-miúda, Mama de pared	
	Serbian: Penjući fikus, Puzajući fikus	
	Burmese: Kyauk kat nyaung nwe	
	Vietnamese: Cây trâu cô'	
	Chinese: Bi li, Man tu luo	
	Japanese: Oo itabi, O-itabi	

TABLE 2.2 (continued)
Ethnographic Data on Various Species of *Ficus*

Ficus pumila L.		References
Habitat and distribution	Disturbed thickets; altitude range: 0–10 m A native of China and Japan, South-East Asia; introduced elsewhere. India, Sri Lanka, China, Japan, Taiwan, Vietnam, Philippines, USA (Florida), West Africa	*GRIN* 1994, Berg 1991, Hanelt et al. 2001, Ghafoor 1985, Wunderlin 1997, Wu et al. 2003, Burkill 1985, Hutchinson and Dalziel 1954, *ALUKA* 2003
Economic importance	Fruit: food; medicinal use Leaf: medicinal use Plant-sap: medicinal use Latex: medicinal use Tree: agri-horticulture: nematode-resistant rootstock for *Ficus carica*; ornamental plant, cultivated or partially tended, mainly for covering walls	Hanelt et al. 2001, *GRIN* 1994, Ghafoor 1985, Wunderlin 1997, Burkill 1985, Hutchinson and Dalziel 1954, *ALUKA* 2003
Local medicinal uses	Parts used: fruit, leaf, latex, plant sap Local uses: generally healing, anodyne, detoxicant, skin problems, carbuncle, parasitic infection (cutaneous, subcutaneous), tumors, cancer, tuberculosis, lactogogue, heart, hernia, thirst, kidney problems, diuretic, diarrhea, dysentery, gonorrhea, spermatorrhea, testicles, virility, fever, parasiticide, vermifuge	*EthnobotDB* 1996, Duke 1993, Burkill 1985, Shih-chen 1973, Hartwell 1967–1971, *Anon.* 1974, *Anon.* 1948–1976

Ficus racemosa L.		
Genus	*Ficus* section: *Sycomorus*	*GRIN* 1994
Synonyms	*Covellia glomerata* (Roxb.) Miq *Ficus glomerata* Roxb. *F. glomerata* var. *elongata* King *F. goolereana* Roxb. *F. goolereea* Roxb. *F. lucescens* Blume *F. mollis* Miq. *F. racemosa* var. *elongata* (King) M. F. Barrett *F. vesca* F. Muell. ex Miq.	*GRIN* 1994, Hanelt et al. 2001, *MMPND* 1995, Ghafoor 1985
Common names	English: Cluster fig, Cluster tree, Country fig, Gular fig, Redwood fig Portuguese: Rumbodo (Brazil) Urdu: Dimiri Sanskrit: Gular, Hemadugdhaka, Jantuphala, Sadaphalah, Udumbar, Udumbara, Udumbarah, Yajnanga Hindi: Ambar (Bombay), Domoor, Doomar, Gular, Jagya dumur, Udumbara, Umar, Umbar Kannada: Alhi, Atti, Atthimara	*MMPND* 1995, *GRIN* 1994, Hanelt et al. 2001, Ghafoor 1985, Wu et al. 2003

(continued on next page)

TABLE 2.2 (continued)
Ethnographic Data on Various Species of *Ficus*

Ficus racemosa L.   References

Common names (continued)	Bengali: Dumur, Hpak-lu, Jagyadumbar, Mayen, Taung tha phan, Thapan, Ye thapan Marathi: Audumbar, Umbar Gujarati: Gular, Umardo Oriya: Dimri Telugu: Arri, Athi, Bodda, Maydi, Paidi, Udumbaramu Malayalam: Athi (Kerala), Athiathial, Atthi Sinhalese: Attikka Tamil: Anai, Athi, Attee marum, Atthi, Atti, Malaiyin munivan, Utumparam, Vellaiatthi Burmese: Hpak-lu, Jagyadumbar, Mayen, Taung tha phan, Thapan, Ye thapan Thai: Duea kliang, Duea nam, Ma duea, Ma duea chumphon, Ma duea utum phon (Central Thailand) Vietnamese: Cây sung, Sung Laotian: Kok dua kieng Malay: Ara, Elo (Indonesia) Nepalese: Dumrii Chinese: Ju guo rong, Yu dan bo luo	
Habitat and distribution	Rainforest, monsoon forest, beside rivers and streams, occasionally in streams; altitude range 100–1700 m. South-East Asia, Malesia, Papua New Guinea and Australia China, India, Nepal, Pakistan, Bangladesh, Sri Lanka, Myanmar, Thailand, Vietnam, Indonesia, Malaysia, Papua New Guinea, Australia, South America	*GRIN* 1994, Hanelt et al. 2001, Ghafoor 1985, Wu et al. 2003, Van Noort and Rasplus 2004
Economic importance	Fruit: food; medicinal use Leaf: food; fodder; soil cover Latex: rubber; water-proof paper; plasticizer for *Hevea* rubber Bark: medicinal use Wood: cart frames, ploughs, box, fittings, match boxes and cheap furniture Tree: agri-horticulture: shade/shelter tree in coffee plantations; ornamental tree; host plant for lac insects; lac; rootstock for *Ficus carica*; social: religious: one of the holy trees of the Hindus	Hanelt et al. 2001, *GRIN* 1994, Ghafoor 1985
Local medicinal uses	Parts used: fruit, bark Local uses: astringent, scabies, boils, wounds, cancer, adenopathy, myalgia, hydrocoele, lactogogue, carminative, stomachic, epididymitis, orchitis, spermatorrhea, menorrhagia, smallpox	Ghafoor 1985, *EthnobotDB* 1996, Duke 1993, Jain and Tarafder 1970, Hartwell 1967–1971

TABLE 2.2 (continued)
Ethnographic Data on Various Species of *Ficus*

Ficus religiosa L.

		References
Genus	*Ficus* section: *Urostigma*	*GRIN* 1994
Synonyms	*Ficus caudata* Stokes	Hanelt et al. 2001,
	F. peepul Griff.	Ghafoor 1985,
	F. rhynchophylla Steud.	Wunderlin 1997, Wu
	F. superstitiosa Link	et al. 2003
	Urostigma affine Miq.	
	U. religiosum (L.) Gasp.	
Common names	English: Bo tree, Bodhi tree, Peepul tree (Nepal), Pipal tree (India), Pippala (India), Po tree (Thailand), Sacred fig, Sacred fig tree	*MMPND* 1995, *GRIN* 1994, Hanelt et al. 2001, Ghafoor 1985, Wunderlin 1997, Wu et al. 2003
	German: Bobaum, Bo-Baum, Bodhi-Baum, Heiliger Feigenbaum, Indischer Pepulbaum, Pappelfeige, Pepul-Baum, Pepulbaum der Inder	
	Danish: Buddhafigen	
	French: Arbre bo de, Arbre de Dieu, Figuier de l'Inde, Figuier des banians, Figuier des pagodes, Figuier indien, Figuier sacré, Figuier sacré de Bodh-Gaya	
	Italian: Fico del diavolo	
	Spanish: Arbol sagrado de la India, Higuera de agua, Higuera de las pagodas, Higuera religiosa de la India, Higuera sagrada de los budistas	
	Portuguese: Figueira-dos-pagodes, Figueira-religiosa	
	Polish: Figowiec swiety	
	Serbian: Gumi-lak smokva, Indijska smokva	
	Arabic: Fikus lisan, Tin al-mu'abid	
	Urdu: Jori, Peepal, Usto	
	Sanskrit: Ashvatha, Ashvathha, Asvattha, Aswattha, Azvattha, Bodhidruma, Bodhivrska, Peepal, Peepul, Piippala, Pipala, Pippala vruksham, Shuchidruma, Vrikshraj, Yajnika	
	Hindi: Pipal, Pipali, Pipli, Pipul	
	Kannada: Arali, Arati, Aswatha	
	Bengali: Pipal, Pippal	
	Marathi: Pimpal	
	Gujarati: Jari, Pipal, Pipers, Piplo	
	Oriya: Jari	
	Telugu: Bodhi, Raavi chettu, Ragichettu, Ravi	
	Malayalam: Arayal (Kerala), Arei al, Ashwatham	
	Sinhalese: Araca maram, Bo, Bodhi	
	Tamil: Aracamaram, Arachu (Ceylon), Araeu, Arali, Arasanaram, Arasi maram, Arasu	
	Burmese: Bawdi nyaung, Lagat, Mai nyawng, Nyaung bawdi	

(continued on next page)

TABLE 2.2 (continued)
Ethnographic Data on Various Species of *Ficus*

Ficus racemosa L.   References

Common names (continued)	Thai: Pho, Pho si maha pho (Central Thailand), Sali (Northern Thailand) Vietnamese: Cây đê Khmer: Pu Laotian: Kok pho Nepalese: Peepul, Pipal, Pippal Chinese: Pu ti shu, Si wei shu Japanese: Indo bodaiju	
Habitat and distribution	Tropics; disturbed thickets Native to Northern India, Nepal, and Pakistan; cultivated throughout the tropics. Bangladesh, India, Sri Lanka, Nepal, Pakistan, Myanmar, Malaysia, Thailand, Vietnam, China, introduced and cultivated in S.E. Asia, Middle East, North Africa (Egypt, Libya), USA (Florida) and elsewhere.	*GRIN* 1994, Berg 1991, Hanelt et al. 2001, Ghafoor 1985, Wunderlin 1997, Wu et al. 2003
Economic importance	Fruit: food; fodder for birds; medicinal use Flower buds: food Leaf buds: food Leaf: fodder for cattle and goats; medicinal use Twigs: fodder for cattle and goats Shoots: medicinal use Latex: varnishes; birdlime Bark: dyeing; tanning Wood: packing cases; sacrificial fires by Hindus Tree: agri-horticulture: ornamental tree with various uses; roadside tree; host tree of lac insects; host tree of silkworms; lac; fodder for silkworms; social: religious: sacred to Hindus and Buddhists	Hanelt et al. 2001, *GRIN* 1994, Ghafoor 1985
Local medicinal uses	Parts used: fruit, leaf, shoots Local uses: astringent, alterative, cooling, refrigerant, bactericide, pimple, carbuncle, sore, skin diseases, tumors, atrophy, cachexia, rheumatism, otitis, stomatitis, gravel, dysuria, laxative, purgative, dysentery, cholera, bowel problems, urogenital problems, gonorrhea, spermatorrhea, fever, smallpox, rinderpest	Hanelt et al. 2001, Ghafoor 1985, *EthnobotDB* 1996, Duke 1993, Al-Rawi 1964, *Anon.* 1948–1976, Jain and Tarafder 1970, Hartwell 1967–1971

Ficus ruficaulis Merr.

Genus	*Ficus* section: *Eriosycea*	*GRIN* 1994
Synonyms	*Ficus antaoensis* Hayata *F. hiiranensis* Hayata *F. ruficaulis* var. *antaoensis* (Hayata) Hatusima & J. C. Liao.	Wu et al. 2003

TABLE 2.2 (continued)
Ethnographic Data on Various Species of *Ficus*

Ficus ruficaulis Merr.

		References
Common names	Chinese: Hong jing rong	Wu et al. 2003
Habitat and distribution	Usually at low elevations. Taiwan, Malaysia, Philippines	Wu et al. 2003
Economic importance	—	
Local medicinal uses	—	

Ficus semicordata Buch.-Ham. ex Sm.

Genus	*Ficus* section: *Hemicardia*	*GRIN* 1994
Synonyms	*Covellia cunia* (Buch.-Ham. ex Roxb) Miq.	*MMPND* 1995, Ghafoor
	Ficus cunia Buch.-Ham. ex Roxb.	1985, Wu et al. 2003
	F. cunia Ham. ex Roxb.	
	F. conglomerata Roxb.	
	F. pretoriae Burtt Davy	
Common names	Hindi: Bhuin gular, Doomar, Khaina	*MMPND* 1995, Ghafoor
	Tamil: Taragadi	1985, Wu et al. 2003
	Vietnamese: Cọ nọt, Đa lá lệch	
	Nepalese: Khanyu	
	Chinese: Ji su zi rong	
Habitat and distribution	Forest margins, valleys, along trails; altitude range 600–1900 (–2800) m.	*GRIN* 1994, Ghafoor 1985, Wu et al. 2003
	Subhimalayan tracts from Pakistan eastwards to Malaya through India, Bhutan and Myanmar.	
	China, Pakistan, Bhutan, India, Nepal, Myanmar, Thailand, Vietnam, Malaysia	
Economic importance	Fruit: food; fodder	Ghafoor 1985, Wu et al. 2003
	Bark: strong fibre	
	Root: medicinal use	
	Tree: agri-horticulture: street or shade tree	
Local medicinal uses	Parts used: root	Ghafoor 1985, Wu et al. 2003
	Local uses: bladder diseases	

Ficus septica Burm. f.

Genus	*Ficus* section: *Sycocarpus*	*GRIN* 1994
Synonyms	*ficus haulii* Blanco	*GRIN* 1994, Wu et al. 2003
	F. kaukauensis Hayata	
	F. leucantatoma Poir.	
	F. oldhamii Hance	
Common names	Chinese: Leng guo rong	*GRIN* 1994

(continued on next page)

TABLE 2.2 (continued)
Ethnographic Data on Various Species of *Ficus*

Ficus septica Burm. f.		References
Habitat and distribution	At low elevations. Japan, Taiwan, Indonesia, Malaysia, Papua New Guinea, Philippines, Australia, Solomon Islands, Vanuatu	*GRIN* 1994, Wu et al. 2003
Economic importance	Medicinal uses	*GRIN* 1994
Local medicinal uses	Local uses: fungoid, tinea, sore, emetic, purgative, antidote, antidote (crab), fish sting	*GRIN* 1994, *EthnobotDB* 1996, Duke 1993, Burkill 1966, Uphof 1968, Hartwell 1967–1971, Altschul 1973

Ficus stenocarpa F.Muell. ex Benth. = *Ficus fraseri* Miq.		
Genus	*Ficus* section: *Sycidium*	Van Noort and Rasplus 2004
Synonyms	*Ficus aspera* var. *subglabra* Benth. *F. fraseri* (Miq.) Miq. *F. stephanocarpa* var. *subglabra* (Benth.) Maiden & Betche *F. subglabra* (Benth.) F.Muell. *Urostigma fraseri* Miq.	Chew 1989, MetaFro Tervuren 1998
Common names	English: Figwood, Watery Fig, White Sandpaper Fig	MetaFro Tervuren 1998, Harden 1999
Habitat and distribution	Drier rainforest, also in littoral and riverine rainforest Australia, New Caledonia, Vanuatu	Chew 1989, Harden 1999
Economic importance	—	
Local medicinal uses	—	

Ficus sur L.		
Genus	*Ficus* section: *Sycomorus*	*GRIN* 1994
Synonyms	*Ficus capensis* Thunb. *F. mallotocarpa* Warb. *F. sur* Forssk.	*MMPND* 1995, *GRIN* 1994
Common names	English: Broom cluster fig, Bush fig, Cape fig, Fig of Heaven, Fire sticks, Kooman, Wild fig German: Kap-Feige French: Petit sycomore Swahili: Mkuyu Kikuyu: Mu-kuyu Hausa: Uwar yara Badyara [Senegal]: Ibé Balanta [Senegal]: Brozi, Bruzi Banyun [Senegal]: Dièntèd, Ki bundal, Si bundèl Basari [Senegal]: A-nak, A-nak o gva	*MMPND* 1995, *GRIN* 1994, Burkill 1985

TABLE 2.2 (continued)
Ethnographic Data on Various Species of *Ficus*

Ficus sur L.		References
Common names (continued)	Bedik [Senegal]: Ba-nák, Ga-nák Crioulo [Senegal]: Défay Diola [Senegal]: Bu rirun, Ku bun	
Habitat and distribution	Widespread in wooded grassland, riverine fringes and evergreen forest; altitude range 0–2300 m. From South Africa to Cape Verde and the Arabian Peninsula. 　Cape Verde, Eritrea, Ethiopia, Cameroon, Zaire, Cote D'Ivoire, Gambia, Ghana, Guinea, Guinea-Bissau, Liberia, Nigeria, Senegal, Sierra Leone, Togo, Bioko, Príncipe, São Tomé, Angola, Malawi, Mozambique, Zambia, Zimbabwe, Botswana, South Africa, Swaziland, Yemen	*GRIN* 1994, Berg 1991, Van Noort and Rasplus 2004
Economic importance	Fruit: food; fodder for wild animals and birds; masticatory; agri-horticulture: increases milk production of cows; pastimes—carving, musical instruments, games, toys, etc.; medicinal use Seed: increases milk production of cows Leaf: food; food: special diets; increases milk production of cows; medicinal use Bark: food; food: special diets; masticatory; tannins; astringents; dyes, stains, inks, tattoos and mordants; fibre; increases milk production of cows; medicinal use Bark-sap: medicinal use Root: increases milk production of cows; medicinal use Root, aerial: food Wood, poles: building materials Wood, sticks: fuel and lighting Tree: agri-horticulture: indicators (soil, water); ornamental, cultivated or partially tended; carpentry and related applications; farming, forestry, hunting and fishing apparatus; household, domestic and personal items; social: religion, superstitions, magic	Burkill 1985
Local medicinal uses	Parts used: fruit, seed, leaf, bark, bark sap, root Local uses: generally healing, pain-killer, astringent, malnutrition, debility, for special diets, parasitic infection (cutaneous, subcutaneous), leprosy, eye treatments, paralysis, epilepsy, convulsions, spasm, yaws, gout, naso-pharyngeal affections, pulmonary troubles, lactation stimulants (incl. veterinary), dropsy, swellings, oedema, kidney problems, diuretic, food poisoning, diarrhea, dysentery, laxative, abortifacient, ecbolic, menstrual cycle, pregnancy, antiabortifacient, venereal diseases, genital stimulants/depressants, febrifuge, antidotes (venomous stings, bites, etc.)	Burkill 1985

FIGURE 2.14 *Ficus religiosa* leaf (19.7.2009, University of Turku, Botanical Garden, Ruissalo, Turku, Finland, by Kaarina Paavilainen).

FIGURE 2.15 *Ficus religiosa* leaves (19.7.2009, University of Turku, Botanical Garden, Ruissalo, Turku, Finland, by Kaarina Paavilainen).

FIGURE 2.16 *Ficus sycomorus* branchlets (29.7.2009, Hebrew University Botanical Garden, Mount Scopus, Jerusalem, Israel, by Krina Brandt-Doekes).

FIGURE 2.17   A lonesome *Ficus sycomorus* (29.7.2009, Hebrew University Botanical Garden, Mount Scopus, Jerusalem, Israel, by Krina Brandt-Doekes).

The following overview (Tables 2.3 and 2.4) of the modern ethnomedical uses of various *Ficus* species is based on the wide scientific and descriptive literature on the subject and, in particular, on the extremely useful databases *NAPRALERT* (*NAPRALERT* 1975), *Dr. Duke's Phytochemical and Ethnobotanical Databases* (Duke 1993), *UK Cropnet*'s *EthnobotDB* (*EthnobotDB* 1996), and Daniel Moerman's *Native American Ethnobotany* (Moerman 2003). The data have been divided into two tables: Table 2.3, which shows the uses suggestive of antineoplastic, anticancerous, and anti-inflammatory action of the plant's chemical compounds, and Table 2.4, which discusses the rest of the indications for which different species of *Ficus* are, or have been, used.

The foregoing data testify to the great importance of figs and the various *Ficus* species all around the world. The present-day ethnomedical uses of the different *Ficus* products may also include practices that deserve wider application. In the following chapters, these potential applications, and others based on historical data, are discussed in more detail.

TABLE 2.3
Contemporary Ethnomedical Uses of *Ficus* spp. Suggestive of Antineoplastic and Anti-Inflammatory Actions

Use	Place	*Ficus* sp.	Plant Part	References
(A) Modern Ethnomedical Uses of *Ficus* spp. against Neoplasia				
Cancer	Ghana, Nigeria	*asperifolia*	Unspecified	Ayensu 1978, Hartwell 1967–1971
Cancers	W. Africa	*asperifolia*	Leaf, latex	Burkill 1985
Cancer	Japan	*carica*	Fruit	Takeuchi et al. 1978
Cancer, orally	United States	*carica*	Fresh fruit	Liebstein 1927
Cancer, externally	Thailand	unspecified	Bark	Wang 1973
Tumors	W. Africa	*asperifolia*	Leaf, latex	Burkill 1985
Tumors	Ghana, Nigeria	*asperifolia*	Unspecified	Ayensu 1978, Hartwell 1967–1971
Tumors	Guinea	*asperifolia/* Unspecified	Latex	Vasileva 1969
Tumors	India	*benghalensis*	Unspecified	Hartwell 1967–1971
Tumor (mixed with *Tamarix dioica* and acetic acid)	India	*carica*	Dried fruit	Siddiqui and Husain 1994
Tumors, breast	Tonga	*pachyrrachis*	Entire plant	Holdsworth 1974a
Tumors: induration of breast associated with redness externally (as part of a drug mixture)	Unspecified	*obliqua*	Leaf	Singh et al. 1984
Warts, externally	India	*benghalensis*	Latex	Reddy et al. 1989
Warts	Iran	*carica*	Fresh latex	Zagari 1992
Warts, externally	Italy	*carica*	Dried latex	Lokar and Poldini 1988
Warts, externally	France	*carica*	Fresh fruit, sap	Novaretti and Lemordant 1990
Warts, verrucas	Italy	*carica*	Latex	De Feo et al. 1992
Warts, externally	Turkey	*carica*	Latex from stem	Yesilada et al. 1995
Warts, externally	Tunisia	*carica*	Fresh latex	Boukef et al. 1982
Warts, externally	Oman	*salicifolia*	Latex	Ghazanfar and Al-Sabahi 1993

(continued on next page)

TABLE 2.3 (continued)
Contemporary Ethnomedical Uses of *Ficus* spp. Suggestive of Antineoplastic and Anti-Inflammatory Actions

Use	Place	*Ficus* sp.	Plant Part	References
(B) Modern Ethnomedical Uses of *Ficus* spp. against Inflammation				
Abscess	Unspecified	*benghalensis*	Unspecified	Duke 1993
Abscess, promotes the maturation of; externally	Turkey	*carica*	Latex	Yesilada et al. 1995
Adenitis	Peru	*carica*	Dried fruit	Ramirez et al. 1988
Antiseptic, externally (in a drug mixture)	Sri Lanka	*glomerata*	Unspecified	Perera et al. 1984
Asthma, water extract, orally	Nepal	*benghalensis*	Stem bark	Bhattarai 1993b
Asthma, decoction, orally	Tanzania	*exasperata*	Dried root bark	Chhabra et al. 1984
Asthma	South Korea	*religiosa*	Unspecified	Han et al. 1984
Asthma, powdered, 2–3 g in warm water infusion, orally	India	Unspecified	Dried fruit	Nisteswar and Kumar 1980
Backache, externally (with potash)	Nigeria	*thonningii*	Fresh leaf	Bhat et al. 1990
Back pain, lower, decoction, orally	Japan	*thunbergii*	Fresh leaf	Kitajima et al. 1994
Bacterial diseases, hot water extract, orally	Papua New Guinea	*septica*	Dried leaf	Baumgartner et al. 1990
Bee sting, externally	Turkey	*carica* subsp. *carica*	Fresh latex	Yazicioglu and Tuzlaci 1996
Boils, externally	Hawaii	*benghalensis*	Dried fruit	Locher et al. 1996
Boils, externally	Papua New Guinea	*botryocarpa*	Fruit juice	Holdsworth 1993
Boils, externally	India	*carica*	Fresh latex	Sebastian and Bhandari 1984
Boils, externally (warm poultice)	Greece	*carica*	Dried fruit	Malamas and Marselos 1992
Boils, externally	Papua New Guinea	*copiosa*	Fresh fruit latex	Holdsworth and Rali 1989
Boils, externally	India	*religiosa*	Stem bark	Singh 1995
Breast, inflamed; externally	Turkey	*carica*	Fruit	Sezik et al. 1997
Bronchitis, orally	Hawaii	*benghalensis*	Dried fruit	Locher et al. 1996
Bronchitis, acute; orally	India	*benghalensis*	Latex	Singh et al. 1996b
Bronchitis, roasted powder/decoction, orally	Iran	*carica*	Dried fruit	Zagari 1992

TABLE 2.3 (continued)
Contemporary Ethnomedical Uses of *Ficus* spp. Suggestive of Antineoplastic and Anti-Inflammatory Actions

Use	Place	*Ficus* sp.	Plant Part	References
Bruise	Unspecified	*benghalensis*	Unspecified	Duke 1993
Bruised finger or toes; externally (in a formula)	Oman	*salicifolia*	Leaf	Ghazanfar and Al-Sabahi 1993
Burns, externally, as a paste	Iran	*carica*	Dried fruit	Zagari 1992
Cardiac edema	South Korea	*religiosa*	Unspecified	Han et al. 1984
Catarrh, roasted powder, orally	Iran	*carica*	Dried fruit	Zagari 1992
Cephalgia, heated, moistened leaves with salt as a hot compress to forehead and other areas, crushed directly on the sore, covered by a similar leaf	Papua New Guinea	*septica*	Fresh leaf	Holdsworth 1992
Chest conditions	Somalia	*sycomorus*	Stem bark	Samuelsson et al. 1992
Cholera	India (Santal)	*benghalensis*	Unspecified	Jain and Tarafder 1970
Colic	Upper Volta	*asperifolia*	Unspecified	Ayensu 1978
Colic, externally	Indonesia	*toxicaria*	Leaf	Duke and Vasquez 1994
Cold, orally (in a drug mixture)	Spain	*carica*	Dried fruit	Martinez-Lirola et al. 1996
Colds, chronic, decoction, orally	Iran	*curica*	Dried fruit	Zagari 1992
Cold, hot water extract, orally	Mexico	*padifolia*	Fresh latex	Dominguez and Alcorn 1985
Cold	Papua New Guinea	*septica*	Leaf	Holdsworth 1984
Cold, common, hot water extract, orally	Papua New Guinea	*septica*	Dried leaf	Baumgartner et al. 1990
Cold, decoction, orally	E. Africa	*vogelii* aff.	Dried bark	Hedberg et al. 1983
Conjunctivitis	Upper Volta	*asperifolia*	Unspecified	Ayensu 1978
Cough (also expectorant), water extract, orally	Nepal	*benghalensis*	Stem bark	Bhattarai 1993b
Cough, hot water extract, orally	S. Korea	*carica*	Unspecified	Han et al. 1984
Cough, orally	Oman	*carica*	Boiled fruit	Ghazanfar and Al-Sabahi 1993

(continued on next page)

TABLE 2.3 (continued)
Contemporary Ethnomedical Uses of *Ficus* spp. Suggestive of Antineoplastic and Anti-Inflammatory Actions

Use	Place	*Ficus* sp.	Plant Part	References
Cough (antitussive), decoction, orally	Italy	*carica*	Fruit	De Feo and Senatore 1993
Cough, orally (taken with *Mallotus* and water)	Buka Island	*pungens*	Leaf	Holdsworth 1980
Cough, orally	Papua New Guinea	*pungens*	Sap from fresh root	Holdsworth and Balun 1992
Cough, externally (for massaging the neck)	Papua New Guinea	*septica*	Leaf	Holdsworth 1984
Cough, orally	Papua New Guinea	*septica*	Fresh sap from leaves	Holdsworth 1992
Cough, orally	Papua New Guinea	*septica*	Fresh root juice	Holdsworth 1984
Cough, orally	New Caledonia	*septica*	Dried fruit	Holdsworth et al. 1983
Cough, juice, orally	Papua New Guinea	*septica*	Young leaf	Holdsworth et al. 1983
Cough, decoction, orally	Somalia	*sycomorus*	Stem bark	Samuelsson et al. 1992
Cough, children's, orally	Indonesia	Unspecified	Fresh latex	Mahyar et al. 1991
Cough, dry, decoction, orally	Guatemala	*carica*	Leaf	Giron et al. 1991
Cough, whooping, powder, orally	Iran	*carica*	Dried fruit	Zagari 1992
Cough, whooping, decoction, orally	India	*hispida*	Bark	Sikarwar and Kaushik 1993
Cough, whooping, orally	Papua New Guinea	*septica*	Fresh root juice	Holdsworth 1984
Cystitis, decoction, orally	Iran	*carica*	Dried fruit	Zagari 1992
Demulcent, orally	United States	*carica*	Dried fruit	*Anon.* 1931
Demulcent, hot water extract, orally	Pakistan	*carica*	Fresh leaf	Ahmed et al. 1990
Dermatitis, externally	Iran	*carica*	Dried fruit	Zagari 1992
Dermatitis, externally	Oman	*carica*	Leaf	Ghazanfar and Al-Sabahi 1993
Dermatitis, externally	Papua New Guinea	*subcuneata*	Fresh latex	Holdsworth and Sakulas 1986
Diarrhea, decoction, orally	Uganda	*anaphalocarpa*	Unspecified	Anokbonggo et al. 1990
Diarrhea, juice, orally (followed by warm saline water)	Nepal	*auriculata*	Stem bark	Bhattarai 1993a

TABLE 2.3 (continued)
Contemporary Ethnomedical Uses of *Ficus* spp. Suggestive of Antineoplastic and Anti-Inflammatory Actions

Use	Place	*Ficus* sp.	Plant Part	References
Diarrhea	India (Santal) Unspecified	*benghalensis*	Unspecified	Jain and Tarafder 1970
Diarrhea, decoction, orally	India	*benghalensis*	Dried bark	Gupta et al. 1993
Diarrhea, decoction, orally	India	*benghalensis*	Shade dried root (adventitious)	Mukherjee et al. 1998
Diarrhea, decoction, orally	Nepal	*benghalensis*	Root, aerial	Bhattarai 1993a
Diarrhea, orally	S. Africa	*capensis*	Unspecified	Simon and Lamla 1991
Diarrhea, decoction (with *Gyrocarpus americanus*), orally	Venda	*capensis*	Dried root	Arnold and Gulumian 1984
Diarrhea, orally	Turkey	*carica*	Fruit	Sezik et al. 1997
Diarrhea, decoction, orally	Iran	*carica*	Fresh bark	Zagari 1992
Diarrhea, water extract, orally	Cameroon	*exasperata*	Leaf	Noumi and Yomi 2001
Diarrhea, hot water extract, orally	Thailand	*glomerata*	Dried bark	Wasuwat 1967
Diarrhea, orally	Papua New Guinea	*pachystemon*	Fresh leaf	Holdsworth 1992
Diarrhea, infusion, orally	Venezuela	*paraensis*	Bark	Boom and Moestl 1990
Diarrhea, orally	Peru	*paraensis*	Latex	Duke and Vasquez 1994
Diarrhea, decoction, orally	India	*racemosa*	Shade dried bark	Mukherjee et al. 1998
Diarrhea, powder, orally, taken for 3 days	India	*racemosa*	Fresh fruit, unripe	Singh and Maheshwari 1994
Diarrhea, juice, orally	Nepal	*religiosa*	Stem bark juice	Bhattarai 1993a
Diarrhea, orally	Papua New Guinea	*septica*	Dried leaf buds	Holdsworth et al. 1989
Diarrhea, water extract, orally	Papua New Guinea	*septica*	Root	Holdsworth 1993
Diarrhea, decoction, orally	Somalia	*sycomorus*	Stem bark	Samuelsson et al. 1992
Diarrhea	Zaire	*sycomorus*	Stem bark	Chifundera et al. 1990
Diarrhea, hot water extract, orally	Zaire	*sycomorus*	Dried leaf, dried root	Chifundera et al. 1990

(continued on next page)

TABLE 2.3 (continued)

Contemporary Ethnomedical Uses of *Ficus* spp. Suggestive of Antineoplastic and Anti-Inflammatory Actions

Use	Place	*Ficus* sp.	Plant Part	References
Diarrhea, orally	Rotuma	*tinctoria*	Fresh leaf	McClatchey 1996
Diarrhea	Peru	*trigona*	Unspecified	Duke and Vasquez 1994
Diarrhea, water extract, orally (in a mixture with salt)	Sierra Leone	Unspecified	Dried leaf	Kargbo 1984
Diarrhea, infusion, orally	Africa	*vogelii*	Bark	Watt and Breyer-Brandwijk 1962
Diarrhea, hot water extract, orally	E. Africa	*vogelii* aff.	Dried bark	Hedberg et al. 1983
Diarrhea, orally	Peru	*yoponensis*	Latex	Duke and Vasquez 1994
Dislocated limbs, leaves burned in hot ash to decolorize, then rubbed on affected parts	Nigeria	*thonningii*	Fresh leaf	Bhat et al. 1990
Dysentery	India (Santal) Unspecified	*benghalensis*	Unspecified	Jain and Tarafder 1970
Dysentery	Turkey	*benghalensis*	Unspecified	Steinmetz 1957
Dysentery, hot water extract, orally	S. Korea	*carica*	Unspecified	Han et al. 1984
Dysentery, hot water extract, orally	Mexico	*padifolia*	Unspecified	Dominguez and Alcorn 1985
Dysentery, orally (taken with sugar)	India	*racemosa*	Latex	Singh and Maheshwari 1994
Dysentery, hot water extract, orally	S. Korea	*religiosa*	Unspecified	Han et al. 1984
Dysentery, hot water extract, orally	India	*religiosa*	Dried seed	Arseculeratne et al. 1985
Dysentery, decoction, orally	Somalia	*sycomorus*	Stem bark	Samuelsson et al. 1992
Dysentery, hot water extract, orally	Rwanda	*thonningii*	Dried leaf	Maikere-Faniyo et al. 1989
Dysentery, infusion, orally	Africa	*vogelii*	Bark	Watt and Breyer-Brandwijk 1962
Dysentery, hot water extract, orally	E. Africa	*vogelii* aff.	Dried bark	Hedberg et al. 1983
Dysentery, orally	Papua New Guinea	*wassa*	Dried bark	Holdsworth 1992

TABLE 2.3 (continued)
Contemporary Ethnomedical Uses of *Ficus* spp. Suggestive of Antineoplastic and Anti-Inflammatory Actions

Use	Place	*Ficus* sp.	Plant Part	References
Eczema, externally (as paste)	Iran	*carica*	Dried fruit	Zagari 1992
Eczema, orally	Turkey	*carica* subsp. *carica*	Dried leaf	Yazicioglu and Tuzlaci 1996
Eczema, infusion, orally	Indonesia	*grossularioides*	Leaf	Grosvenor et al. 1995a
Eruptions, externally	India	*carica*	Fresh latex	Sebastian and Bhandari 1984
Eruptive skin disease, externally (in a drug mixture)	Sierra Leone	*exasperata*	Dried leaf	Macfoy and Sama 1983
Eye inflammation, locally in the eye	India	*glomerata*	Latex	Singh et al. 1996b
Eye injury, infusion, locally in the eye	Rotuma	*scabra*	Fresh bark	McClatchey 1996
Eye irritation, infusion, locally in the eye	Rotuma	*scabra*	Fresh bark	McClatchey 1996
Eyes, redness of, used locally in the eye	India	*glomerata*	Latex	Singh et al. 1996b
Eczema, orally	Turkey	*carica*	Dried leaf	Yazicioglu and Tuzlaci 1996
Fever	W. Africa	*asperifolia*	Leaf	Burkill 1985, Ayensu 1978
Fever	India (Santal)	*benghalensis*	Unspecified	Jain and Tarafder 1970
Fever	Thailand	*glomerata*	Dried root	Mokkhasmit et al. 1971, Wasuwat 1967
Fever, decoction, orally	Indonesia	*grossularioides*	Dried leaf	Mahyar et al. 1991
Fever, infusion, orally	Brazil	*maxima*	Dried leaf	Diaz et al. 1997
Fever, orally	Bangladesh	*religiosa*	Dried fruit	Khanom et al. 2000
Fever, hot water extract, orally	India	*religiosa*	Dried seed	Arseculeratne et al. 1985
Fever, hot water extract, orally	Papua New Guinea	*septica*	Dried leaf	Baumgartner et al. 1990
Fever, preventing it, decoction, orally	Papua New Guinea	*septica*	Dried leaf and stem	Holdsworth 1975
Fever, juice, orally	New Ireland	Unspecified	Fresh leaf	Holdsworth et al. 1980

(continued on next page)

TABLE 2.3 (continued)
Contemporary Ethnomedical Uses of *Ficus* spp. Suggestive of Antineoplastic and Anti-Inflammatory Actions

Use	Place	*Ficus* sp.	Plant Part	References
Fever, infusion, orally	Brazil	Unspecified	Bark	Milliken 1997
Fistulas, externally (in a drug mixture)	Sri Lanka	*glomerata*	Unspecified	Perera et al. 1984
Food	India	*auriculata*	Ripe fresh fruit	Bennet 1983
Food	China	*auriculata, pubigera*	Fruit	Pei 1985
Food (jelly), water extract	Taiwan	*awkeotsang*	Seed	Suzuno et al. 1997
Food	China	*callosa, glomerata*	Immature leaf, fresh	Pei 1985
Food	India	*hispida*	Ripe and unripe fruit	Maikhuri and Gangwar 1993
Food	Thailand	*lacor*	Fresh leaf	Murakami et al. 1995
Food	Andaman Islands	*nervosa*	Fruit	Awasthi 1991
Food	Tanzania	*populifolia*	Gum	Johns et al. 1996
Food	Egypt	*pseudo-sycomorus*	Fruit	Goodman and Hobbs 1988
Food	India	*racemosa*	Fruit	Ramachandran and Nair 1981a, 1981b
Food	Tanzania	*sycomorus*	Fruit	Johns et al. 1996
Food	Tanzania	*wakefieldii*	Fruit; gum	Johns et al. 1996
Foot problems	Unspecified	*benghalensis*	Unspecified	Uphof 1968
Gastralgia (gripping), decoction, orally	India	*racemosa*	Fruit	Sharma et al. 1992b
Gastralgia, seawater infusion, orally	Buka Island	*septica*	Fresh leaf	Holdsworth 1980
Gastritis, decoction, orally	South Korea	*religiosa*	Unspecified	Han et al. 1984
Gastritis, acute, hot water extract, orally	S. Korea	*religiosa*	Unspecified	Han et al. 1984
Gastroenteritis, leaf buds, orally	Papua New Guinea	*septica*	Dried buds	Holdsworth et al. 1989
Gastroenteritis, orally	Indonesia	*toxicaria*	Leaf	Grosvenor et al. 1995a
Gastrointestinal ulcers, infusion, orally	Rodrigues Islands	*reflexa*	Bark and leaf	Gurib-Fakim et al. 1996
Gingivitis, decoction made in milk, as a gargle	Iran	*carica*	Dried fruit	Zagari 1992

TABLE 2.3 (continued)
Contemporary Ethnomedical Uses of *Ficus* spp. Suggestive of Antineoplastic and Anti-Inflammatory Actions

Use	Place	*Ficus* sp.	Plant Part	References
Gingivitis, decoction, orally	Honduras	*maxima*	Dried leaf and stem	Lentz et al. 1998
Gingivitis, decoction, orally	Honduras	*maxima*	Sap, wood	Lentz 1993
Gonorrhea	Upper Volta	*asperifolia*	Unspecified	Ayensu 1978
Gonorrhea, vaginally	India	*racemosa*	Latex from twigs	Diddiqui and Husain 1993
Gonorrhea, powder, externally	Fiji	*religiosa*	Dried bark	Singh 1986
Gonorrhea, powder, 5 g/2 × per day, orally	India	*religiosa*	Dried bark	Diddiqui and Husain 1993
Gout	W. Africa	*asperifolia*	Latex	Burkill 1985
Gout, infusion, orally	Rodrigues Islands	*reflexa*	Bark and leaf	Gurib-Fakim et al. 1996
Headache	Upper Volta	*asperifolia*	Unspecified	Ayensu 1978
Headache, decoction, orally	Mexico	*obtusifolia*	Fresh latex, fresh leaf	Dominguez and Alcorn 1985
Headache, hot water extract, externally	Mexico	*padifolia*	Fresh latex	Dominguez and Alcorn 1985
Headache, leaf buds, orally	Papua New Guinea	*septica*	Dried buds	Holdsworth et al. 1989
Headache, orally	Papua	*septica*	Fresh leaf and stem	Holdsworth 1974b
Headaches, leaves tied onto ears	Papua New Guinea	Unspecified	Leaf	Holdsworth 1989
Hemorrhoids	Upper Volta	*asperifolia*	Unspecified	Ayensu 1978
Hemorrhoids	W. Africa	*asperifolia*	Root, fruit	Burkill 1985
Hemorrhoids, externally	Turkey	*carica*	Latex from stem	Yesilada et al. 1995
Hemorrhoids, orally	Tunisia	*carica*	Dried fruit	Boukef et al. 1982
Hemorrhoids, externally	Iran	*carica*	Dried leaf	Zagari 1992
Hemorrhoids, hot water extract, orally	S. Korea	*carica*	Unspecified	Han et al. 1984
Hemorrhoids, orally	Turkey	*carica* subsp. *carica*	Dried leaf	Yazicioglu and Tuzlaci 1996
Hemorrhoids, rectally	Bangladesh	*religiosa*	Dried fruit	Khanom et al. 2000
Inflammations, external applications of decoction for	Iran	*carica*	Dried leaf	Zagari 1992

(continued on next page)

TABLE 2.3 (continued)
Contemporary Ethnomedical Uses of *Ficus* spp. Suggestive of Antineoplastic and Anti-Inflammatory Actions

Use	Place	*Ficus* sp.	Plant Part	References
Inflammations, reduces them, orally	Iran	carica	Dried fruit	Zagari 1992
Inflammations caused by common cold, decoction made in milk, as a gargle	Iran	carica	Dried fruit	Zagari 1992
Leucorrhea, bloody, 5 g of ash with *Saraca asoca* bark, multivitamin tablet, and erythromycin	India	racemosa	Bark ash	Vedavathyk and Rao 1995
Lumbago	Unspecified	benghalensis	Unspecified	Duke 1993
Lumbago, four to six leaves with a pinch of potash ground together and applied to affected part after making nine incisions	Nigeria	thonningii	Fresh leaf	Bhat et al. 1990
Lumbago, decoction, orally	Japan	thunbergii	Fresh leaf	Kitajima et al. 1994
Malaria, orally	Papua New Guinea	adenosperma	Root	Holdsworth 1977
Malaria	Nigeria	ingens, ovata	Dried bark	Etkin 1997
Malaria, decoction, orally	Togo, Nigeria	polita	Dried leaf	Gbeassor et al. 1990, Etkin 1997
Malaria, hot water extract, orally	Madagascar	pyrifolia	Dried leaf	Ratismamanga-Urverg et al. 1991
Malaria, pills (with solidified sugar cane juices), orally	India	religiosa	Dried root	Anis and Iqbal 1986
Malarial fever, tablets, orally	India	religiosa	Roots, aerial	Singh and Ali 1994
Malaria, hot water extract, orally	Indonesia	ribes	Bark and leaf	Dragendorff 1898
Malaria, juice, orally	New Ireland	Unspecified	Fresh leaf	Holdsworth et al. 1980
Malaria, infusion, orally	Brazil	Unspecified	Bark	Milliken 1997
Mastitis	Turkey	carica	Fruit	Sezik et al. 1997
Menstrual pain, decoction, orally	Honduras	insipida	Dried leaf and stem	Lentz et al. 1998

TABLE 2.3 (continued)
Contemporary Ethnomedical Uses of *Ficus* spp. Suggestive of Antineoplastic and Anti-Inflammatory Actions

Use	Place	*Ficus* sp.	Plant Part	References
Menstrual pains, decoction, orally	Madagascar, Sierra Leone	Unspecified	Dried bark	Quansah 1988
Mouth sores of a baby, orally	Papua New Guinea	*ampelos*	Young leaf	Holdsworth et al. 1989
Nasopharyngeal afflictions	W. Africa	*asperifolia*	Root	Burkill 1985
Nausea	Unspecified	*benghalensis*	Unspecified	Duke 1993
Nephritis, decoction, orally	Iran	*carica*	Dried fruit	Zagari 1992
Neuralgia, hot water extract, orally	S. Korea	*carica*	Unspecified	Han et al. 1984
Ophthalmia	India (Santal)	*benghalensis*	Unspecified	Jain and Tarafder 1970
Ophthalmia, externally	India	*glomerata*	Latex	Singh et al. 1996a
Ophthalmia, infusion, externally	Rotuma (Fiji)	*scabra*	Fresh bark	McClatchey 1996
Oral disinfectant, decoction, orally	Italy	*carica*	Dried fruit	Lokar and Poldini 1988
Pains, externally	Turkey	*carica*	Latex from stem	Fujita et al. 1995
Pains, externally (as paste)	Iran	*carica*	Dried fruit	Zagari 1992
Pains and aches, decoction, orally	Honduras	*insipida*	Dried leaf and stem	Lentz et al. 1998
Pains and aches, decoction, orally/ externally	Nicaragua	*insipida*	Sap	Coe and Anderson 1996a, 1996b
Pains in body, orally (in a drug mixture)	Nepal	*benghalensis*	Dried stem bark	Bhattarai 1991
Pains in body, heated leaves applied externally	Papua New Guinea	*pungens, septica*	Leaf	Holdsworth 1993
Pains, muscular, externally	India	*benghalensis*	Fresh latex	Shah and Gopal 1985
Pains, somatic, heated, moistened leaves with salt as a hot compress to forehead and other areas, crushed directly to sores, covered by a similar leaf, externally	Papua New Guinea	*septica*	Fresh leaf	Holdsworth 1992
Pain, scorpion bite, externally	Turkey	*carica*	Latex from stem	Fujita et al. 1995

(continued on next page)

TABLE 2.3 (continued)

Contemporary Ethnomedical Uses of *Ficus* spp. Suggestive of Antineoplastic and Anti-Inflammatory Actions

Use	Place	*Ficus* sp.	Plant Part	References
Paralysis, orally	Bangladesh	*religiosa*	Dried fruit	Khanom et al. 2000
Pharyngeal problems, infusion, orally	Canary Islands	*carica*	Dried shoots	Darias et al. 1986
Pharyngitis, orally	Canary Islands	*carica*	Dried shoots	Darias et al. 1986
Pleurisy, roasted powder, orally	Iran	*carica*	Dried fruit	Zagari 1992
Pneumonia, orally	India	*racemosa*	Fresh leaf juice	Sebastian and Bhandari 1984
Pruritis	Malaysia	*aurantiacea*	Leaf	Ahmad and Holdsworth 1994
Pulmonary catarrh, decoction, orally	Iran	*carica*	Dried fruit	Zagari 1992
Respiratory disease	Peru	Unspecified	Dried leaf	Desmarchelier et al. 1997
Rheumatalgia, decoction, orally	Japan	*thunbergii*	Fresh leaf	Kitajima et al. 1994
Rheumatic inflammations, externally	Peru	*insipida* var. *insipida*	Latex	Duke and Vasquez 1994
Rheumatism	Unspecified	*benghalensis*	Unspecified	Duke 1993
Rheumatism, externally	India	*benghalensis*	Fresh latex	Shah and Gopal 1985
Rheumatism, externally	Peru	*insipida, maxima*	Latex	Duke and Vasquez 1994
Rheumatism, orally, infusion	Brazil	*maxima*	Dried leaf	Diaz et al. 1997
Rheumatism, decoction, orally	Japan	*thunbergii*	Fresh leaf	Kitajima et al. 1994
Scrofula, orally	United States	*carica*	Fresh fruit	Liebstein 1927
Scrofula, decoction, orally	Somalia	*sycomorus*	Stem bark	Samuelsson et al. 1992
Skin rash, externally, used for a week	Papua New Guinea	*subcuneata*	Fresh sap	Holdsworth and Sakulas 1986
Sores	Africa	*asperifolia*	Unspecified	Ayensu 1978
Sores	India (Santal)	*benghalensis*	Unspecified	Jain and Tarafder 1970
Sores, externally	Papua New Guinea	*calopilina*	Fresh leaf and fresh fruit	Holdsworth and Rali 1989
Sores, hot water extract, orally	Mexico	*padifolia*	Fresh leaf	Dominguez and Alcorn 1985

TABLE 2.3 (continued)
Contemporary Ethnomedical Uses of *Ficus* spp. Suggestive of Antineoplastic and Anti-Inflammatory Actions

Use	Place	*Ficus* sp.	Plant Part	References
Sores, externally	Papua New Guinea	*septica*	Fresh leaf	Holdsworth et al. 1983
Sores, externally	Papua New Guinea	*sublimata*	Fresh latex from leaf	Holdsworth and Sakulas 1986
Stomachache, orally	Papua New Guinea	*copiosa*	Fresh, unripe fruit	Holdsworth and Balun 1992
Stomachache, orally/ externally (massaging stomach with crushed leaves)	Papua New Guinea	*copiosa*	Fresh leaf and root	Holdsworth and Balun 1992, Holdsworth 1992
Stomachache, decoction, orally	S. Africa	*craterostoma*	Leaf	Bhat and Jacobs 1995
Stomachache, decoction, orally	Nicaragua	*insipida*	Dried sap	Coe and Anderson 1996b
Stomachache, orally	Indonesia	*pandana*	Dried leaf	Grosvenor et al. 1995b
Stomachache, orally (taken with turmeric)	India	*racemosa*	Dried fruit	Sharma et al. 1992a
Stomachache, externally	Peru	*radula*	Latex	Duke and Vasquez 1994
Stomachache, water extract (made with seawater), orally	Buka Island	*septica*	Fresh leaf	Holdsworth 1980
Stomachache, externally	Indonesia	*toxicaria*	Leaf	Duke and Vasquez 1994
Stomachache	E. Africa	*vogelii* aff.	Dried bark	Hedberg et al. 1983
Stomach pain, gripping, decoction, orally	India	*racemosa*	Fruit	Sharma et al. 1992b
Stomach problems	W. Africa	*asperifolia*	Bark	Burkill 1985
Stomach problems	Guinea	*asperifolia*	Unspecified	Ayensu 1978
Stomach problems, boiled figs mixed with curd and salt, orally	India	*sarmentosa* var. *ludduca*	Fresh fruit	Chandra and Pandey 1983
Stomach problems (upset stomach)	Papua New Guinea	*septica*	Dried leaf buds	Holdsworth et al. 1989
Stomachic, orally	India	*racemosa*	Root juice, sap	Mukherjee and Namhata 1990
Stomachic, infusion, orally	Africa	*vogelii*	Bark	Watt and Breyer-Brandwijk 1962

(continued on next page)

TABLE 2.3 (continued)

Contemporary Ethnomedical Uses of *Ficus* spp. Suggestive of Antineoplastic and Anti-Inflammatory Actions

Use	Place	*Ficus* sp.	Plant Part	References
Stomatitis	India (Santal)	*benghalensis*	Unspecified	Jain and Tarafder 1970
Stomatitis, orally	Hawaii	*benghalensis*	Dried fruit	Locher et al. 1996
Stomatitis, decoction made in milk, as a gargle	Iran	*carica*	Dried fruit	Zagari 1992
Stomatitis, orally	Canary Islands	*carica*	Dried shoots	Darias et al. 1986
Swelling, reduces it, externally (as paste)	Iran	*carica*	Dried fruit	Zagari 1992
Swellings, externally	India	*hispida*	Shade dried bark	Nyman et al. 1998
Swelling of lymph nodes, externally (mixed with *Tamarix dioica* and acetic acid)	India	*carica*	Dried fruit	Siddiqui and Husain 1994
Throat, inflammation of, locally (rubbing with the leaf)	Tanzania	*exasperata*	Fresh leaf	Chhabra et al. 1984
Throat pain, orally	Hawaii	*benghalensis*	Dried fruit	Locher et al. 1996
Throat pains, externally (spread on a cloth placed on the neck)	India	*benghalensis*	Latex	Nagaraju and Rao 1990
Throat pain, decoction made in milk, as a gargle	Iran	*carica*	Dried fruit	Zagari 1992
Throat pain, orally	Tanzania	*exasperata*	Dried flowers	Chhabra et al. 1984
Throat pain, infusion, orally	Somalia	*sycomorus*	Stem bark	Samuelsson et al. 1992
Tonsils, inflammation of, locally (rubbing with the leaf)	Tanzania	*exasperata*	Fresh leaf	Chhabra et al. 1984
Tooth, abscess in, decoction	Italy	*carica*	Dried fruit	Lokar and Poldini 1988
Tooth, caries in, orally	India	*auriculata*	Fresh latex	Jain and Puri 1984
Tooth, caries in, orally	Peru	*radula*	Latex	Duke and Vasquez 1994
Tooth, caries in	India	*retusa*	Dried root	Nagaraju and Rao 1990

TABLE 2.3 (continued)

Contemporary Ethnomedical Uses of *Ficus* spp. Suggestive of Antineoplastic and Anti-Inflammatory Actions

Use	Place	*Ficus* sp.	Plant Part	References
Toothache	Unspecified	*benghalensis*	Unspecified	Duke 1993
Toothache, decoction, directly in tooth cavity	Madagascar	*cocculifolia*	Dried bark	Kaufmann and Elvin-Lewis 1995
Toothache, hot water extract, orally	Senegal	*dicranostyla*	Dried bark	Le Grand 1989
Toothache, hot water extract, orally	Senegal	*dicranostyla*	Fresh latex	Le Grand and Wondergem 1987
Toothache, decoction, orally	Mexico	*obtusifolia*	Fresh latex	Dominguez and Alcorn 1985
Toothache, direct to cavity	Indonesia	*toxicaria*	Fresh latex	Mahyar et al. 1991
Toothache, direct to cavity to calm down the pain	Indonesia	*toxicaria*	Latex	Mahyar et al. 1991
Toothache	Unspecified	Unspecified	Fresh latex	Lewis and Elvin-Lewis 1984
Toothache, directly	Mexico	Unspecified	Latex	Zamora-Martinez and Pola 1992
Toothache, directly	Upper Amazon Basin	Unspecified	Latex	Lewis and Elvin-Lewis 1984
Toothache, directly	Mexico	Unspecified	Latex	Zamora-Martinez and Pola 1992
Toothbrush, used as	India	*benghalensis*	Fresh branches	John 1984
Toothbrush, used as	India	*benghalensis* var. *retusa*	Root, aerial	Ramachandran and Nair 1981b
Toothbrush, used as	Fiji	*religiosa*	Fresh twig	Singh 1986
Traumatic injury, decoction, orally	Mexico	*obtusifolia*	Fresh latex, fresh bark, leaf	Dominguez and Alcorn 1985
Traumatic injury, decoction, orally	Mexico	*padifolia*	Fresh bark and/or root	Dominguez and Alcorn 1985
Tuberculosis, infusion, orally (mixture of *F. capensis* and *F. sycomorus*)	Venda	*capensis, sycomorus*	Dried fruit	Arnold and Gulumian 1984
Tuberculosis, smoked	Turkey	*carica* subsp. *carica*	Dried leaf	Yazicioglu and Tuzlaci 1996

(continued on next page)

TABLE 2.3 (continued)

Contemporary Ethnomedical Uses of *Ficus* spp. Suggestive of Antineoplastic and Anti-Inflammatory Actions

Use	Place	*Ficus* sp.	Plant Part	References
Tuberculosis, orally	Bangladesh	*religiosa*	Dried fruit	Khanom et al. 2000
Tuberculosis, orally	India	*religiosa*	Dried leaf	Jain 1989
Ulcers, cleaning them, decoction, externally	India	*benghalensis*	Dried bark	John 1984
Ulcers, mixed with egg yolk or vegetable oil for, externally	Iran	*carica*	Fresh latex	Zagari 1992
Ulcers, orally	Nigeria	*exasperata*	Dried leaf	Akah et al. 1997
Ulcers, externally (in a drug mixture)	Sri Lanka	*glomerata*	Unspecified	Perera et al. 1984
Ulcers, infusion, orally	India	*infectoria*	Dried bark	Siddiqi et al. 1989
Ulcers, decoction, orally	Nicaragua	*insipida*	Dried sap	Coe and Anderson 1996b
Ulcer, externally	Papua New Guinea	*pachyrrachis*	Sap	Holdsworth 1993
Ulcers, infusion, orally	Rodrigues Islands	*reflexa*	Bark and leaf	Gurib-Fakim et al. 1996
Ulcers, leprous, infusion, orally	E. Africa	*vogelii* aff.	Dried bark	Hedberg et al. 1983
Ulcers, leprous, infusion, externally	Africa	*vogelii*	Bark	Watt and Breyer-Brandwijk 1962
Ulcers, stomach, infusion, orally	Nigeria	*exasperata*	Dried leaf	Akah et al. 1998
Ulcers, venereal, decoction, orally (in a drug mixture)	India	*benghalensis, racemosa, religiosa, talbotii*	Dried bark	Kumar and Prabhakar 1987
Unconsciousness	Fiji	*religiosa*	Dried leaf juice	Singh 1986
Urinary tract inflammation, water extract, orally (with coconut water)	Papua New Guinea	*septica*	Root	Holdsworth 1993
Vagina, promoting healing post partum, decoction, externally	India	*benghalensis*	Dried bark	John 1984
Vaginal rash, powder, vaginally	Sierra Leone	*exasperata*	Dried leaf	Macfoy and Sama 1983

TABLE 2.3 (continued)
Contemporary Ethnomedical Uses of *Ficus* spp. Suggestive of Antineoplastic and Anti-Inflammatory Actions

Use	Place	*Ficus* sp.	Plant Part	References
Voice, loss of, infusion, orally	Madagascar	*megapoda*	Leaf	Novy 1997
Vulnerary after abortion, ash, orally	India	*benghalensis*	Ash from fresh leaf	Reddy et al. 1988
Vulnerary for post partum bleeding, 5 g of ash with *Saraca asoca* bark and multivitamin tablet	India	*racemosa*	Bark ash	Vedavathyk and Rao 1995
Vulnerary after childbirth, decoction, orally	Madagascar, Sierra Leone	Unspecified	Dried bark	Quansah 1988
Wounds, cleaning them, decoction, externally	India	*benghalensis*	Dried bark	John 1984
Wounds, mixed with egg yolk or vegetable oil for, externally	Iran	*carica*	Fresh latex	Zagari 1992
Wounds, hot water extract, externally	Mexico	*padifolia*	Fresh latex	Dominguez and Alcorn 1985
Wounds, externally	Peru	*radula*	Latex	Duke and Vasquez 1994
Wounds, infected, externally	Peru	*citrifolia, paraensis*	Latex	Duke and Vasquez 1994
Wounds from poisoned arrows, decoction, externally	India	Unspecified	Dried bark	Bisset and Mazars 1984
Wounds of head, externally	Turkey	*carica* subsp. *carica*	Ash from stem and twigs	Yazicioglu and Tuzlaci 1996

TABLE 2.4
Other Contemporary Ethnomedical Uses of *Ficus* spp.

Use	Place	*Ficus* sp.	Plant Part	References
(A) Modern Ethnomedical Uses of *Ficus* spp., General				
Alterative	Turkey	*benghalensis*	Unspecified	Steinmetz 1957
Anodyne	W. Africa	*asperifolia*	Bark	Burkill 1985
Anodyne	Unspecified	*benghalensis*	Unspecified	Duke 1993
Astringent	Unspecified	*benghalensis*	Unspecified	Duke 1993
Astringent	Turkey	*benghalensis*	Unspecified	Steinmetz 1957
Astringent, externally (in a drug mixture)	Sri Lanka	*glomerata*	Unspecified	Perera et al. 1984
Astringent, hot water extract, externally	India	*religiosa*	Dried seed	Arseculeratne et al. 1985
Astringent, infusion, orally and externally	E. Africa	*vogelii* aff.	Dried bark	Hedberg et al. 1983
Astringent, infusion, externallyly	Africa	*vogelii*	Bark	Watt and Breyer-Brandwijk 1962
Atrophy	India (Santal)	*benghalensis*	Unspecified	Jain and Tarafder 1970
Cachexia	India (Santal)	*benghalensis*	Unspecified	Jain and Tarafder 1970
Debility	W. Africa	*asperifolia*	Seed	Burkill 1985
Healing (generally)	W. Africa	*asperifolia*	Leaf, latex, bark, root	Burkill 1985
Infants, ointment for, ground scrapings put in boiled palm oil	Sierra Leone	Unspecified	Dried bark	Kargbo 1984
Malnutrition	W. Africa	*asperifolia*	Seed	Burkill 1985
Medicine, used as, hot water extract, orally	Senegal	*capensis*	Dried stem bark	Laurens et al. 1985
Medicine, used as, hot water extract, orally	Peru	*urbaniana*	Dried leaf	Ramirez et al. 1988
Medicine, used as, hot water extract, orally	Yemen	*vasta*	Unspecified	Fleurentin et al. 1983
Rejuvenating, decoction, orally	Malaysia	*deltoidea*	Leaf	Ahmad and Holdsworth 1994
Tonic	Africa	*asperifolia*	Unspecified	Ayensu 1978
Tonic	W. Africa	*asperifolia*	Leaf, latex, bark, root	Burkill 1985
Tonic	Unspecified	*benghalensis*	Unspecified	Duke 1993
Tonic	Turkey	*benghalensis*	Unspecified	Steinmetz 1957
Tonic, powder, orally	India	*benghalensis*	Dried root bark	Reddy et al. 1988

TABLE 2.4 (continued)
Other Contemporary Ethnomedical Uses of *Ficus* spp.

Use	Place	*Ficus* sp.	Plant Part	References
Tonic, decoction, orally	Malaysia	*lepicarpa*	Root	Ahmad and Holdsworth 1995
Tonic, orally	Thailand	*semicordata*	Leaf, boiled	Anderson 1986b
Weakness of legs, hot water extract, orally	Thailand	*semicordata*	Dried leaf	Anderson 1986a

(B) Modern Ethnomedical Uses of *Ficus* spp., Skin Problems

Use	Place	*Ficus* sp.	Plant Part	References
Emollient, hot water extract, externally	Pakistan	*carica*	Fresh leaf	Ahmed et al. 1990
Fungal diseases, hot water extract, orally	Papua New Guinea	*septica*	Dried leaf	Baumgartner et al. 1990
Leucoderma, externally	India	*benjamina*	Fresh latex	Singh et al. 1996a
Parasitic infection, cutaneous, subcutaneous	W. Africa	*asperifolia*	Leaf, bark, root bark	Burkill 1985
Ringworm	Africa	*asperifolia*	Unspecified	Ayensu 1978
Ringworm, externally	S. Korea	*carica*	Unspecified	Han et al. 1984
Scabies, externally	Malaysia	*aurantiacea*	Leaf	Ahmad and Holdsworth 1994
Scabies	India (Santal)	*benghalensis*	Unspecified	Jain and Tarafder 1970
Scabies, infusion, externally	Indonesia	*grossularioides*	Leaf	Grosvenor et al. 1995a
Scabies, infusion, orally	India	*religiosa*	Bark	Bajpai et al. 1995
Scabies, externally, used for a week	Papua New Guinea	*subcuneata*	Fresh sap	Holdsworth and Sakulas 1986
Skin diseases, infusion, externally/hot water extract, orally	Senegal	*dicranostyla*	Dried bark and leaf	Le Grand and Wondergem 1987, Le Grand 1989
Skin itch, externally	Malaysia	*aurantiacea*	Leaf	Ahmad and Holdsworth 1994
Skin problems, externally/infusion, orally	Indonesia	*grossularioides*	Leaf	Grosvenor et al. 1995a, 1995b
Skin treatment, externally	Oman	*carica*	Leaf	Ghazanfar and Al-Sabahi 1993

(continued on next page)

TABLE 2.4 (continued)
Other Contemporary Ethnomedical Uses of *Ficus* spp.

Use	Place	*Ficus* sp.	Plant Part	References
(C) Modern Ethnomedical Uses of *Ficus* spp., Eye Problems				
Eye problems	India (Santal)	*benghalensis*	Unspecified	Jain and Tarafder 1970
Eye problems, water extract, locally in the eye	Tanzania	*exasperata*	Fresh leaf	Chhabra et al. 1984
Ophthalmic antiseptic, locally in the eye	Yemen	*salicifolia*	Sap	Fleurentin and Pelt 1982
(D) Modern Ethnomedical Uses of *Ficus* spp., Mental Problems				
Depression, orally	Rotuma	*tinctoria*	Fresh leaf	McClatchey 1996
Hallucinogen, orally	Peru	Unspecified	Latex	Luna 1984a
Hallucinogenic, used during shamanic training, decoction, orally (in a drug mixture)	Peru	*insipida,* unspecified	Unspecified	Luna 1984b
Hysteria, decoction, orally	Somalia	*bussei*	Stem bark	Samuelsson et al. 1992
Madness	India (Santal)	*benghalensis*	Unspecified	Jain and Tarafder 1970
Tranquilizer, externally	India	*racemosa*	Unspecified	Mukherjee and Namhata 1990
(E) Modern Ethnomedical Uses of *Ficus* spp., Neurological Problems				
Convulsions, infusion, orally	Rodrigues Islands	*rubra*	Bark	Gurib-Fakim et al. 1996
Hiccups, infusion, orally	Somalia	*bussei*	Stem bark	Samuelsson et al. 1992
Incontinence, hot water extract, orally	S. Korea	*carica*	Unspecified	Han et al. 1984
Neurotonic, hot water extract, orally	Thailand	*hirta*	Dried root	Wasuwat 1967
(F) Modern Ethnomedical Uses of *Ficus* spp., Heart and Blood Circulation				
Anemia, hot water extract, orally	Peru	*anthelmintica*	Dried latex	Ramirez et al. 1988
Anemia, hot water extract, orally	S. Korea	*carica*	Unspecified	Han et al. 1984
Anemia, hot water extract, orally	India	*hispida*	Dried bark	Acharya and Kumar 1984
Anemia, infusion, orally	Brazil	*maxima*	Dried leaf	Diaz et al. 1997
Anemia, as food	India	*racemosa*	Fruit	Mukherjee and Namhata 1990

TABLE 2.4 (continued)
Other Contemporary Ethnomedical Uses of *Ficus* spp.

Use	Place	*Ficus* sp.	Plant Part	References
Cardiac disorders, infusion, orally	Fiji	*religiosa*	Dried leaf	Singh 1986
Cardiac pain, decoction, orally	Somalia	*bussei*	Stem bark	Samuelsson et al. 1992
Cardiotonic, hot water extract, orally	Thailand	*hirta*	Dried root	Wasuwat 1967
Cardiotonic, orally	Saudi Arabia	*salicifolia*	Dried entire plant	Tariq et al. 1983
Dropsy, swellings, oedema	W. Africa	*asperifolia*	Latex	Burkill 1985
Dropsy, externally (rubbed on the body)	Bangladesh	*pumila*	Plant ash	Alam 1992
Edema, decoction, orally	Iran	*carica*	Dried branchlets	Zagari 1992
Epistaxis, nasally	Nepal	*benghalensis*	Stem bark	Bhattarai 1993b
Hematemesis, hot water extract, orally	S. Korea	*carica*	Unspecified	Han et al. 1984
Hemorrhage, decoction, orally	Iran	*carica*	Fresh bark	Zagari 1992
Hemostat	Turkey	*benghalensis*	Unspecified	Steinmetz 1957
Hemostatic, externally	Ivory Coast	*exasperata*	Fresh leaf	Kone-Bamba et al. 1987
Hemostatic, externally	S. Korea	*religiosa*	Unspecified	Han et al. 1984

(G) Modern Ethnomedical Uses of *Ficus* spp., Gastrointestinal

Use	Place	*Ficus* sp.	Plant Part	References
Acidant, infusion, orally	India	*carica*	Dried leaf	Ahmad and Beg 2001
Ankylostomiasis, orally	Paraguay	*eximia*	Latex	Dragendorff 1898
Anthelmintic, orally	Peru	*anthelmintica, glabrata*	Dried latex	Hansson et al. 1986, Ramirez et al. 1988
Anthelmintic, hot water extract, orally	Pakistan	*carica*	Fresh leaf	Ahmed et al. 1990
Anthelmintic, hot water extract, orally	Brazil	*doliaria*	Leaf	Peckolt 1942
Anthelmintic, orally	Tanzania	*exasperata*	Fresh leaf	Chhabra et al. 1984
Anthelmintic, orally	Paraguay	*eximia*	Latex	Dragendorff 1898
Anthelmintic, orally	Peru	*insipida*	Fresh latex	Phillips 1990
Anthelmintic, infusion, orally	Brazil	*maxima, vermifuga*	Dried leaf	Diaz et al. 1997, Peckolt 1942
Anthelmintic, orally	Indonesia	*toxicaria*	Latex	Dragendorff 1898

(continued on next page)

TABLE 2.4 (continued)

Other Contemporary Ethnomedical Uses of *Ficus* spp.

Use	Place	*Ficus* sp.	Plant Part	References
Anthelmintic, orally	Venezuela	Unspecified	Fresh latex	Jaffe 1943
Antiemetic, hot water extract, orally	Peru	*nitida*	Dried leaf	Ramirez et al. 1988
Antiparasitic purgative, orally	Colombia	Unspecified	Latex	Laferriere 1994
Appetite restorer, infusion, orally	Madagascar	*megapoda*	Leaf	Novy 1997
Ascaricide, orally	Tanzania	*exasperata*	Dried flowers	Chhabra et al. 1984
Constipation, orally	United States	*carica*	Dried fruit	Giordano and Levine 1989
Cramping, abdominal, due to sluggish bowel, orally	United States	*carica*	Dried fruit	Giordano and Levine 1989
Digestant, orally	Iran	*carica*	Dried fruit	Zagari 1992
Digestive, decoction, orally	Nicaragua	*insipida*	Sap	Coe and Anderson 1996a
Digestive aid, orally	Peru	*insipida*	Fresh latex	Phillips 1990
Gastrointestinal disorders, orally	United States	*carica*	Dried fruit	Giordano and Levine 1989
Laxative, orally	United States	*carica*	Dried fruit	Anon. 1931
Laxative, hot water extract, orally	Peru	*carica*	Dried fruit	Ramirez et al. 1988
Laxative, decoction, orally	Italy	*carica*	Dried fruit	Amico 1972
Laxative, orally	Iran	*carica*	Dried fruit, baked dried leaf, and latex	Zagari 1992
Laxative, orally	Morocco	*carica*	Fruit	Bellakhdar et al. 1991
Laxative, hot water extract, orally	Saudi Arabia	*carica*	Dried entire plant	Ashy and El-Tawil 1981
Laxative in connection with serious diseases, oil from frying aerial root tips with coconut cream, orally	Cook Islands	*prolixa*	Aerial root tips	Whistler 1985
Laxative, orally	India	*religiosa*	Dried fruit	Shah 1982
Parasites, intestinal, decoction, orally	Honduras	*cotinifolia*	Dried flower, leaf, and stem	Lentz et al. 1998
Parasites, intestinal, decoction, orally	Honduras	*insipida*	Dried leaf and stem	Lentz et al. 1998
Purgative, orally	Hawaii	*benghalensis*	Dried fruit	Locher et al. 1996

TABLE 2.4 (continued)
Other Contemporary Ethnomedical Uses of *Ficus* spp.

Use	Place	*Ficus* sp.	Plant Part	References
Purgative, wine extract, orally	Ivory Coast	*capensis*	Fruit	Bouquet and Debray 1974
Purgative, orally	Argentina	*insipida*	Latex from stem	Desmarchelier et al. 1996
Purgative, infusion, orally	India	*religiosa*	Dried leaf	Ahmad and Beg 2001
Polydipsia, ash mixed with water, orally	India	*religiosa*	Stem bark ash	Reddy et al. 1989
Roundworms, orally (taken mixed with honey)	India	*elastica*	Latex	Nagaraju and Rao 1990
Tapeworms, orally (taken mixed with honey)	India	*elastica*	Latex	Nagaraju and Rao 1990
Vermicide, water extract, orally	Nepal	*benghalensis*	Leaf juice	Bhattarai 1992a
Vermifuge, orally	Peru	*citrifolia, insipida* var. *insipida*	Latex	Duke and Vasquez 1994
Vermifuge, orally (taken with black pepper and water)	India	*religiosa*	Leaf	Singh et al. 1996a
Worms, decoction, orally, for 7 days	India	*benghalensis*	Root	Reddy et al. 1989
Worms, decoction, orally	Honduras	*cotinifolia*	Dried flower, leaf, and stem	Lentz et al. 1998
Worms, decoction, orally	Honduras	*insipida*	Dried leaf and stem	Lentz et al. 1998
Worms	Peru	*paraensis*	Latex	Duke and Vasquez 1994
Worms, intestinal	Peru	*yoponensis*	Latex	Duke and Vasquez 1994
Worms, in stomach, orally (taken mixed with honey)	India	*elastica*	Latex	Nagaraju and Rao 1990

(H) Modern Ethnomedical Uses of *Ficus* spp., Liver

Use	Place	*Ficus* sp.	Plant Part	References
Cholagogue, hot water extract, orally	Niger	*gnaphalocarpa*	Leaf	Baoua et al. 1976
Jaundice, hot water extract, orally	India	*hispida*	Dried bark	Acharya and Kumar 1984
Jaundice, infusion, orally	Madagascar	*megapoda*	Leaf	Novy 1997
Jaundice, infusion, orally	Fiji	*religiosa*	Dried bark, leaf and seed	Singh 1986

(continued on next page)

TABLE 2.4 (continued)
Other Contemporary Ethnomedical Uses of *Ficus* spp.

Use	Place	*Ficus* sp.	Plant Part	References
Jaundice, orally	Rotuma	*tinctoria*	Fresh leaf	McClatchey 1996
Liver conditions, orally	United States	*carica*	Fresh fruit	Liebstein 1927
Liver disease, hot water extract, orally	Taiwan	*awkeotsang, beecheyana, retusa*	Dried root	Yang et al. 1987
Liver problems	W. Africa	*asperifolia*	Bark	Burkill 1985
Megalospleny	Guinea	*asperifolia*	Unspecified	Ayensu 1978
(I) Modern Ethnomedical Uses of *Ficus* spp., Kidney				
Diuretic	W. Africa	*asperifolia*	Leaf, bark, root	Burkill 1985
Diuretic, orally (root paste mixed with salt), for 8 days	India	*benghalensis*	Aerial parts	Singh and Maheshwari 1994
Diuretic, orally	Guinea	*capensis*	Latex	Vasileva 1969
Diuretic, decoction, orally	Italy	*carica*	Latex	De Feo et al. 1992
Diuretic, hot water extract, orally	Peru	*carica*	Dried fruit	Ramirez et al. 1988
Diuretic, hot water extract, orally	Pakistan	*carica*	Fresh leaf	Ahmed et al. 1990
Diuretic, orally	India	*hispida*	Unspecified	Nyman et al. 1998
Kidney problems	W. Africa	*asperifolia*	Leaf, bark, root	Burkill 1985
(J) Modern Ethnomedical Uses of *Ficus* spp., Pancreas				
Diabetes	Unspecified	*benghalensis*	Unspecified	Duke 1993
Diabetes	Turkey	*benghalensis*	Unspecified	Steinmetz 1957
Diabetes, orally (root paste mixed with salt), for 8 days	India	*benghalensis*	Aerial parts	Singh and Maheshwari 1994
Diabetes, orally	India	*benghalensis*	Unspecified	Cherian and Augusti 1993
Diabetes, water extract, orally	India	*benghalensis*	Dried bark and juice	Jain and Sharma 1967, Geetha et al. 1994, Augusti et al. 1994, Deshmukh et al. 1960
Diabetes, decoction, orally (with *Terminalia bellerica*)	Rodrigues Islands	*benghalensis*	Root	Gurib-Fakim et al. 1996

TABLE 2.4 (continued)
Other Contemporary Ethnomedical Uses of *Ficus* spp.

Use	Place	*Ficus* sp.	Plant Part	References
Diabetes, decoction, orally	India	*benghalensis*	Dried root	Shah and Gopal 1985
Diabetes, decoction, orally	India	*benghalensis*	Stem bark	Kar et al. 1999
Diabetes, orally	India	*benghalensis*	Dried stem bark, dried aerial root	Khan and Singh 1996
Diabetes	Spain	*carica*	Leaf	Dominguez et al. 1996, Perez et al. 1996
Diabetes, decoction, orally	Taiwan	*erecta* var. *beecheyana*	Dried stem parenchyma	Lin 1992
Diabetes, orally	Pakistan	*glomerata*	Dried fruit	Akhtar 1992
Diabetes, decoction, orally	India	*glomerata*	Root	Kar et al. 1999
Diabetes, hot water extract, orally	India	*racemosa*	Dried bark and fruit	Jain and Sharma 1967
Diabetes, hot water extract, orally	India	*racemosa*	Dried fruit and dried bark	Jain and Verma 1981
Diabetes, hot water extract, orally	S. Korea	*religiosa*	Unspecified	Han et al. 1984
Diabetes, hot water extract, orally	Thailand	*religiosa*	Dried root	Mueller-Oerlinghausen et al. 1971
Diabetes, hot water extract, orally	India	*religiosa*	Dried bark	Jain and Sharma 1967
Diabetes, hot water extract, orally	India	*religiosa*	Trunk bark	Dragendorff 1898
Diabetes, dried and powdered fruits 10 g/2 × per day; or fresh ripe fruit 25 g/3 × per day, orally	India	*virens*	Dried fruit	Alam et al. 1990

(K) Modern Ethnomedical Uses of *Ficus* spp., Gynecological/Urogenital Problems

Abortifacient	W. Africa	*asperifolia*	Leaf	Burkill 1985, Ayensu 1978
Abortifacient, orally (with *Xylopia* sp. seeds)	Guinea	*asperifolia*	Leaf	Vasileva 1969
Abortifacient, water extract, orally (in a drug mixture)	India	*benghalensis*	Fresh root	Hemadri and Sasibhushana Rao 1983
Abortifacient, infusion, orally	Sierra Leone	*exasperata*	Dried leaf	Macfoy and Sama 1983

(continued on next page)

TABLE 2.4 (continued)
Other Contemporary Ethnomedical Uses of *Ficus* spp.

Use	Place	*Ficus* sp.	Plant Part	References
Abortifacient, orally (with *Embelia ribes* and borax with cow's milk)	India	*religiosa*	Dried leaf	Lal and Lata 1980
Abortifacient, juice of macerated leaves, orally	Gabon	*vogeliana*	Leaf	Raponda-Walker and Sillans 1961
Abortifacient, water extract, orally	Guinea	*vogeliana*	Leaf	Vasileva 1969
Antifertility agent, female, hot water extract, orally	Papua New Guinea	*benjamina, subnervosa*	Unspecified	Womersley 1974
Aphrodisiac, wine extract, orally	Ivory Coast	*capensis*	Fruit	Bouquet and Debray 1974
Aphrodisiac, orally	Guinea	*capensis*	Latex	Vasileva 1969
Aphrodisiac, hot water extract, orally	Ivory Coast	*capensis*	Unspecified	Kerharo and Bouquet 1950
Birth, decoction, orally	Honduras	*insipida*	Dried leaf and stem	Lentz et al. 1998
Birth, to ease delivery, decoction, orally	Peru	*carica*	Unspecified	Duke and Vasquez 1994
Birth, for stimulating contractions, water extract, orally/externally (rubbed on abdomen)	Sierra Leone	*exasperata*	Dried leaf	Kargbo 1984
Birth, given after, decoction, orally	Malaysia	*fistulosa*	Root	Burkill 1966
Birth, given after, hot water extract, orally	Malaysia	*hispida*	Leaf	Burkill 1966
Conception, helping induce it	India	*religiosa*	Unspecified	Sharma et al. 1992b
Contraceptive, male/female, prevents conception permanently, decoction, orally	India	*racemosa*	Fresh stem	Nisteswar 1988
Contraceptive, female, prevents conception permanently, hot water extract, orally	India	*racemosa*	Fresh stem bark	Billore and Audichya 1978
Contraceptive, hot water extract, orally	Papua New Guinea	*wassa*	Root	Holdsworth 1977
Ecbolic, hot water extract, orally	Niger	*gnaphalocarpa*	Leaf	Baoua et al. 1976
Ecbolic	Somalia	*sycomorus*	Root stem bark	Samuelsson et al. 1992
Ecbolic	Uganda	*sycomorus*	Root	Mitchell 1938

TABLE 2.4 (continued)
Other Contemporary Ethnomedical Uses of *Ficus* spp.

Use	Place	*Ficus* sp.	Plant Part	References
Emmenagogue	Africa	*asperifolia*	Unspecified	Ayensu 1978
Emmenagogue	Europe	*carica*	Unspecified	Jöchle 1974
Emmenagogue, hot water extract, orally/as a pessary, vaginal	Arabic countries	*carica*	Unspecified	Razzack 1980
Emmenagogue, hot water extract, orally	Reunion Island	*laterifolia*	Dried entire plant	Vera et al. 1990
Fertility, increases it, orally (with traditional salt)	Papua New Guinea	*nasuta*	Dried leaf	Holdsworth and Sakulas 1986
Galactagogue, externally	E. Africa	*capensis*	Bark	Bally 1937
Galactagogue, hot water extract, orally	Mozambique	*capensis*	Bark	Amico 1977
Galactagogue, wine extract, orally	Ivory Coast	*capensis*	Fruit	Bouquet and Debray 1974
Galactagogue, hot water extract, orally	Mozambique	*ingens*	Unspecified	Amico 1977
Galactagogue, orally	Tanganyika	*natalensis*	Unspecified	Watt and Breyer-Brandwijk 1962
Galactagogue, hot water extract, orally	Malaysia	*obpyramidata*	Entire plant	Burkill 1966
Galactagogue, juice, orally (pressed with a *Zingiber* leaf)	Papua New Guinea	*obpyramidata*	Dried stem bark	Holdsworth et al. 1983
Hemorrhage (in a female), decoction, orally	Honduras	*insipida*	Dried leaf and stem	Lentz et al. 1998
Hemorrhage, menstrual, orally (taken with rice water)	Bangladesh	*hispida*	Dried root	Alam 1992
Impotence, decoction, orally	Somalia	*bussei*	Stem bark	Samuelsson et al. 1992
Lactation, in failure of, hot water decoction, orally	India	*glomerata*	Unspecified	Jain and Tarafder 1970
Lactation, increases milk flow, orally	Vietnam	*hispida*	Fruit	Petelot 1954
Lactation, increases milk flow, hot water extract, orally	E. Africa	*stuhlmannii*	Root	Kokwaro 1976
Lactation, induces it, decoction, orally	Somalia	*sycomorus*	Stem bark	Samuelsson et al. 1992
Lactation, induces it, water extract, orally	E. Africa	*thonningii*	Bark, root	Kokwaro 1976

(continued on next page)

TABLE 2.4 (continued)
Other Contemporary Ethnomedical Uses of *Ficus* spp.

Use	Place	*Ficus* sp.	Plant Part	References
Lactation, stimulates it, hot water extract, orally	Tanganyika	*thonningii*	Root	Watt and Breyer-Brandwijk 1962
Menorrhagia, orally (in a drug mixture)	Vanuatu	*adenosperma*	Latex	Bourdy and Walter 1992
Menorrhagia, orally	India	*benghalensis*	Dried root	Joshi et al. 1980
Menorrhagia, hot water extract, orally	India	*glomerata*	Unspecified	Jain and Tarafder 1970
Menorrhagia, 5 g of ash with *Saraca asoca* bark and multivitamin tablet	India	*racemosa*	Bark ash	Vedavathyk and Rao 1995
Menstrual cycle, regulating it	W. Africa	*asperifolia*	Leaf	Burkill 1985
Menstruation: for dysmenorrhea, decoction/powder, orally	Tanganyika	*mucuso*	Root	Haerdi 1964
Menstruation, regulating it, juice, orally	Nepal	*religiosa*	Stem bark juice	Dangol and Gurung 1991
Menstruation, stimulating it in a pregnant woman [= abortifacient], orally (with *Xylopia* sp. seeds)	Guinea	*asperifolia*	Leaf	Vasileva 1969
Placenta, draws the fragments of, 10 crushed fruits in water as a drink, orally	Vanuatu	*septica* var. *cauliflora*	Fruit	Bourdy and Walter 1992
Placental retention, decoction, orally	Somalia	*bussei*	Stem bark	Samuelsson et al. 1992
Sexual potency, for the increasing of, male, orally (taken with *Cocos nucifera* fruit)	India	*benghalensis*	Latex	Lal and Yadav 1983
Spermatorrhea, orally (with Gurhal flowers)	India	*benghalensis*	Aerial parts	Singh and Maheshwari 1994
Spermatorrhea, orally	India	*benghalensis*	Latex	Singh and Ali 1992, Singh and Maheshwari 1994
Spermatorrhea	India	*glomerata, religiosa*	Unspecified	Jain and Tarafder 1970
Sterility, hot water extract, orally	Senegal	*capensis*	Dried stem bark	Tignokpa et al. 1986

TABLE 2.4 (continued)
Other Contemporary Ethnomedical Uses of *Ficus* spp.

Use	Place	*Ficus* sp.	Plant Part	References
Sterility, female, orally	Ivory Coast	*capensis*	Latex	Vasileva 1969
Sterility, female, hot water extract, orally	Ivory Coast	*capensis*	Unspecified	Kerharo and Bouquet 1950
Sterility, female, orally	Zimbabwe	*pretoriae*	Root	Watt and Breyer-Brandwijk 1962
Sterility, female, hot water extract, orally	East Africa	*vogelii* aff.	Dried bark	Hedberg et al. 1983
Sterility-inducing, orally (with *Cuminum cyminum* seed)	India	*arnottiana*	Dried leaf	Lal and Lata 1980
Syphilis, hot water extract, orally	India	*religiosa*	Trunk bark	Dragendorff 1898
Urogenital problems	Africa	*asperifolia*	Unspecified	Ayensu 1978
Uterine tonic, infusion, orally	India	*religiosa*	Dried fruit	Kakrani and Saluja 1993
Venereal diseases	West Africa	*asperifolia*	Fruit	Burkill 1985
Venereal disease, juice, orally, for 48 days	India	*racemosa*	Root exudates	Ramachandran and Nair 1981b

(L) Modern Ethnomedical Uses of *Ficus* spp., Infectious Diseases

Use	Place	*Ficus* sp.	Plant Part	References
Filarial swellings, hot water extract, externally	Cook Islands	*prolixa*	Root, aerial	Whistler 1985, Comley 1990
Leprosy, hot water extract	Nepal	*cunia*	Fruit and bark	Suwal 1970
Leprosy, hot water extract, orally	India	*cunia*	Dried fruit and bark	Chopra 1933, Kapur 1983
Leprosy, hot water extract, orally	India	*hispida*	Dried bark	Acharya and Kumar 1984
Leprosy, hot water extract, orally	India	*hispida*	Fruit	Dragendorff 1898
Leprosy, hot water extract, externally	India	*racemosa*	Dried fruit and dried bark	Jain and Verma 1981
Leprosy, decoction, externally	Fiji	*religiosa*	Dried bark and leaf	Singh 1986
Leprosy, hot water extract	India	*talbotii*	Bark	Chopra 1933
Leprotic lesions, externally	Iran	*carica*	Fresh latex	Zagari 1992
Measles, decoction, orally	Iran	*carica*	Dried fruit	Zagari 1992
Measles, German, decoction, orally	Iran	*carica*	Dried fruit	Zagari 1992

(continued on next page)

TABLE 2.4 (continued)
Other Contemporary Ethnomedical Uses of *Ficus* spp.

Use	Place	*Ficus* sp.	Plant Part	References
Sleeping sickness, externally (in a drug mixture as part of complex therapy)	Ivory Coast	*exasperata*	Leaf	Kerharo 1974
Sleeping sickness, hot water extract, orally and externally as a vapor bath (taken with *Afzelia africana* and *Tamarindus indica* bark and leaves)	Ivory Coast	Unspecified	Bark and leaf	Kerharo 1974
Sleeping sickness, hot water extract, as an inhalation and a bath (mixed with *Cussonia barteri* leaves)	Ivory Coast	*vallis-choudae*	Leaf	Kerharo 1974
Sleeping sickness, externally, as a poultice (in a drug mixture as part of complex therapy)	Ivory Coast	*vallis-choudae*	Latex	Kerharo 1974
Smallpox, decoction, orally	Iran	*carica*	Dried fruit	Zagari 1992
Smallpox, prevention of, hot water extract, orally	India	*racemosa*	Dried fruit and dried bark	Jain and Verma 1981
Typhoid, orally (taken with *Diospyros melanoxylon* gum and *Balanites roxburghii* seeds)	India	*microcarpa*	Dried root	Anis and Iqbal 1986

(M) Modern Ethnomedical Uses of *Ficus* spp., Poison-Related

Use	Place	*Ficus* sp.	Plant Part	References
Antidote	Upper Volta	*asperifolia*	Unspecified	Ayensu 1978
Antidote, orally	Papua New Guinea	*copiosa*	Fresh leaf	Holdsworth and Balun 1992
Antidote, orally	Papua	*septica*	Fresh root	Holdsworth 1974b
Antivenin	India	*natalensis*	Root	Selvanayahgam et al. 1994
Antivenin, juice, orally	India	*religiosa*	Bark	Selvanayahgam et al. 1994
Arrow poison	W. Africa	*asperifolia*	Leaf	Burkill 1985
Poison, used as	Malaysia	*aurantiacea*	Fruit	Ahmad and Holdsworth 1994

TABLE 2.4 (continued)
Other Contemporary Ethnomedical Uses of *Ficus* spp.

Use	Place	*Ficus* sp.	Plant Part	References
Scorpion sting, externally	Turkey	*carica*	Latex	Honda et al. 1996, Yesilada et al. 1995, Fujita et al. 1995
Snake bite, externally	Peru	*glabrata*	Dried latex	Hansson et al. 1986
Snake bite, infusion, externally/decoction, orally	Colombia	*nymphaeifolia*	Aerial parts	Otero et al. 2000a, 2000b
Sting ray punctures, externally	Peru	*glabrata*	Dried latex	Hansson et al. 1986
(N) Modern Ethnomedical Uses of *Ficus* spp., Veterinary				
Digestive for domestic animals, decoction, orally (mixed in the feed)	Italy	*carica*	Latex	De Feo et al. 1992
Fodder for cattle and goats	India	*sarmentosa* var. *ludduca*	Fresh leaf	Chandra and Pandey 1983
Fodder for pigs	Spain	*carica*	Dried fruit	Martinez-Lirola et al. 1996
Increases flow of milk when fed to cows	Zaire	*sycomorus*	Leaf, fruit	Watt and Breyer-Brandwijk 1962
Increases flow of milk when fed to cows	S. Africa	*capensis*	Bark and leaf	Watt and Breyer-Brandwijk 1962
Increases flow of milk when fed to cows	S. Africa	*ingens*	Unspecified	Watt and Breyer-Brandwijk 1962
Placenta, removes it in cattle, orally (given with *Trachyspermum ammi* crude sugar)	Nepal	*lacor*	Stem bark	Bhattarai 1992b
Rinderpest	India (Santal)	*benghalensis*	Unspecified	Jain and Tarafder 1970
(O) Modern Ethnomedical Uses of *Ficus* spp., Other				
Bones, broken, hastens the repair of, externally	India	*tsiela*	Dried leaf	Mitra et al. 1978
Bone fractures, externally, tied on point of the fracture	Andaman Islands	Unspecified	Root, aerial	Awasthi 1991

(continued on next page)

TABLE 2.4 (continued)
Other Contemporary Ethnomedical Uses of *Ficus* spp.

Use	Place	*Ficus* sp.	Plant Part	References
Buccal problems, infusion, orally	Canary Islands	*carica*	Dried shoots	Darias et al. 1986
Fontanel, promotes the closing of, externally (as a poultice with clay and water)	Sierra Leone	*capensis*	Dried leaf	Macfoy and Sama 1983
Frostbite, externally	S. Korea	*carica*	Unspecified	Han et al. 1984
Gangrene, orally	United States	*carica*	Fresh fruit	Liebstein 1927
Goiter, decoction, orally	Fiji	*religiosa*	Dried root	Singh 1986
Hair, causing it to grow long, paste/powder, externally	India	*benghalensis*	Dried root	Tiwari et al. 1979
Hernia, hot water extract, orally (in a drug mixture)	Tanzania	*vogelii* aff.	Dried root	Hedberg et al. 1983
Parasites, orally	Nicaragua	*maxima*	Sap	Barrett 1994
Scurvy, orally	United States	*carica*	Fresh fruit	Liebstein 1927
Thorns, removing them, externally	Oman	*carica*	Latex	Ghazanfar and Al-Sabahi 1993
Thorns, removing them from flesh, externally	Nepal	*pseudo-sycomorus*	Latex	Manandhar 1995

REFERENCES

Abdullahi, J.J. 2006. Herbal extract with antiviral properties. USPTO Application #: 20060024386, Class: 424725000 (USPTO).

Acharya, B.M., and K.A. Kumar. 1984. Chemical examination of the bark of *Ficus hispida* Linn. *Curr Sci.* 53: 1034–5.

Adhya, M., B. Singha, and B.P. Chatterjee. 2006. *Ficus cunia* agglutinin for recognition of bacteria. *Indian J Biochem Biophys.* 43: 94–7.

Ahmad, I., and A.Z. Beg. 2001. Antimicrobial and phytochemical studies on 45 Indian medicinal plants against multi-drug resistant human pathogens. *J Ethnopharmacol.* 74: 113–23.

Ahmad, F.B., and D.K. Holdsworth. 1994. Medicinal plants of Sabah, Malaysia. Part II. The Muruts. *Int J Pharmacog.* 32: 378–83.

Ahmed, W., Z. Ahmed, and A. Malik. 1990. Triterpenes from the leaves of *Ficus carica*. *Fitoterapia* 61: 373.

Ajose, F.O. 2007. Some Nigerian plants of dermatologic importance. *Int J Dermatolog.* 46(Suppl. 1): 48–55.

Akah, P.A., K.S. Gamaniel, C.N. Wambebe, A. Shittu, S.D. Kapu, and O.O. Kunle. 1997. Studies on the gastrointestinal properties of *Ficus exasperata*. *Fitoterapia* 68: 17–20.

Akah, P.A., O.E. Orisakwe, K.S. Gamaniel, and A. Shittu. 1998. Evaluation of Nigerian traditional medicines. II. Effects of some Nigerian folk remedies on peptic ulcer. *J Ethnopharmacol.* 62: 123–7.

Akhtar, M.S. 1992. Hypoglycaemic activities of some indigenous medicinal plants traditionally used as antidiabetic drugs. *J Pak Med Assoc.* 42: 271–7.

Alam, M.K. 1992. Medical ethnobotany of the Marma tribe of Bangladesh. *Econ Bot.* 46: 330–5.

Alam, M.M., M.B. Siddiqui, and W. Husain. 1990. Treatment of diabetes through herbal drugs in rural India. *Fitoterapia* 61: 240–2.

Al-Rawi, A. 1964. *Medicinal Plants of Iraq.* Baghdad: Ministry of Agriculture Technology, Bull. No. 15.

Altschul, S. Von Reis. 1973. *Drugs and Foods from Little-Known Plants.* Cambridge, MA: Harvard University Press.

ALUKA. 2003. [Online Database]. Princeton, NJ 08540. URL: http://www.aluka.org/ [accessed July 23, 2009].

Amico, A. 1972. *Piante medicinali.* Bari: Adriantica Editrice.

Amico, A. 1977. Medicinal plants of Southern Zambesia. *Fitoterapia* 48: 101–39.

Amos, S., L. Binda, B. Chindo et al. 2001. Evaluation of methanolic extract of *Ficus platyphylla* on gastrointestinal activity. *Indian J Exp Biol.* 39: 63–7.

Anderson, E.F. 1986a. Ethnobotany of hill tribes of Northern Thailand. I. Medicinal plants of Akha. *Econ Bot.* 40: 38–53.

Anderson, E.F. 1986b. Ethnobotany of hill tribes of Northern Thailand. II. Lahu medicinal plants. *Econ Bot.* 40: 442–50.

Anis, M., and M. Iqbal. 1986. Antipyretic utility of some Indian plants in traditional medicine. *Fitoterapia* 57: 52–5.

Annan, K., and P.J. Houghton. 2008. Antibacterial, antioxidant and fibroblast growth stimulation of aqueous extracts of *Ficus asperifolia* Miq. and *Gossypium arboreum* L., wound-healing plants of Ghana. *J Ethnopharmacol.* 119: 141–4.

Anokbonggo, W.W., R. Odoi-Adome, and P.M. Oluju. 1990. Traditional methods in management of diarrheal diseases in Uganda. *Bull World Health Organ.* 68: 359–63.

Anon. 1974. *A Barefoot Doctor's Manual.* DHEW Publication No. (NIH): 75-695.

Anon. 1978. *List of Plants.* Kyoto Herbal Garden, Pharmacognostic Research Lab., Central Research Division, Takeda Chem. Industries, Ltd., Ichijoji, Sakyoku, Kyoto, Japan.

Anon. 1931. *The Herbalist.* Hammond: Hammond Book Company.

Anon. 1948–1976. *The Wealth of India Raw Materials.* 11 Vols. New Delhi: Publications and Information Directorate, CSIR.

Arnold, H.J., and M. Gulumian. 1984. Pharmacopoeia of traditional medicine in Venda. *J Ethnopharmacol.* 12: 35–74.

Arseculeratne, S.N., A.A.L. Gunatilaka, and R.G. Panabokke. 1985. Studies on medicinal plants of Sri Lanka. Part 14. Toxicity of some traditional medicinal herbs. *J Ethnopharmacol.* 13: 323–35.

Ashy, M.A., and B.A.H. El-Tawil. 1981. Constituents of local plants. Part 7. The coumarin constituents of *Ficus carica* L. and *Convolvulus aeyranisis* L. *Pharmazie* 36: 297.

Augusti, K.T., R.S. Daniel, S. Cherian, C.G. Sheela, and C.R.S. Nair. 1994. Effect of leucopelargonin derivative from *Ficus bengalensis* Linn. on diabetic dogs. *Indian J Med Res.* 99: 82–6.

Awasthi, A.K. 1991. Ethnobotanical studies on the Negrito islanders of Andaman Islands, India—The Great Andamanese. *Econ Bot.* 45: 274–80.

Ayensu, E.S. 1978. *Medicinal Plants of West Africa.* Algonac: Reference Publications.

Azevedo, M.P. 1949. [Mechanism of anti-coagulant action of the latex of the *Ficus glabrata* H.B.K.] *Mem Inst Butantan* 22: 25–30.

Bafor, E.E., and O. Igbinuwen. 2009. Acute toxicity studies of the leaf extract of *Ficus exasperata* on haematological parameters, body weight and body temperature. *J Ethnopharmacol.* 123: 302–7.

Bajpai, A., J.K. Ojha, and H.R. Sant. 1995. Medicobotany of the Varanasi District. *Int J Pharmacog.* 33: 172–6.

Bally, P.R.O. 1937. Native medicinal and poisonous plants of East Africa. *Bull Misc Information Roy Bot Gard.* 1: 10–26.

Baoua, M., J. Fayn, and J. Bassiere. 1976. Preliminary phytochemical testing of some medical plants of Niger. *Plant Med Phytother.* 10: 251–66.

Barrett, B. 1994. Medicinal plants of Nicaragua's Atlantic coast. *Econ Bot.* 48: 8–20.

Basudan, O.A., M. Ilyas, M. Parveen, H.M.H. Muhisen, and R. Kumar. 2005. A new chromone from *Ficus lyrata. J Asian Nat Prod Res.* 7: 81–5.

Baumgartner, B., C.A.J. Erdelmeier, A.D. Wright, T. Rali, and O. Sticher. 1990. An antimicrobial alkaloid from *Ficus septica. Phytochemistry* 29: 3327–30.

Bellakhdar, J., R. Claisse, J. Fleurentin, and C. Younos. 1991. Repertory of standard herbal drugs in the Moroccan pharmacopoeia. *J Ethnopharmacol.* 35: 123–43.

Bennet, S.S.R. 1983. Ethnobotanical studies in Sikkim. *Indian For.* 109: 477–81.

Berg, C.C. 1991. Moraceae. In *Flora Zambesiaca*, Vol. 9, Part 6. London: Crown Agents for Oversea Governments and Administrations.

Bhat, R.B., E.O. Eterjere, and V.T. Oladipo. 1990. Ethnobotanical studies from Central Nigeria. *Econ Bot.* 44: 382–90.

Bhat, R.B., and T.V. Jacobs. 1995. Traditional herbal medicine in Transkei. *J Ethnopharmacol.* 48: 7–12.

Bhattarai, N.K. 1992a. Folk anthelmintic drugs of Central Nepal. *Int J Pharmacog.* 30: 145–50.

Bhattarai, N.K. 1993a. Folk herbal remedies for diarrhoea and dysentery in Central Nepal. *Fitoterapia* 64: 243–50.

Bhattarai, N.K. 1991. Folk herbal medicines of Makawanpur District, Nepal. *Int J Pharmacog.* 29: 284–95.

Bhattarai, N.K. 1992b. Folk use of plants in veterinary medicine in Central Nepal. *Fitoterapia* 63: 497–506.

Bhattarai, N.K. 1993b. Medical ethnobotany in the Rapti Zone, Nepal. *Fitoterapia* 64: 483–93.

Billore, K.V., and K.C. Audichya. 1978. Some oral contraceptives—family planning tribal way. *J Res Indian Med Yoga Homeopath.* 13: 104–9.

Bisset, N.G., and G. Mazars. 1984. Arrow poisons in South Asia. Part 1. Arrow poisons in ancient India. *J Ethnopharmacol.* 12: 1–24.

Boom, B.M., and S. Moestl. 1990. Ethnobotanical notes of Jose M. Cruxent from the Franco-Venezuelan expedition to the headwaters of the Orinoco river, 1951–1952. *Econ Bot.* 44: 416–19.

Boukef, K., H.R. Souissi, and G. Balansard. 1982. Contribution to the study on plants used in traditional medicine in Tunisia. *Plant Med Phytother.* 16: 260–79.

Bouquet, A., and M. Debray. 1974. Plantes médicinales de la Côte d'Ivoire. Paris: Grstom.

Bourdy, G., and A. Walter. 1992. Maternity and medicinal plants in Vanuatu. I. The cycle of reproduction. *J Ethnopharmacol.* 37: 179–96.

Bown, D. 1995. *Encyclopaedia of Herbs and Their Uses.* London: Dorling Kindersley.

Brutus, T.C., and A.V. Pierce-Noel. 1960. *Les Plantes et les Legumes d'Haiti qui Guerissent.* Port-Au-Prince: Imprimerie De L'Etat.

Burkill, H.M. 1985. *The Useful Plants of West Tropical Africa.* Vol. 4. Richmond: Royal Botanic Gardens, Kew.

Burkill, J.D. 1966. *A Dictionary of the Economic Products of the Malay Peninsula.* Kuala Lumpur: Art Printing Works.

Cai, Q.Y., H.B. Chen, S.Q. Cai et al. 2007. [Effect of roots of *Ficus hirta* on cocaine-induced hepatotoxicity and active components.] *Zhongguo Zhong Yao Za Zhi.* 32: 1190–3.

Chandra, K., and H.C. Pandey. 1983. Collection of plants around Agora-Dodital in Uttarkashi District of Uttar Pradesh, with medicinal values and folk-lore claims. *Int J Crude Drug Res.* 21: 21–8.

Chang, M.S., Y.C. Yang, Y.C. Kuo et al. 2005. Furocoumarin glycosides from the leaves of *Ficus ruficaulis* Merr. var. *antaoensis*. *J Nat Prod.* 68: 11–3, Erratum 68: 634.

Channabasavaraj, K.P., S. Badami, and S. Bhojraj. 2008. Hepatoprotective and antioxidant activity of methanol extract of *Ficus glomerata*. *Nat Med (Tokyo)* 62: 379–83.

Cherian, S., and K.T. Augusti. 1993. Antidiabetic effects of a glycoside of leucopelargonidin isolated from *Ficus bengalensis* Linn. *Indian J Exp Biol.* 31: 26–9.

Chew, W.-L. 1989. *Ficus*. In *Flora of Australia Online*. [Online Database.] Australian Biological Resources Study, Canberra. URL: http://www.environment.gov.au/biodiversity/abrs/online-resources/flora/main/index.html [accessed July 24, 2008].

Chhabra, S.C., F.C. Uiso, and E.N. Mshiu. 1984. Phytochemical screening of Tanzanian medicinal plants. I. *J Ethnopharmacol.* 11: 157–79.

Chiang, Y.M., J.Y. Chang, C.C. Kuo, C.Y. Chang, and Y.H. Kuo. 2005. Cytotoxic triterpenes from the aerial roots of *Ficus microcarpa*. *Phytochemistry* 66: 495–501.

Chifundera, K., W.M. Mbuyi, and B. Kizungu. 1990. Phytochemical and antibacterial screening of extracts of *Ficus sycomorus*. *Fitoterapia* 61: 535–9.

Chindo, B.A., S. Amos, A.A. Odutola et al. 2003. Central nervous system activity of the methanol extract of *Ficus platyphylla* stem bark. *J Ethnopharmacol.* 85: 131–7.

Chindo, B.A., J.A. Anuka, L. McNeil et al. 2009. Anticonvulsant properties of saponins from *Ficus platyphylla* stem bark. *Brain Res Bull.* 78: 276–82.

Chopra, R.N. 1933. *Indigenous Drugs of India: their Medical and Economic Aspects*. Calcutta: The Art Press.

Chua, A.C.N., W.M. Chou, C.L. Chyan, and J.T.C. Tzen. 2007. Purification, cloning, and identification of two thaumatin-like protein isoforms in jelly fig (*Ficus awkeotsang*) achenes. *J Agric Food Chem.* 55: 7602–8.

Chua, A.C., E.S. Hsiao, Y.C. Yang, L.J. Lin, W.M. Chou, and J.T. Tzen. 2008. Gene families encoding 11S globulin and 2S albumin isoforms of jelly fig (*Ficus awkeotsang*) achenes. *Biosci Biotechnol Biochem.* 72: 506–13.

Coe, F.G., and G.J. Anderson. 1996a. Ethnobotany of the Garifuna of Eastern Nicaragua. *Econ Bot.* 50: 71–107.

Coe, F.G., and G.J. Anderson. 1996b. Screening of medicinal plants used by the Garifuna of Eastern Nicaragua for bioactive compounds. *J Ethnopharmacol.* 53: 29–50.

Comley, J.C.W. 1990. New macrofilaricidal leads from plants? *Trop Med Parasitol.* 41: 1–9.

Damu, A.G., P.C. Kuo, L.S. Shi et al. 2005. Phenanthroindolizidine alkaloids from the stems of *Ficus septica*. *J Nat Prod.* 68: 1071–5.

Damu, A.G., P.C. Kuo, L.S. Shi, C.Y. Li, C.R. Su, and T.S. Wu. 2009. Cytotoxic phenanthroindolizidine alkaloids from the roots of *Ficus septica*. *Planta Med*. March 18, 2009. [Epub ahead of print]

Dangol, D.R., and S.B. Gurung. 1991. Ethnobotany of the Tharu tribe of Chitwan District, Nepal. *Int J Pharmacog.* 29: 203–9.

Darbour, N., C. Bayet, S. Rodin-Bercion et al. 2007. Isoflavones from *Ficus nymphaefolia*. *Nat Prod Res.* 21: 461–4.

Darias, V., L. Bravo, E. Barquin, D.M. Herrera, and C. Fraile. 1986. Contribution to the ethno-pharmacological study of the Canary Islands. *J Ethnopharmacol.* 15: 169–93.

De Feo, V., R. Aquino, A. Menghini, E. Ramundo, and F. Senatore. 1992. Traditional phytotherapy in the Peninsula Sorrentina, Campania, Southern Italy. *J Ethnopharmacol.* 36: 113–25.

De Feo, V., and F. Senatore. 1993. Medicinal plants and phytotherapy in the Amalfitan Coast, Salerno Province, Campania, Southern Italy. *J Ethnopharmacol.* 39: 39–51.

Deshmukh, V.K., D.S. Shrotri, and R. Aiman. 1960. Isolation of a hypoglycemic principle from the bark of *Ficus bengalensis* Linn. *Indian J Physiol Pharmacol*. 4: 182–5.

Desmarchelier, C., A. Gurni, G. Ciccia, and A.M. Giulietti. 1996. Ritual and medicinal plants of the Ese'ejas of the Amazonian rainforest (Madre de Dios, Peru). *J Ethnopharmacol*. 52: 45–51.

Desmarchelier, C., M. Repetto, J. Coussio, S. Llesuy, and G. Ciccia. 1997. Total reactive antioxidant potential (TRAP) and total antioxidant reactivity (TAR) of medicinal plants used in Southwest Amazona (Bolivia and Peru). *Int J Pharmacog*. 35: 288–96.

Devaraj, K.B., P.R Kumar, and V. Prakash. 2009. Characterization of acid-induced molten globule like state of ficin. *Int J Biol Macromol*. 45: 248–54.

Diaz, G., A.C. Arruda, M.S.P. Arruda, and A.H. Muller. 1997. Methoxyflavones from *Ficus maxima*. *Phytochemistry* 45: 1697–9.

Diddiqui, M.B., and W. Husain. 1993. Traditional treatment of gonorrhoea through herbal drugs in the Province of Central Uttar Pradesh, India. *Fitoterapia* 64: 399–403.

Dominguez, X.A., and J.B. Alcorn. 1985. Screening of medicinal plants used by Huastec Mayans of Northeastern Mexico. *J Ethnopharmacol*. 13: 139–56.

Dominguez, E., J.R. Canal, M.D. Torres, J.E. Campillo, and C. Perez. 1996. Hypolipidaemic activity of *Ficus carica* leaf extract in streptozotocin-diabetic rats. *Phytother Res*. 10: 526–8.

Dragendorff, G. 1898. *Die Heilpflanzen der verschiedenen Volker und Zeiten*. Stuttgart: F. Enke.

Duke, J.A. 1993. *Dr. Duke's Phytochemical and Ethnobotanical Databases*. [Online Database]. USDA—ARS—NGRL, Beltsville Agricultural Research Center, Beltsville, Maryland. URL: http://www.ars-grin.gov/duke/ethnobot.html [accessed May 12, 2009].

Duke, J.A. 1975. Ethnobotanical observations on the Cuna Indians. *Econ Bot*. 29: 278–93.

Duke, J.A. 1972. *Isthmian Ethnobotanical Dictionary*. Fulton, MD.

Duke, J.A., and R. Vasquez. 1994. *Amazonian Ethnobotanical Dictionary*. Boca Raton: CRC Press.

EthnobotDB. *EthnobotDB—Worldwide Plant Uses*. 1996. [Online Database]. National Agricultural Library, Agricultural Genome Information System, Washington DC. URL: http://ukcrop.net/perl/ace/search/EthnobotDB [accessed July 24, 2009].

Etkin, N.L. 1997. Antimalarial plants used by Hausa in Northern Nigeria. *Trop Doct*. 27: 12–16.

Fazliana, M.S., H. Muhajir, H. Hazilawati, K. Shafii, and M. Mazleha. 2008. Effects of *Ficus deltoidea* aqueous extract on hematological and biochemical parameters in rats. *Med J Malaysia* 63 (Suppl. A): 103–4.

Fleurentin, J., G. Mazars, and J.M. Pelt. 1983. Additional information on the cultural background of drugs and medicinal plants of Yemen. *J Ethnopharmacol*. 8: 335–44.

Fleurentin, J., and J.M. Pelt. 1982. Repertory of drugs and medicinal plants of Yemen. *J Ethnopharmacol*. 6: 85–108.

Font Query, P. 1979. *Plantas Medicinales el Dioscorides Renovado*. 5th ed. Barcelona: Editorial Labor, S.A.

Friis, I. 1999. *Flora Somalia*, Vol. 2 [updated by M. Thulin 2008]. Royal Botanic Gardens, Kew.

Fujita, T., E. Sezik, M. Tabata et al. 1995. Traditional medicine in Turkey. VII. Folk medicine in Middle and West Black Sea regions. *Econ Bot*. 49: 406–22.

Gbeassor, M., A.Y., Kedjagni, K. Koumaglo et al. 1990. In vitro antimalarial activity of six medicinal plants. *Phytother Res*. 4: 115–7.

Geetha, B.S., B.C. Mathew, and K.T. Augusti. 1994. Hypoglycemic effects of leucodelphindin derivative isolated from *Ficus bengalensis* (Linn.). *Indian J Physiol Pharmacol*. 38: 220–2.

Ghafoor, A. 1985. Moraceae. In *Flora of Pakistan, eFloras*. Missouri Botanical Garden, St. Louis, MO & Harvard University Herbaria, Cambridge, MA. URL: http://www.efloras.org [accessed July 24, 2009].

Ghazanfar, S.A., and M.A. Al-Sabahi. 1993. Medicinal plants of Northern and Central Oman (Arabia). *Econ Bot.* 47: 89–98.

Gilani, A.H., M.H. Mehmood, K.H. Janbaz, A. Khan, and S.A. Saeed. 2008. Ethnopharmacological studies on antispasmodic and antiplatelet activities of *Ficus carica*. *J Ethnopharmacol.* 119: 1–5.

Giordano, J., and P.J. Levine. 1989. Botanical preparations used in Italian folk medicine: Possible pharmacological and chemical basis of effect. *Social Pharmacol.* 3: 83–110.

Giron, L.M., V. Freire, A. Alonzo, and A. Caceres. 1991. Ethnobotanical survey of the medicinal flora used by the Caribs of Guatemala. *J Ethnopharmacol.* 34: 173–87.

Goodman, S.M., and J.J. Hobbs. 1988. The ethnobotany of the Egyptian eastern desert: A comparison of common plant usage between two culturally distinct Bedouin groups. *J Ethnopharmacol.* 23: 73–89.

GRIN. Germplasm Resources Information Network. 1994. [Online Database]. USDA, ARS, National Genetic Resources Program. National Germplasm Resources Laboratory, Beltsville, Maryland. URL: http://www.ars-grin.gov/npgs/ [accessed July 22, 2009].

Grosvenor, P.W., P.K. Gothard, N.C. McWilliam, A. Supriono, and D.O. Gray. 1995a. Medicinal plants from Riau Province, Sumatra, Indonesia. Part 1. Uses. *J Ethnopharmacol.* 45: 75–95.

Grosvenor, P.W., A. Supriono, and D.O. Gray. 1995b. Medicinal plants from Riau Province, Sumatra, Indonesia. Part 2. Antibacterial and antifungal activity. *J Ethnopharmacol.* 45: 97–111.

Gupta, S., J.N.S. Yadava, and J.S. Tandon. 1993. Antisecretory (antidiarrhoeal) activity of Indian medicinal plants against *Escherichia coli* enterotoxin-induced secretion in rabbit and guinea pig ileal loop models. *Int J Pharmacog.* 31: 198–204.

Gurib-Fakim, A., M.D. Sweraj, J. Gueho, and E. Dulloo. 1996. Medicinal plants of Rodrigues. *Int J Pharmacog.* 34: 2–14.

Haerdi, F. 1964. Native medicinal plants of Ulanga District of Tanganyika (East Africa). Dissertation (University of Basel). Basel: Verlag fur Recht und Gesellschaft AG.

Han, D.S., S.J. Lee, and H.K. Lee. 1984. Ethnobotanical survey in Korea. In *Proc Fifth Asian Symp Med Plants Spices Seoul Korea August 20–24 1984*, ed. B.H. Han, D.S. Han, Y.N. Han, and W.S. Woo, 125–144.

Hanelt, P., R. Mansfeld, and R. Büttner, (ed.). 2001. *Mansfeld's Encyclopedia of Agricultural and Horticultural Crops.* Berlin/Heidelberg: Springer-Verlag.

Hansson, A., G. Veliz, C. Naquira, M. Amren, M. Arroyo, and G. Arevalo. 1986. Preclinical and clinical studies with latex from *Ficus glabrata* HBK, a traditional intestinal anthelminthic in the Amazonian area. *J Ethnopharmacol.* 17: 105–38.

Harden, G.J. 1999. *New South Wales Flora Online.* [Online Database]. National Herbarium of New South Wales, Royal Botanic Gardens & Domain Trust, Sydney Australia. URL: http://plantnet.rbgsyd.nsw.gov.au/ [accessed July 22, 2009].

Hartwell, J.L. 1967–1971. Plants used against cancer: A survey. *Lloydia* 30–34.

Hedberg, I., O. Hedberg, P.J. Madati, K.E. Mshigeni, E.N. Mshiu, and G. Samuelsson. 1983. Inventory of plants used in traditional medicine in Tanzania. II. Plants of the families Dilleniaceae-Opiliaceae. *J Ethnopharmacol.* 9: 105–27.

Hemadri, K., and S. Sasibhushana Rao. 1983. Antifertility, abortifacient and fertility promoting drugs from Dandakaranya. *Ancient Sci Life* 3: 103–107.

Hirschhorn, H.H. 1983. Botanical Remedies of the Former Dutch East Indies (Indonesia). Part I. Eumycetes, Pteridophyta, Gymnospermae, Angiospermae (Monocotyledons Only). *J Ethnopharmacol.* 7: 123–56.

Holdsworth, D.K. 1974a. A phytochemical survey of medicinal plants in Papua New Guinea. I. *Sci New Guinea* 2: 142–54.

Holdsworth, D.K. 1974b. A phytochemical survey of medicinal plants of the D'Entrecasteaux Islands, Papua. *Sci New Guinea* 2: 164–71.

Holdsworth, D. 1989. High altitude medicinal plants of Papua New Guinea. *Int J Crude Drug Res.* 27: 95–100.

Holdsworth, D.K. 1977. Medicinal Plants of Papua-New Guinea. Technical Paper No.175. Noumea, New Caledonia: South Pacific Commission.

Holdsworth, D. 1992. Medicinal plants of the Gazelle Penisula, New Britain Island, Papua New Guinea. Part I. *Int J Pharmacog.* 30: 185–90.

Holdsworth, D. 1984. Phytomedicine of the Madang Province, Papua New Guinea. Part I. Karkar Island. *Int J Crude Drug Res.* 22: 111–19.

Holdsworth, D. 1993. Medicinal plants of the Oro (Northern) Province of Papua New Guinea. *Int J Pharmacog.* 31: 23–28.

Holdsworth, D.K. 1980. Traditional medicinal plants of the North Solomons Province, Papua New Guinea. *Q J Crude Drug Res.* 18: 33–44.

Holdsworth, D. 1975. Traditional medicinal plants used in the treatment of malaria and fevers in Papua New Guinea. *P N G Med J* 18: 142–8.

Holdsworth, D., and L. Balun. 1992. Medicinal plants of the East and West Sepik Provinces, Papua New Guinea. *Int J Pharmacog.* 30: 218–22.

Holdsworth, D., O. Gideon, and B. Pilokos. 1989. Traditional medicine of New Ireland, Papua New Guinea. Part III. Konos, Central New Ireland. *Int J Crude Drug Res.* 27: 55–61.

Holdsworth, D.K., C.L. Hurley, and S.E. Rayner. 1980. Traditional medicinal plants of New Ireland, Papua New Guinea. *Q J Crude Drug Res.* 18: 131–9.

Holdsworth, D., B. Pilokos, and P. Lambes. 1983. Traditional medicinal plants of New Ireland, Papua New Guinea. *Int J Crude Drug Res.* 21: 161–8.

Holdsworth, D., and T. Rali. 1989. A survey of medicinal plants of the Southern Highlands, Papua New Guinea. *Int J Crude Drug Res.* 27: 1–8.

Holdsworth, D., and H. Sakulas. 1986. Medicinal plants of the Morobe Province. Part II. The Aseki Valley. *Int J Crude Drug Res.* 24: 31–40.

Honda, G., E. Yesilada, M. Tabata et al. 1996. Traditional medicine in Turkey. VI. Folk medicine in West Anatolia: Afyon, Kutahya, Denizli, Mugla, Aydin Provinces. *J Ethnopharmacol.* 53: 75–87.

Hsiao, E.S., J.C. Chen, H.Y. Tsai, K.H. Khoo, S.T. Chen, and J.T. Tzen. 2009. Determination of N-glycosylation site and glycan structures of pectin methylesterase in jelly fig (*Ficus awkeotsang*) achenes. *J Agric Food Chem.* July 13, 2009 [Epub ahead of print].

Hutchinson, J. 1925. *Ficus*, Linn. In *Flora Capensis: Being a Systematic Description of the Plants of the Cape Colony, Caffraria, & Port Natal (and neighbouring territories)*, ed. W.H. Harvey, O.W. Sonder, and W.T. Thiselton-Dyer, Vol. 5, Part 2, 523–541. Kent, etc.: L. Reeve and Co.

Hutchinson, J. 1917. *Ficus*, Linn. In *Flora of Tropical Africa*, ed. D. Oliver et al., Vol. 6, Part 2, 78–215. London: L. Reeve and Co.

Hutchinson, J., and J.M. Dalziel. 1958. *Flora of West Tropical Africa; The British West African Territories, Liberia, the French and Portuguese Territories South of Latitude 18 N. to Lake Chad, and Fernando Po.* Vol. 1, Part 2. London: Crown Agents for Oversea Governments and Administrations.

Jaffe, W. 1943. The sap of fig trees (as an anthelmintic). *Trop Dis Bull.* 40: 612.

Jahan, I.A., N. Nahar, M. Mosihuzzaman et al. 2009. Hypoglycaemic and antioxidant activities of *Ficus racemosa* Linn. fruits. *Nat Prod Res.* 23: 399–408.

Jain, S.P. 1989. Tribal remedies from Saranda Forest, Bihar, India. I. *Int J Crude Drug Res.* 27: 29–32.

Jain, S.P., and H.S. Puri. 1984. Ethnomedicinal plants of Jaunsar-Bawar Hills, Uttar Pradesh, India. *J Ethnopharmacol.* 12: 213–22.

Jain, S.R., and S.N. Sharma. 1967. Hypoglycaemic drugs of Indian indigenous origin. *Planta Med.* 15: 439–42.

Jain, S.K., and C.R. Tarafder. 1970. Medicinal plant-lore of the Santals. *Econ Bot.* 24: 241–78.

Jain, S.P., and D.M. Verma. 1981. Medicinal plants in the folk-lore of North-East Haryana. *Natl Acad Sci Lett (India)* 4: 269–71.

Jöchle, W. 1974. Menses-inducing drugs: Their role in antique, medieval and renaissance gynecology and birth control. *Contraception* 10: 425–39.

John, D. 1984. One hundred useful raw drugs of the Kani tribes of Trivandrum Forest Division, Kerala, India. *Int J Crude Drug Res.* 22: 17–39

Johns, T., E.B. Mhoro, and P. Sanaya. 1996. Food plants and masticants of the Batemi of Ngorongoro District, Tanzania. *Econ Bot.* 50: 115–21.

Jørgensen, P. 2008. Bolivia Checklist. In *eFloras*. [Online Database]. Missouri Botanical Garden, St. Louis, MO & Harvard University Herbaria, Cambridge, MA. URL: http://www.efloras.org/flora_page.aspx?flora_id=40 [accessed May 18, 2009].

Joshi, M.C., M.B. Patel, and P.J. Mehta. 1980. Some folk medicines of Dangs, Gujarat State. *Bull Med Ethnobot Res.* 1: 8–24.

Jung, H.W., H.Y. Son, C.V. Minh, Y.H. Kim, and Y.K. Park. 2008. Methanol extract of *Ficus* leaf inhibits the production of nitric oxide and pro-inflammatory cytokines in LPS-stimulated microglia via the MAPK pathway. *Phytother Res.* 22: 1064–9.

Kakrani, H.K., and A.K. Saluja. 1993. Traditional treatment through herbal drugs in Kutch District, Gujarat State, India. Part I. Uterine disorders. *Fitoterapia* 65: 463–5.

Kapur, S.K. 1983. Medico-botanic survey of medicinal and aromatic plants of Mawphlang (Shillong). *Indian Drugs* 21: 1–5.

Kar, A., B.K. Choudhary, and N.G. Bandyopadhyay. 2003. Comparative evaluation of hypoglycaemic activity of some Indian medicinal plants in alloxan diabetic rats. *J Ethnopharmacol.* 84: 105–8.

Kar, A., B.K. Choudhary, and N.G. Bandyopadhyay. 1999. Preliminary studies on the inorganic constituents of some indigenous hypoglycaemic herbs on oral glucose tolerance test. *J Exp Bot.* 64: 179–84.

Kargbo, T.K. 1984. *Traditional practices affecting the health of women and children in Africa.* Unpublished manuscript.

Kaufmann, J.C., and M. Elvin-Lewis. 1995. Towards a logic of ethnodentistry at Antongobe, Southwestern Madagascar. *Econ Bot.* 49: 213–22.

Kerharo, J. 1974. Historic and ethnopharmacognosic review on the belief and traditional practices in the treatment of sleeping sickness in West Africa. *Bull Soc Med Afr Noire Lang Fr* 19: 400–20.

Kerharo, J., and A. Bouquet. 1950. *Plantes médicinales et toxiques de la Côte-d'Ivoire— Haute-Volta.* Paris: Editions Vigot Frères.

Khan, M.A., and V.K. Singh. 1996. A folklore survey of some plants of Bhopal District Forests, Madhya Pradesh, India, described as antidiabetics. *Fitoterapia* 67: 416–21.

Khanom, F., H. Kayahara, and K. Tadasa. 2000. Superoxide-scavenging and prolyl endopeptidase inhibitory activities of Bangladeshi indigenous medicinal plants. *Biosci Biotechnol Biochem.* 64: 837–40.

Kitajima, J., M. Arai, and Y. Tanaka. 1994. Triterpenoid constituents of *Ficus thunbergii*. *Chem Pharm Bull (Tokyo)* 42: 608–10.

Kitajima, J., K. Kimizuka, and Y. Tanaka. 2000. Three new sesquiterpenoid glucosides of *Ficus pumila* fruit. *Chem Pharm Bull (Tokyo)* 48: 77–80.

Kokwaro, J.O. 1976. *Medicinal Plants of East Africa.* Nairobi: East African Literature Bureau.

Koné, W.M., K.K. Atindehou, C. Terreaux, K. Hostettmann, D. Traoré, and M. Dosso. 2007. Traditional medicine in north Côte-d'Ivoire: screening of 50 medicinal plants for antibacterial activity. *J Ethnopharmacol.* 93: 43–9.

Kone-Bamba, D., Y. Pelissier, Z.F. Ozoukou, and D. Kouao. 1987. Hemostatic activity of 216 plants used in traditional medicine in the Ivory Coast. *Plant Med Phytother.* 21: 122–30.

Konno, K., C. Hirayama, M. Nakamura et al. 2004. Papain protects papaya trees from herbivorous insects: role of cysteine protease in latex. *Plant J.* 37: 370–8.

Kuete, V., F. Nana, B. Ngameni, A.T. Mbaveng, F. Keumedjio, and B.T. Ngadjui. 2009. Antimicrobial activity of the crude extract, fractions and compounds from stem bark of *Ficus ovata* (Moraceae). *J Ethnopharmacol.* 124: 556–61.

Kuete, V., B. Ngameni, C.C. Simo et al. 2008. Antimicrobial activity of the crude extracts and compounds from *Ficus chlamydocarpa* and *Ficus cordata* (Moraceae). *J Ethnopharmacol.* 120: 17–24.

Kumar, D.S., and Y.S. Prabhakar. 1987. On the ethnomedical significance of the Arjun tree, *Terminalia arjuna* (Roxb.) Wight & Arnot. *J Ethnopharmacol.* 20: 173–90.

Laferriere, J.E. 1994. Medicinal plants of the Lowland Inga people of Colombia. *Int J Pharmacog.* 32: 90–94.

Lal, S.D., and K. Lata. 1980. Plants used by the Bhat community for regulating fertility. *Econ Bot.* 34: 273–275.

Lal, S.D., and B.K. Yadav. 1983. Folk medicine of Kurukshetra District (Haryana), India. *Econ Bot.* 37: 299–305.

Latorre, D.L., and F. A. Latorre. 1977. Plants Used by the Mexican Kickapoo Indians. *Econ Bot.* 31: 340–357.

Laurens, A., S. Mboup, M. Tignokpa, O. Sylla, and J. Masquelier. 1985. Antimicrobial activity of some medicinal species of Dakar markets. *Pharmazie* 40: 482–5.

Le Grand, A., 1989. Anti-infectious phytotherapy of the tree-savannah, Senegal (Western Africa). III. A review of the phytochemical substances and anti-microbial activity of 43 species. *J Ethnopharmacol.* 25: 315–38.

Le Grand, A., and P.A. Wondergem. 1987. Antiinfective phytotherapy of the Savannah forests of Senegal (East Africa). I. An inventory. *J Ethnopharmacol.* 21: 109–125.

Lentz, D.L. 1993. Medicinal and other economic plants of the Paya of Honduras. *Econ Bot.* 47: 358–70.

Lentz, D.L., A.M. Clark, C.D. Hufford et al. 1998. Antimicrobial properties of Honduran medicinal plants. *J Ethnopharmacol.* 63: 253–63.

Lewis, W.H., and M.P.F. Elvin-Lewis. 1984. Plants and dental care among the Jivaro of the Upper Amazon Basin. In *Ethnobotany in the Neotropics*, ed. G.T. Prance, and J.A. Kallunki (Advances in Economic Botany 1), 53–61. New York: New York Botanical Garden.

Li, Y., J. Duan, T. Guo et al. 2009. In vivo pharmacokinetics comparisons of icariin, emodin and psoralen from Gan-kang granules and extracts of *Herba Epimedii*, Nepal dock root, *Ficus hirta* Vahl. *J Ethnopharmacol.* May 18, 2009. [Epub ahead of print].

Li, Y.C., Y.C. Yang, J.S. Hsu, D.J. Wu, H.H. Wu, and J.T. Tzen. 2005. Cloning and immuno-localization of an antifungal chitinase in jelly fig (*Ficus awkeotsang*) achenes. *Phytochemistry.* 66: 879–86.

Liebstein, A.M. 1927. Therapeutic effects of various food articles. *Am Med.* 33: 33–38.

Lin, C.C. 1992. Crude drugs used for the treatment of diabetes mellitus in Taiwan. *Am J Chin Med.* 20: 269–79.

Liogier, A.H. 1974. *Diccionario botanico de nombres vulgares de la espanola.* Santo Domingo: Universidad Nacional Pedro Henriquez Urena.

Locher, C.P., M. Witvrouw, M.P. De Bethune et al. 1996. Antiviral activity of Hawaiian medicinal plants against human immunodeficiency virus type-1 (HIV-1). *Phytomedicine* 2: 259–64.

Lokar, L.C., and L. Poldini. 1988. Herbal remedies in the traditional medicine of the Venezia Giulia region (North East Italy). *J Ethnopharmacol.* 22: 231–9.

Luna, L.E. 1984b. The concept of plants as teachers among four mestizo shamans of Iquitos, northeastern Peru. *J Ethnopharmacol.* 11: 135–56.

Luna, L.E. 1984a. The healing practices of a Peruvian shaman. *J Ethnopharmacol.* 11: 123–33.

Lv, Y.J., F.L. Jia, M. Ruan, and B.X. Zhang. 2008. [The hepatoprotective effect of aqueous extracts from *Ficus hirta* on N, N-dimethylformamide induced acute liver injury in mice]. *Zhong Yao Cai* 31: 1364–8.

Macfoy, C.A., and A.M. Sama. 1983. Medicinal plants in Pujehun District of Sierra Leone. *J Ethnopharmacol.* 8: 215–23.

Mahyar, U.W., J.S. Burley, C. Gyllenhaal, and D.D. Soejarto. 1991. Medicinal plants of Seberida (Riau Province, Sumatra, Indonesia). *J Ethnopharmacol.* 31: 217–37.

Maikere-Faniyo, R., L. Van Puyvelde, A. Mutwewingabo, and F.X. Habiyaremye. 1989. Study of Rwandese medicinal plants used in the treatment of diarrhoea. I. *J Ethnopharmacol.* 26: 101–9.

Maikhuri, R.K., and A.K. Gangwar. 1993. Ethnobiological notes on the Khasi and Garo tribes of Meghalaya, Northeast India. *Econ Bot.* 47: 345–57.

Malamas, M., and M. Marselos. 1992. The tradition of medicinal plants in Zagori, Epirus (northwestern Greece). *J Ethnopharmacol.* 37: 197–203.

Manandhar, N.P. 1995. A survey of medicinal plants of Jajarkot District, Nepal. *J Ethnopharmacol.* 48: 1–6.

Mandal, S.C., and Ashok Kumar, C.K. 2002. Studies on anti-diarrhoeal activity of *Ficus hispida* Linn. leaf extract in rats. *Fitoterapia* 73: 663–7.

Mandal, S.C., B. Saraswathi, C.K. Kumar, S. Mohana Lakshmi, and B.C. Maiti. 2000. Protective effect of leaf extract of *Ficus hispida* Linn. against paracetamol-induced hepatotoxicity in rats. *Phytother Res.* 14: 457–9.

Maregesi, S.M., L. Pieters, O.D. Ngassapa et al. 2008. Screening of some Tanzanian medicinal plants from Bunda district for antibacterial, antifungal and antiviral activities. *J Ethnopharmacol.* 119: 58–66.

Martinez, M. 1969. *Las plantas medinales de Mexico.* Mexico: La Impresora Aztecas, Ediciones Botas.

Martinez-Lirola, M.J., M.R. Gonzalez-Tejero, and J. Molero-Mesa. 1996. Ethnobotanical resources in the Province of Almeria, Spain: Campos de Nijar. *Econ Bot.* 50: 40–56.

McClatchey, W. 1996. The ethnopharmacopoeia of Rotuma. *J Ethnopharmacol.* 50: 147–56.

Metafro Tervuren. *Metafro Infosys: Tervuren Xylarium Wood Database.* (1998) [Online Database]. Royal Museum for Central Africa (RMCA), Tervuren, Belgium. URL: http://www.metafro.be/xylarium [accessed July 22, 2009].

Milliken, W. 1997. Traditional anti-malarial medicine in Roraima, Brazil. *Econ Bot.* 51: 212–37.

Mitchell, J.P. 1938. On the causes of obstructed labour in Uganda. *East Afr Med J.* 15: 177.

Mitra, R., S. Mehrotra, and L.D. Kapoor. 1978. Pharmacognostical study of Plaksha. II. Leaf of *Ficus tsiela* Roxb. *J Res Indian Med Yoga Homeopath.* 13: 74–83.

Miyazaki, Y., S. Yakou, and K. Takayama. 2004. Study on jelly fig extract as a potential hydrophilic matrix for controlled drug delivery. *Int J Pharm.* 287: 39–46.

MMPND. Multilingual Multiscript Plant Name Database (M.M.P.N.D.). 1995. [Online Database]. Ed. M.H. Porcher. School of Agriculture and Food Systems, Faculty of Land & Food Resources, The University of Melbourne, Australia. URL: http://www.plant-names.unimelb.edu.au/ [accessed July 22, 2009].

Moerman, D. 2003. *Native American Ethnobotany.* [Online database]. University of Michigan—Dearborn. Dearborn, Michigan, USA. URL: http://herb.umd.umich.edu./ [accessed June 21, 2008].

Mokkhasmit, M., W. Ngarmwathana, K. Sawasdimongkol, and U. Permphiphat. 1971. Pharmacological evaluation of Thai medicinal plants. *J Med Assoc Thai.* 54: 490–504.

Mpiana, P.T., V. Mudogo, D.S. Tshibangu et al. 2008. Antisickling activity of anthocyanins from *Bombax pentadrum, Ficus capensis* and *Ziziphus mucronata*: Photodegradation effect. *J Ethnopharmacol.* 120: 413–8.

Mueller-Oerlinghausen, B., W. Ngamwathana, and P. Kanchanapee. 1971. Investigation into Thai medicinal plants said to cure diabetes. *J Med Assoc Thai.* 54: 105–11.

Mukherjee, A., and D. Namhata. 1990. Some medicinal plants of Sundargarh District, Orissa. *Int J Crude Drug Res.* 28: 177–82.

Mukherjee, P.K., K. Saha, T. Murugesan, S.C. Mandal, M. Pal, and B.P. Saha. 1998. Screening of anti-diarrhoeal profile of some plant extracts of a specific region of West Bengal, India. *J Ethnopharmacol.* 60: 85–89.

Murakami, A., S. Jiwajiinda, K. Koshimizu, and H. Ohigashi. 1995. Screening for in vitro anti-tumor promoting activities of edible plants from Thailand. *Cancer Lett.* 95: 137–46.

Muregi, F.W., S.C. Chhabra, E.N. Njagi et al. 2003. In vitro antiplasmodial activity of some plants used in Kisii, Kenya against malaria and their chloroquine potentiation effects. *J Ethnopharmacol.* 84: 235–9.

Muregi, F.W., A. Ishih, T. Miyase et al. 2007. Antimalarial activity of methanolic extracts from plants used in Kenyan ethnomedicine and their interactions with chloro-quine (CQ) against a CQ-tolerant rodent parasite, in mice. *J Ethnopharmacol.* 111: 190–5.

Musabayane, C.T., M. Gondwe, D.R. Kamadyaapa, A.A. Chuturgoon, and J.A. Ojewole. 2007. Effects of *Ficus thonningii* (Blume) [Moraceae] stem-bark ethanolic extract on blood glucose, cardiovascular and kidney functions of rats, and on kidney cell lines of the proximal (LLC-PK1) and distal tubules (MDBK). *Ren Fail.* 29: 389–97.

Nagaraju, N., and K.N. Rao. 1990. A survey of plant crude drugs of Rayalaseema, Andhra Pradesh, India. *J Ethnopharmacol.* 29: 137–58.

*NAPRALERT*SM. *Natural Products Alert Database.* 1975. [Online Database]. The Program for Collaborative Research in the Pharmaceutical Sciences (PCRPS), College of Pharmacy, University of Illinois at Chicago. Chicago, Illinois, United States. URL: http://www. napralert.org [accessed June 24, 2009].

Nisteswar, K. 1988. Review of certain indigenous antifertility agents. *Deerghayu International* 4: 4–7.

Nisteswar, K., and K.A. Kumar. 1980. Utilitarian values of medical-lore of Rampa Agency (A.P) in primary health care. *Sacitra Ayurveda* 1980: 210–211.

Noumi, E., and A. Yomi. 2001. Medicinal plants used for intestinal diseases in Mbalmayo Region, Central Province, Cameroon. *Fitoterapia* 72: 246–54.

Novaretti, R., and D. Lemordant. 1990. Plants in the traditional medicine of the Ubaye Valley. *J Ethnopharmacol.* 30: 1–34.

Novy, J.W. 1997. Medicinal plants of the Eastern Region of Madagascar. *J Ethnopharmacol.* 55: 119–26.

Nyman, U., P. Joshi, L.B. Madsen et al. 1998. Ethnomedical information and in vitro screening for angiotensin-coverting enzyme inhibition of plants utilized as traditional medicines in Gujarat, Rajasthan and Kerala (India). *J Ethnopharmacol.* 60: 247–263

Ong, C.Y., S.K. Ling, R.M. Ali et al. 2009. Systematic analysis of in vitro photo-cytotoxic activity in extracts from terrestrial plants in Peninsula Malaysia for photodynamic ther-apy. *J Photochem Photobiol B.* July 1, 2009. [Epub ahead of print.]

Otero, R., R. Fonnegra, S.L. Jimenez et al. 2000b. Snakebites and ethnobotany in the northwest region of Colombia. Part I. Traditional use of plants. *J Ethnopharmacol.* 71: 493–504.

Otero, R., V. Nunez, J. Barona et al. 2000a. Snakebites and ethnobotany in the northwest region of Colombia. Part III. Neutralization of the haemorrhagic effect of Bothrops atrox venom. *J Ethnopharmacol.* 73: 233–41.

Ouyang, M.A., P.Q. Chen, and S.B. Wang. 2007. Water-soluble phenylpropanoid constituents from aerial roots of *Ficus microcarpa*. *Nat Prod Res.* 21: 769–74.

Owolabi, O.J., Z.A. Nworgu, A. Falodun, B.A. Ayinde, and C.N. Nwako. 2009. Evaluation of tocolytic activity of ethanol extract of the stem bark of *Ficus capensis* Thunb. (Moraceae). *Acta Pol Pharm.* 66: 293–6.

Pandey, S.K., B.D. Tripathi, S.K. Prajapati et al. 2005. Magnetic properties of vehicle-derived particulates and amelioration by *Ficus infectoria*: A keystone species. *Ambio* 34: 645–6.

Parveen, M., R.M. Ghalib, S.H. Mehdi, S.Z. Rehman, and M. Ali. 2009. A new triterpenoid from the leaves of *Ficus benjamina* (var. *comosa*). *Nat Prod Res.* 23: 729–36.

Peckolt, G. 1942. Brazilian anthelmintic plants. *Rev Flora Med.* 9: 333.

Pei, S.-J. 1985. Preliminary study of ethnobotany in Xishuang Banna, People's Republic of China. *J Ethnopharmacol.* 13: 121–37.

Peraza-Sánchez, S.R., H.B. Chai, Y.G. Shin et al. 2002. Constituents of the leaves and twigs of *Ficus hispida*. *Planta Med.* 68: 186–8.

Perera, P., D. Kanjanapoothi, F. Sandberg, and R. Verpoorte. 1984. Screening for biological activity of different plant parts of *Tabernaemontana dichotoma*, known as Divikaduru in Sri Lanka. *J Ethnopharmacol.* 11: 233–41.

Perez, C., E. Dominguez, J.M. Ramiro, A. Romero, J.E. Campillo, and M.D. Torres. 1996. A study on the glycaemic balance in streptozotocin-diabetic rat treated with an aqueous extract of *Ficus carica* (fig tree) leaves. *Phytother Res.* 10: 82–3.

Petelot, A. 1954. *Les plantes medicinales du Cambodge, du Laos et du Vietnam*. Vols. 1–4. *Archives des recherches agronomiques et pastorales au Vietnam*, No. 23.

Phillips, O. 1990. *Ficus insipida* (Moraceae): Ethnobotany and ecology of an Amazonian anthelmintic. *Econ Bot.* 44: 534–6.

Pittier, H. 1926. *Manual de las plantas usuales de Venezuela*. Caracas: Litografia del Comercio.

Plants For A Future. Plants For A Future Database. 1996. [Online Database]. Plants For A Future, Lostwithiel, Cornwall, England. URL: http://www.pfaf.org/ [accessed July 24, 2009].

Plants of Hawaii. 2009. [Online Database]. Starr, F., and K. Starr. 149 Hawea Place, Makawao, Maui, Hawaii, 96768. URL: http://www.hear.org/starr/plants/images/species/ [accessed October 10, 2009].

Polunin, O., and A. Huxley. 1987. *Flowers of the Mediterranean*. London: Hogarth Press.

Poumale, H.M.P., R.T. Kengap, J.C. Tchouankeu, F. Keumedjio, H. Laatsch, and B.T. Ngadjui. 2008. Pentacyclic triterpenes and other constituents from *Ficus cordata* (Moraceae). *Z Naturforsch.* 63b: 1335–8.

Prajapati, S.K., and B.D. Tripathi. 2008. Management of hazardous road derived respirable particulates using magnetic properties of tree leaves. *Environ Monit Assess* 139: 351–4.

Prasad, P.V., P.K. Subhaktha, A. Narayana, and M.M. Rao. 2006. Medico-historical study of "asvattha" (sacred fig tree). *Bull Indian Inst Hist Med Hyderabad.* 36: 1–20.

Quansah, N. 1988. Ethnomedicine in the Maroantsetra Region of Madagascar. *Econ Bot.* 42: 370–5.

Ragasa, C.Y., E. Juan, and J.A. Rideout. 1999. A triterpene from *Ficus pumila*. *J Asian Nat Prod Res.* 1: 269–75.

Rahuman, A.A., P. Venkatesan, K. Geetha, G. Gopalakrishnan, A. Bagavan, and C. Kamaraj. 2007. Mosquito larvicidal activity of gluanol acetate, a tetracyclic triterpenes derived from *Ficus racemosa* Linn. *Parasitol Res.* 103: 333–9.

Ramachandran, V.S., and N.C. Nair. 1981b. Ethnobotanical observations on Irulars of Tamil Nadu (India). *J Econ Taxon Bot.* 2: 183–90.

Ramachandran, V.S., and V.J. Nair. 1981a. Ethnobotanical studies in Cannanore District, Kerala State (India). *J Econ Taxon Bot.* 2: 65–72.

Ramirez, V.R., L.J. Mostacero, A.E. Garcia et al. 1988. *Vegetales empleados en medicina tradicional norperuana.* Trujillo: Banco Agrario del Peru & Universidad Nacional de Trujillo.

Rao, Ch.V., A.R. Verma, M. Vijayakumar, and S. Rastogi. 2008. Gastroprotective effect of standardized extract of *Ficus glomerata* fruit on experimental gastric ulcers in rats. *J Ethnopharmacol.* 115: 323–6.

Raponda-Walker, A., and R. Sillans. 1961. *Plants Used in Gabon.* Paris: Encyclopedie Biologique.

Ratismamanga-Urverg, S., P. Rasoanaivo, A. Rakoto-Ratsimamanga, J. Le Bras, O. Ramiliarisoa, and J. Savel. 1991. Antimalarial activity and cytotoxicity of *Ficus pyrifolia* and *Rhus (= Baronia) aratana* leaf extracts. *Phytother Res.* 5: 32–4.

Razzack, H.M.A. 1980. *The Concept of Birth Control in Unani Medical Literature.* Unpublished manuscript.

Reddy, M.B., K.R. Reddy, and M.N. Reddy. 1988. A survey of medicinal plants of Chenchu tribes of Andhra Pradesh, India. *Int J Crude Drug Res.* 26: 189–96.

Reddy, M.B., K.R. Reddy, and M.N. Reddy. 1989. A survey of plant crude drugs of Anantapur District, Andhra Pradesh, India. *Int J Crude Drug Res.* 27: 145–55.

Riotte, L. 1978. *Companion Planting for Successful Gardening.* Vermont: Garden Way.

Sackeyfio, A.C., and O.M. Lugeleka. 1986. The anti-inflammatory effect of a crude aqueous extract of the root bark of "*Ficus elastica*" in the rat. *Arch Int Pharmacodyn Ther.* 281: 169–76.

Samuelsson, G., M.H. Farah, P. Claeson et al. 1992. Inventory of plants used in traditional medicine in Somalia. III. Plants of the families Lauraceae-Papilionaceae. *J Ethnopharmacol.* 37: 93–112.

Sandabe, U.K., P.A. Onyeyili, and G.A. Chibuzo. 2006. Phytochemical screening and effect of aqueous extract of *Ficus sycomorus* L. (Moraceae) stem bark on muscular activity in laboratory animals. *J Ethnopharmacol.* 104: 283–5.

Sanon, S., E. Ollivier, N. Azas et al. 2003. Ethnobotanical survey and in vitro antiplasmodial activity of plants used in traditional medicine in Burkina Faso. *J Ethnopharmacol.* 86: 143–7.

Sebastian, M.K., and M.M. Bhandari. 1984. Medico-ethnobotany of Mount Abu, Rajasthan, India. *J Ethnopharmacol.* 12: 223–230.

Selvanayahgam, Z.E., S.G. Gnanevendhan, K. Balakrishna, and R.B. Rao. 1994. Antisnake venom botanicals from ethnomedicine. *J Herbs Spices Med Plants* 2: 45–100.

Sezik, E., E. Yesilada, M. Tabata et al. 1997. Traditional medicine in Turkey. VIII. Folk medicine in East Anatolia: Erzurum, Erzincan, Agri, Kars, Igdir Provinces. *Econ Bot.* 51: 195–211.

Shah, N.C. 1982. Herbal folk medicines in Northern India. *J Ethnopharmacol.* 6: 293–301.

Shah, G.L., and G.V. Gopal. 1985. Ethnomedical notes from the tribal inhabitants of the North Gujarat (India). *J Econ Taxon Botany* 6: 193–201.

Shah, N.C., and M.C. Joshi. 1972. An ethnobotanical study of the Kumaon region of India. *Econ Bot.* 25: 414–22.

Sharma, M.P., J. Ahmad, A. Hussain, and S. Khan. 1992a. Folklore medicinal plants of Metwat (Gurgaon District), Haryana, India. *Int J Pharmacog.* 30: 129–34.

Sharma, M.P., J. Ahmad, A. Hussain, and S. Khan. 1992b. Folklore medicinal plants of Mewat (Gurgaon District), Haryana, India. *Int J Pharmacog.* 30: 135–7.

Siddiqi, T.O., J. Ahmed, K. Javed, and M.S.Y. Khan. 1989. Pharmacognostical studies of the bark of *Ficus infectoria* Roxb. *Indian Drugs* 26: 205–10.

Siddiqui, M.B., and W. Husain. 1994. Medicinal plants of wide use in India with special reference to Sitapur District (Uttar Pradesh). *Fitoterapia* 65: 3–6.

Sikarwar, R.L.S., and J.P. Kaushik. 1993. Folk medicines of the Morena District, Madhya Pradesh, India. *Int J Pharmacog.* 31: 283–7.

Simon, P.N., A. Chaboud, N. Darbour et al. 2001. Modulation of cancer cell multidrug resistance by an extract of *Ficus citrifolia*. *Anticancer Res.* 21: 1023–7.

Simon, C., and M. Lamla. 1991. Merging pharmacopoeia: Understanding the historical origins of incorporative pharmacopoeial processes among Xhosa healers in Southern Africa. *J Ethnopharmacol.* 33: 237–42.

Singh, V. 1995. Herbal folk remedies of Morni Hills (Haryana), India. *Fitoterapia* 66: 425–30.

Singh, Y.N. 1986. Traditional medicine in Fiji: Some herbal folk cures used by Fiji Indians. *J Ethnopharmacol.* 15: 57–88.

Singh, V.K., and Z.A. Ali. 1992. A contribution to the ethnopharmacological study of the Udaipur Forests of Rajasthan, India. *Fitoterapia* 63: 136–44.

Singh, V.K., and Z.A. Ali. 1994. Folk medicines in primary health care: Common plants used for the treatment of fevers in India. *Fitoterapia* 65: 68–74.

Singh, V.K., Z.A. Ali, and M.K. Siddiqui. 1996a. Ethnomedicines in the Bahraich District of Uttar Pradesh, India. *Fitoterapia* 67: 65–76.

Singh, V.K., Z.A. Ali, S.T.H. Zaidi, and M.K. Siddiqui. 1996b. Ethnomedicinal uses of plants of Gonda District Forests of Uttar Pradesh, India. *Fitoterapia* 67: 129–39.

Singh, D., and R.K. Goel. 2009. Anticonvulsant effect of *Ficus religiosa*: Role of serotonergic pathways. *J. Ethnopharmacol.* 123: 330–4.

Singh, Y.N., T. Ikahihifo, M. Panuve, and C. Slatter. 1984. Folk medicine in Tonga. A study on the use of herbal medicines for obstetric and gynaecological conditions and disorders. *J Ethnopharmacol.* 12: 305–29.

Singh, K.K., and J.K. Maheshwari. 1994. Traditional phytotherapy of some medicinal plants used by the Tharus of the Nainital District, Uttar Pradesh, India. *Int J Pharmacog.* 32: 51–8.

Song, B., Y.Q. Peng, J.M. Guan, P. Yang, and D.R. Yang. 2008. [Sex ratio adjustment of a non-pollinating fig wasp species on *Ficus semicordata* in Xishuangbanna]. *Ying Yong Sheng Tai Xue Bao* 19: 588–92.

Standley, P.C., and J.A. Steyermark. 1952. *Flora of Guatemala.* Fieldiana Botany, Vol. 24, Part 3. Chicago: Field Museum Press.

Steinmetz, E.F. 1957. *Codex Vegetabilis.* Amsterdam: Steinmetz.

Stone, K.J., A.R. Wellburn, F.W. Hemming, and J.F. Pennock. 1967. The characterization of ficaprenol-10, -11 and -12 from the leaves of *Ficus elastica* (decorative rubber plant). *Biochem J.* 102: 325–30.

Subhaktha, P.K., R. Rajasekaran, and A. Narayana. 2007. Udumbara (*Ficus glomerata* Roxb.): A medico-historical review. *Bull Indian Inst Hist Med Hyderabad.* 37: 29–44.

Sulaiman, M.R., M.K. Hussain, Z.A. Zakaria et al. 2008. Evaluation of the antinociceptive activity of *Ficus deltoidea* aqueous extract. *Fitoterapia* 79: 557–61.

Suwal, P.N. 1970. *Medicinal Plants of Nepal.* Kathmandu: Ministry of Forests, Department of Medicinal Plants.

Suzuno, H., S. Kinugasa, H. Nakahara, and A. Kawabata. 1997. Molecular characteristics of water-soluble polysaccharide extracted from jelly fig (*Ficus awkeotsang* Makino) seeds. *Biosci Biotechnol Biochem.* 61: 1491–4.

Taira, T., A. Ohdomari, N. Nakama, M. Shimoji, and M. Ishihara. 2005. Characterization and antifungal activity of gazyumaru (*Ficus microcarpa*) latex chitinases: Both the chitin-binding and the antifungal activities of class I chitinase are reinforced with increasing ionic strength. *Biosci Biotechnol Biochem.* 69: 811–18.

Takeuchi, S., M. Kochi, K. Sakaguchi, K. Nakagawa, and T. Mizutani. 1978. Benzaldehyde as a carcinostatic principle in figs. *Agric Biol Chem.* 42: 1449–51.

Tambunan, P., S. Baba, A. Kuniyoshi et al. 2006. Isoprene emission from tropical trees in Okinawa Island, Japan. *Chemosphere* 65: 2138–44.

Tariq, M., T.S. Moss, I.A. Al-Meshal, and M.A. Al-Yahya. 1983. Studies on cardiotonic plants of Saudi Arabia. *43rd International Congress of Pharmaceutical Sciences,* Fip 83, Montreux, Switzerland.

Teklehaymanot, T., and M. Giday. 2007. Ethnobotanical study of medicinal plants used by people in Zegie Peninsula, Northwestern Ethiopia. *J Ethnobiol Ethnomed.* 3: 12.

Tignokpa, M., A. Laurens, S. Mboup, and O. Sylla. 1986. Popular medicinal plants of the markets of Dakar (Senegal). *Int J Crude Drug Res.* 24: 75–80.

Tiwari, K.C., R. Majumder, and S. Bhattacharjee. 1979. Folklore medicines from Assam and Arunachal Pradesh (District Tirap). *Int J Crude Drug Res.* 17: 61–67.

Uphof, J.C.T. 1968. *Dictionary of Economic Plants.* 2nd ed. Port Jervis: Lubrecht & Cramer, Ltd.

Van Noort, S., and J.-Y. Rasplus, *Figweb.* 2004. [Online database]. Iziko Museums of Cape Town. Cape Town, South Africa. URL: http://www.figweb.org [accessed June 23, 2008].

Vasileva, B. 1969. *Plantes medicinales de Guinee.* Conakry.

Vedavathyk, S., and D.N. Rao. 1995. Herbal folk medicine of Tirumala and Tirupati Region of Chittoor District, Andhra Pradesh. *Fitoterapia* 66: 167–171.

Veerapur, V.P., K.R. Prabhakar, V.K. Parihar et al. 2009. *Ficus racemosa* stem bark extract: A potent antioxidant and a probable natural radioprotector. *Evid Based Complement Alternat Med.* 6: 317–24.

Vera, R., J. Smadja, and J.Y. Conan. 1990. Preliminary assay of some plants with alkaloids from Reunion Island. *Plant Med Phytother.* 24: 50–65.

Waheed, S., and N. Siddique. 2009. Evaluation of dietary status with respect to trace element intake from dry fruits consumed in Pakistan: A study using instrumental neutron activation analysis. *Int J Food Sci Nutr.* 60: 333–43.

Wang, S.I. 1973. Isolation and characterization of some constituents of a species of *Ficus.* Diss Abstr Int B 33: 2675.

Wasuwat, S. 1967. *A List of Thai Medicinal Plants.* Report no.1 on Res. Project 17. Bangkok, A.S.R.C.T.

Watcho, P., E. Ngadjui, N.P. Alango, N.T. Benoît, and A. Kamanyi. 2009. Reproductive effects of *Ficus asperifolia* (Moraceae) in female rats. *Afr Health Sci.* 9: 49–53.

Watt, J.M., and M.G. Breyer-Brandwijk. 1962. *The Medicinal and Poisonous Plants of Southern and Eastern Africa.* 2nd ed. London: E. & S. Livingstone, Ltd.

Whistler, W.A. 1985. Traditional and herbal medicine in the Cook Islands. *J Ethnopharmacol.* 13: 239–80.

Williams, D.C., V.C. Sgarbieri, and J.R. Whitaker. 1968. Proteolytic activity in the genus *Ficus. Plant Physiol.* 43: 1083–8.

Womersley, J.S. 1974. Botanical validification in medicinal plant investigations. In *Report of Regional Technical Meeting on Medicinal Plants (Papeete, Tahiti, 12–17 November 1973),* 117. Noumea, New Caledonia: South Pacific Commission.

Wong, W. 1976. Some folk medicinal plants from Trinidad. *Econ Bot.* 30: 103–42.

Wu, J.S.B., M.C. Wu, C.M. Jiang, Y.P. Hwang, S.C. Shen, and H.M. Chang. 2005. Pectinesterase inhibitor from jelly-fig (*Ficus awkeotsang* Makino) achenes reduces methanol content in carambola wine. *J Agric Food Chem.* 53: 9506–11.

Wu, Z., Z.K. Zhou, and M.G. Gilbert. 2003. Moraceae. In *Flora of China* Vol. 5, *eFloras.* [Online Database.] Missouri Botanical Garden, St. Louis, MO & Harvard University Herbaria, Cambridge, MA. URL: http://www.efloras.org [accessed May 12, 2009].

Wunderlin, R.P. 1997. *Ficus.* In *Flora of North America North of Mexico,* Vol. 3. Ed. Editorial Committee. New York and Oxford.

Yang, L.L., K.Y. Yen, Y. Kiso, and H. Kikino. 1987. Antihepatotoxic actions of Formosan plant drugs. *J Ethnopharmacol.* 19: 103–110.

Yang, X.M., W. Yu, Z.P. Ou, H.L. Ma, W.M. Liu, and X.L. Ji. 2009. Antioxidant and immunity activity of water extract and crude polysaccharide from *Ficus carica* L. fruit. *Plant Foods Hum Nutr.* 64: 167–73.

Yateem, A., T. Al-Sharrah, and A. Bin-Haji. 2008. Investigation of microbes in the rhizosphere of selected trees for the rhizoremediation of hydrocarbon-contaminated soils. *Int J Phytoremediation* 10: 311–24.

Yazicioglu, A., and E. Tuzlaci. 1996. Folk medicinal plants of Trabzon (Turkey). *Fitoterapia* 67: 307–18.

Yesilada, E., G. Honda, E. Sezik et al. 1995. Traditional medicine in Turkey. V. Folk medicine in the Inner Taurus Mountains. *J Ethnopharmacol.* 46: 133–52.

Yoon, W.J., H.J. Lee, G.J. Kang, H.K. Kang, and E.S. Yoo. 2007. Inhibitory effects of *Ficus erecta* leaves on osteoporotic factors in vitro. *Arch Pharm Res.* 30: 43–9.

Zagari, A. 1992. *Medicinal Plants*. Vol. 4. 5th ed. Tehran: Tehran University Publications.

Zamora-Martinez, M.C., and C.N.P. Pola. 1992. Medicinal plants used in some rural populations of Oaxaca, Puebla and Veracruz, Mexico. *J Ethnopharmacol.* 35: 229–57.

Zhang, H.J., P.A. Tamez, Z. Aydogmus et al. 2002. Antimalarial agents from plants. III. Trichothecenes from *Ficus fistulosa* and *Rhaphidophora decursiva*. *Planta Med.* 68: 1088–91.

ZipcodeZoo. ZipcodeZoo.com. 2004. [Online Database]. BayScience Foundation. URL: http://zipcodezoo.com [accessed July 22, 2009].

3 Fruits

The Chinese term for fig, "no flower fruit," gives a hint as to its convoluted form. Similar to pomegranates, figs contain numerous seeds, each technically representing a tiny fruit. However, unlike pomegranate, which issues its tough leathery shell from a bright orange, red, or pink trumpet-shaped flower heralding its emergence, the more modest and inwardly feminine fig hides its flowers within its fruit, while the pomegranate hides its fruit within its flower. Flowers of the fig are not visible to the uninitiated, but they do exist. In fact, the flowers (achenes) are inside the fig fruit; they are out of sight and do not ordinarily have access to the outside. Some tens of species of wasps, both true fig wasps and "counterfeit" fig wasps (Chapter 8), possess the chemical sensory equipment to sniff out fig trees and to use the fruits for their amorous preparation and ovular incubation.

All *Ficus* species produce figs, that is, fig fruits. These fruits have thin and delicate skins (which toughen a bit on drying) and contain the flowers in an inverted form; that is, the flowers grow into the interior of the fruits. Fertilization occurs exclusively by true fig wasps that burrow into the fruits from their exteriors, mate, and leave their eggs to hatch. Males in some species also enlarge the entrance holes to be used later as exit holes for their mates, who may go on to find other figs.

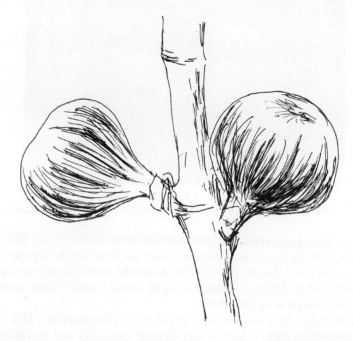

FIGURE 3.1 *Ficus carica* fruits on the tree in midsummer (pen and ink by Zipora Lansky).

FIGURE 3.2 *Ficus carica*, ripe fruit (by Kaarina Paavilainen).

FIGURE 3.3 *Ficus carica*, ripe fruit with focus on the osiole (by Kaarina Paavilainen).

When the fruits go to seed, they do so within these sacs of many flowers, and subsequently there are many seeds per fruit. They are thus similar to pomegranates, in having a multiplicity of seeds per fruit, but dissimilar in that the pomegranate fruit develops from a single flower, but hundreds of fig flowers actually occur within each single fig fruit, a single seed per flower.

The chemistry of fig fruits has only begun to be characterized. This is rather startling, considering their extensive employment medically and nutritionally for thousands of years. Unripe fruits contain latex, the chemistry of which will be taken

FIGURE 3.4 *Ficus carica* fruit stem (by Kaarina Paavilainen).

FIGURE 3.5 Fig from *Ficus carica*, cut to expose the seeds and vestigial flowers (photo by Zipora Lansky).

up in Chapter 5. Phytosterols are part of this composition and occur in the ripe and dried figs, along with amino acids, salicylates, and sugars. Secondary metabolites that have been characterized include anthocyanins with strong antioxidant activity.

In a study by Anat Solomon and colleagues from several Israeli institutions published in the *Journal of Agricultural and Food Chemistry* in 2006, cyanidin-3-*O*-rhamnoglucoside (cyanidin-3-*O*-rutinoside; C3R) is revealed by proton and carbon

FIGURE 3.6   *Ficus carica*, unripe fruit (29.7.2009, Hebrew University Botanical Garden, Mount Scopus, Jerusalem, Israel, by Krina Brandt-Doekes).

NMR to be the most prominent anthocyanin in figs, while cyanidin is the major aglycon product of hydrolysis. All figs used in the study were of *Ficus carica* and included varieties ranging from the lightest-skinned fruits to the darkest, namely, Mission figs. The concentration of anthocyanins was much higher in the skins than in the inner pulp and directly proportional to the darkness of the fruit skins, with the anthocyanin fraction contributing 28% of the total antioxidant effect in red Brown-Turkey varieties, for example, and 36% in the black missions. In our earlier studies of pomegranate peel extracts versus pomegranate fermented juice extracts in inhibiting angiogenesis (the creation of new blood vessels from existing ones that is associated with both wound healing and cancer metastasis), pomegranate juice fermented and extracted with ethyl acetate exerted a topically visible in vivo antiangiogenic effect in a chick chorioallantoic membrane (CAM), while pomegranate peel extract (unfermented) did not (Toi et al. 2003). An important difference chemically between pomegranate peel and pomegranate juice is the presence of more and varied anthocyanins in the latter. In the same way, the antiangiogenic action of fig fruits from

Mission   Chechik   Brown Turkey

Bursa   Brunswick   Kadota

FIGURE 3.7   (See color insert after page 128.) *Ficus carica* varieties examined for anthocyanin content and antioxidant activity in the study by Solomon et al. 2006 (Solomon et al. 2006. Antioxidant activities and anthocyanin content of fresh fruits of common fig (*Ficus carica* L.). *J Agric Food Chem.* 54: 7717–23. Reprinted with permission). The darkest varieties possess the highest amounts of anthocyanins. (Reprinted with permission of the American Chemical Society)

anthocyanins is potentially synergistic with the antiantiogenic action described elsewhere for immature fruit, leaf, or stem latex.

Figs have a number of characteristics that make them highly suitable for fermentation into a fruit wine. Yeasts that may prove suitable for ongoing fig fermentations have been isolated from dried figs (Van der Walt and Tscheuschner 1956). Conversely, dried figs that have been exhaustively extracted with water provide suitable "rafts" for immobilizing yeasts in the fermentation of glucose (Bekatorou et al. 2002). The method of immobilizing yeast allows for a smoother and more even fermentation. Used for beer fermentation, yeast immobilization by the fig rafts produced a richer, more full-bodied brew than obtained by immobilization with other natural materials. According to one Chinese patent (Fang 2007), fermentation mixtures containing

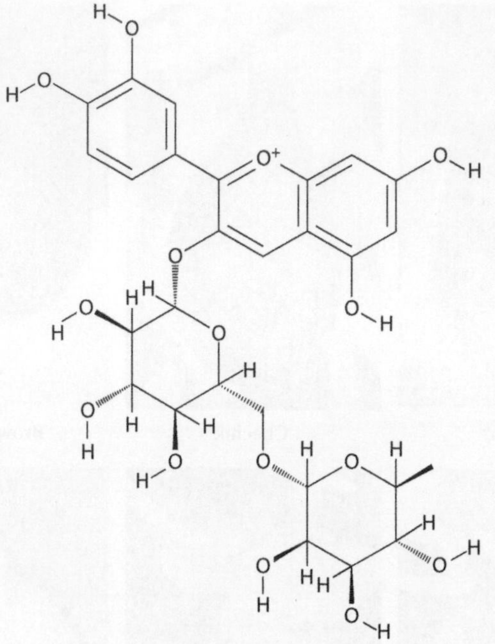

FIGURE 3.8 Cyanidin-3-*O*-rhamnoglucoside (cyanidin-3-*O*-rutinoside; C3R), the principal anthocyanin in *F. carica* fruits. (*PubChem.* The PubChem Project. 2001. [Online Database]. National Center for Biotechnology Information. U.S. National Library of Medicine. 8600 Rockville Pike, Bethesda, MD 20894. URL: http://pubchem.ncbi.nlm.nih.gov/ [accessed August 16, 2009].)

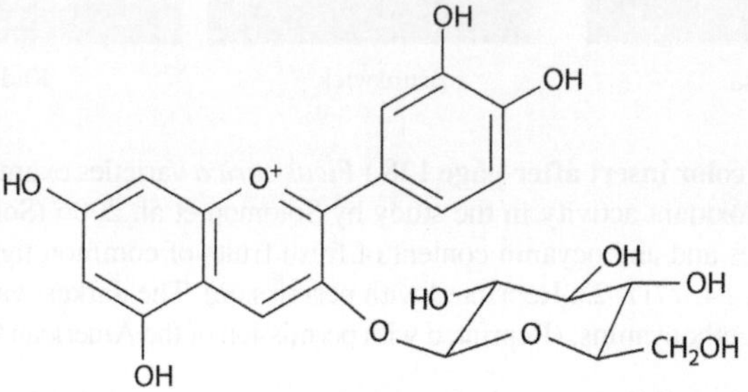

FIGURE 3.9 Cyanidin-3-*O*-glucoside (another anthocyanin in *F. carica* fruits). (Adapted from Solomon et al. 2006. Antioxidant activities and anthocyanin content of fresh fruits of common fig (*Ficus carica* L.). *J Agric Food Chem.* 54: 7717–23. Reprinted with permission of the American Chemical Society)

approximately equal parts of grape juice and fig juice yield a wine with "good health promoting effect, rich nutrition, and nice color, fragrance, and taste. It has the effects of invigorating spleen, promoting digestion, relieving hemorrhoids, relieving constipation, delaying aging, preventing and treating cancer, and enhancing immunity."

Other innovations in fig wine production have also been described. Thus, Wang and Yue (2005) demonstrated that the addition of sulfur dioxide (SO_2; 60–90 mg/L)

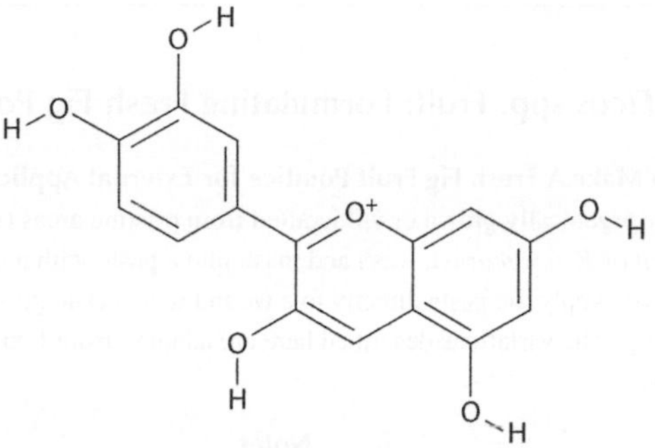

FIGURE 3.10 Cyanidin (the principal hydrolysis product of anthocyanidin in figs of *F. carica* according to Solomon et al. 2006. Antioxidant activities and anthocyanin content of fresh fruits of common fig (*Ficus carica* L.). *J Agric Food Chem.* 54: 7717–23; figure from *PubChem.* The PubChem Project. 2001. [Online Database]. National Center for Biotechnology Information. U.S. National Library of Medicine. 8600 Rockville Pike, Bethesda, MD 20894. URL: http://pubchem.ncbi.nlm.nih.gov/ [accessed August 16, 2009]. Reprinted with permission of the American Chemical Society)

and active dry wine yeast (5–7 g/L !) to clarified fig juice containing 20%–22% sugar produced a high-quality product of 11.2% alcohol (volume/volume) when allowed to ferment for 14 days at 20°C. Wu (2005) described a different method whereby the fig juice is cooked along with rice, either common *Oryza sativa* or sweet *Oryza glutinosa*, that has soaked in water for some time, then been allowed to cool, and active yeast added. The resultant wine prepared in this manner was said to be nutritious, to contain amino acids and microelements essential to the human body, and to possess appetite-promoting, health-promoting, spleen-invigorating, and immunity-strengthening effects, and to be endowed with a soft, sweet, and rich flavor. Finally, adding the powder of the dried medicinal mushroom, *Ganoderma lucidum*, 5%–10% to fig juice prior to sterilization and subsequent yeast fermentation at 35°C–42°C for 8–45 days yields, after pressing, filtration, and addition of water, a medicinal wine with "good antitumor effects, and suitable for treatment of various malignant tumors and leukemia" (Sun 2008).

Lectins, for example, *F. benghalensis* agglutinin (FBA) (Singha et al. 2007) isolated from the seeds of the fruits of the banyan tree, *F. benghalensis*, are sugar-binding proteins specific to the sugars they bind. Lectins are found both in plants and animals. They are an important alternative to the classical complement pathway of activating the complement system in the blood. *Complement* is a term coined by Dr Paul Erlich in the 1890s to indicate a component of the blood that is *complementary* to the immune system's ability to neutralize pathogens. Lectins are essential in allowing the complement system to "get a good fit" on the cell surface of a pathogen, by allowing "opsono" (they do the shopping while the phagocytes do the eating) phagocytic cells to attach. The "pathogen" may be either a classical microorganism, such as a bacterium, or from the body's own cells, such as a neoplasm.

Ficolins are a class of lectins that regulate specific stages in the *lectin pathway* of the complement system, which is an alternative to its classical component. The complement system is part of the *human innate immune system*. The innate immune

TABLE 3.1
Historical Uses of *Ficus* spp. Fruit: Formulating Fresh Fig Poultices

How to Make A Fresh Fig Fruit Poultice for External Applications

Take fresh figs, preferably organically grown or wildcrafted from pristine areas (usually but not necessarily of *Ficus carica* or *F. sycomorus*), wash and mash into a paste with a sterile metal fork into a sterile stainless steel bowl. Apply the paste directly to a wound with sterile gloves or instruments and cover with a sterile dressing. The variations described here are adapted from Lansky et al. 2008, and from the data in Table 3.7.

Variations	Notes
Mix the mashed fruit with wax	Use as an ointment for tumors, swellings, furuncles, abscesses, paronychia
Mix the mashed fruit with barley flour	Use as a poultice for hard tumors, dermatitis
Aqueous decoction of whole figs or mash	As a poultice or ointment for swellings, swollen glands, hard or inflamed tumors and boils
Mix the mashed fruit with gum ammoniac U.S.P., (i.e., the "tears" of gum-resin of the plant *Dorema ammoniacum*)	Poultice, for splenomegaly or chronic recalcitrant low output respiratory conditions. Gum ammoniac possesses antispasmodic and expectorant properties.
Mix mashed fruit with vitriol (ferrous or copper sulfate or other similar metal salts of sulfuric acid)	For malignant, exuding ulcers, especially of leg or groin, vitriol in medicine has the effect of "chemical debridement," as well as being a purgative.
Mix mashed fruit with pounded or water-decocted pomegranate peels	For paronychia
Mix mashed fruit with crushed lily (*Lilium* sp.) roots and wheat flour	As poultice for groin buboes (swollen lymph glands) to facilitate ripening
Mix mashed fruit with wheat flour, fenugreek powder (*Trigonella foenum graecum*), flax seed (*Linum usitatissimum*), and marshmallow (*Althaea*) roots	As a warm poultice for hot swellings and tumors behind the ears (= swollen lymph glands)
Mash pulpiest fruit and mix with verdigris (the green coating formed on copper, bronze, or brass following exposure to air or water, usually copper carbonate; near the sea, copper chloride; in contact with vinegar, copper acetate)	Poultice for leg ulcers; for abscesses of parotid glands
Mashed fresh fruit mixed with honey	As poultice for ulcers with honey-like exudates; for the bite of a rabid dog.
Mix mashed fresh ripe fruit with natron, a naturally occurring mixture of sodium carbonate decahydrate ($Na_2CO_3 \cdot 10H_2O$, a form of soda ash) and about 17% sodium bicarbonate ($NaHCO_3$) with traces of sodium chloride and sodium sulfate, and vinegar	As a poultice for pustules on the head
Mash fresh fruit with fenugreek flour and vinegar	Poultice for gout
Mash fresh fruit and mix with vinegar	As a poultice on swollen pancreas; causes it to shrink

TABLE 3.1 (continued)

Historical Uses of *Ficus* spp. Fruit: Formulating Fresh Fig Poultices

Variations	Notes
Mash fresh fruit with vinegar and salt	As a poultice for pimples and ulcers on the head
Mash fresh fruit and mix with sourdough and salt	Poultice for ulcers and specifically plague ulcers
Mash fresh fruit and mix with natron and flour	As a poultice on warts
Mash fresh fruit and mix with salt	As a poultice on scorpion sting; itch
Mash fresh fruit with leaves of wild poppy	As a poultice removes splinters of bones
Mash fresh fruit with lentils and wine	As a poultice for the bite of centipedes
Grind sharp-tasting mustard seeds and mix the powder with water. Add finely chopped unripe figs	Used as ear drops/inserted into ears as a paste for tinnitus and itching of ears
Decoct fresh figs in water till they reach a honey-like consistency, and add to the decoction froth from mustard seeds boiled in water	As ear drops for tinnitus; an ointment for itch

system is opposed to its adaptive immune system, the latter having a complex cognitive-like ability to remember and adapt to changes in its environment through learning. The complement system does not learn as such, but is sure and consistent in defense. It is also important in repair, particularly of the cardiovascular system following myocardial infarct (Bjerre et al. 2008). Plants have immune systems (Jones and Dangl 2006) similar to the human complement system, and both the plant immune system and the complement system use proteins to bind biodiverse sugars in their recognition of not-self and in a reflexive defense. Both the mammalian complement system, which renders blood inherently a natural antiseptic and anticancer solution, and the plant immune system, which also mounts biochemical defenses against both microbes and cancer as well as animal predators, do recognize pathogenic forces and respond biochemically. The mechanism is related to facilitating phagocytosis by macrophages via the lectin pathway, allowing macrophages to secure a better grip on their prokaryotic or eukaryotic cellular prey. The lectin pathway is modulated by ficolins. Metaphorically, the innate immune system is similar to the brain stem, whereas the regular immune system is similar to the cerebral cortex.

Lectins that specifically agglutinate leukemia cells but not normal blood cells were isolated from the seeds of *F. racemosa* (Agrawal and Agarwal 1990), the giant cluster tree. The possibility that leukemia-specific agglutinins derived from fig seed lectins could be used in therapy of leukemia was raised and enhances considerably the anticancer profile of *Ficus*. Details about a specific lectin in the seeds of *F. cunia* were determined with chromatography and immunodiffusion. It was shown to be a monomeric metalloprotein with a molecular weight between 3200 and 3500 (Ray et al. 1993). This lectin was later shown to bind to the bacteria *Escherichia coli*, *Pseudomonas aeruginosa*, *Klebsiella pneumoniae*, *Bacillus subtilis*, and *Staphylococcus aureus*, and is important both as a laboratory tool and as a potential adjuvant to antibiotic therapy (Adhya et al. 2006).

The ficolins and the collectins are related classes of lectins, oligomeric proteins present in animal plasma and mucosal surfaces that include mannan-binding lectin

TABLE 3.2
Historical Uses of *Ficus* spp. Fruit: Formulating Dry Fig Poultices

How to Make A Dry Fig Poultice for External Applications

Take dried figs, and cover with water in a stainless steel or ceramic pot. Boil over low heat for about 2 h until a mash forms. Allow to cool and apply to affected area as per the fresh fig poultice. The mash may be substituted for fresh fig mash in any of the variations described for fresh fig poultice. In addition, some specific variations for dried fig poultice are described here. Alternatively, the dried figs may be decocted as earlier but in wine instead of water. The alcohol in the wine helps yield a product with somewhat different chemical properties in that somewhat less polar constituents are also extracted. Unless otherwise specified, the variations here are assumed to be carried out with water decoctions. In general, the poultices are applied to swellings, both of cancerous and infectious etiologies, and hard tumors.

Variations	Notes
Decoct dried figs with rhizomes of blue flag (*Iris versicolor*), natron, and quicklime (CaO)	Poultice for groin or axilla buboes, pustules, hardenings; hot swellings
Dried figs decocted in wine	Poultice for tumors of soft flesh and swelling of parotid glands; particularly emollient to the tumors and for furuncles (boils); as a suppository for stomach pain
Dried figs decocted in water with sugar	Poultice for paronychia
Decocted or simply chopped dried fig mash mixed with cooked or ground fenugreek and barley seeds (or their flour)	Poultice for tumors, especially hard tumors, cutaneous ulcers, and abscesses; swelling of parotid glands; as a poultice for uterine problems
Soak the dried figs in red vinegar for 9 days, then mash	Poultice for tumors, especially hard tumors, and cutaneous ulcers; splenomegaly
Dried fruit aqueous decoction mixed with a combination of barley, wheat, and fenugreek flour (or barley flour and wheat flour alone or barley flour and fenugreek flour alone)	Poultice to ripen and dissolve hard tumors, or poultice over liver or spleen for hepatomegaly or splenomegaly
Dried fruit aqueous decoction or chopped dried fruits mixed with thyme, black pepper, ginger, and mint or hyssop (*Hyssopus* sp.)	Poultice for hard tumors, splenomegaly, or hepatomegaly
Decoction of dried fig and cabbage or blue flag	Poultice for hard tumors, splenomegaly, or hepatomegaly
Decoction of dried fig mixed with shoemaker's blackening (copper sulfate or similar vitriol)	Poultice for running or malignant ulcers
Decoction of dried fig mixed with barley flour and wormwood (*Artemisia absinthium*)	Poultice for dropsy
Decoction of dried fig mixed with violet, lily, or marshmallow root	Poultice for groin buboes

(MBL), surfactant proteins A and D (SP-A and SP-D) (the collectins), and the structurally similar ficolins: L-ficolin, M-ficolin, and H-ficolin. The oligomers are composed of carbohydrate-recognition domains (CRDs) attached to collagenous regions. The collectins and ficolins are essentially identical, except that they make use of different CRD structures, namely, C-type lectin domains for the collectins (collagen-like

How to Make Fig Fruit Wine

1. **From fresh figs**. Obtain several kilograms of fresh, ripe figs. Soak first in water for an hour to remove debris and pesticides if not organically grown. Rinse well, drain the water, and chop the figs into quarters. Add the chopped figs to a vessel such as a 5 gal (19 L) spring water bottle until about 2/3 full. (Alternatively, the chopped figs can be blended in a blender with a little spring water and the mash added to the bottle in batches. This works very well!) Add spring water just to cover the figs and 5 g of wine yeast suitable for white wine. Affix a water lock to the bottle neck or a sterile surgical glove turned inside out secured with a rubber band. The glove should be tight enough to prevent the entry of air, but loose enough to allow gases to exit. If using the glove method, the glove will stand erect when fermentation is underway. Shake the bottle gently from side to side once a day. Fermentation is complete when gases stop being emitted (evidenced by a quiet water valve or flaccid glove), in about 10 days.

2. **From dried figs**. Inspect the figs carefully and discard any that appear diseased or moldy. Place in a large stainless steel pot with spring water, about 1:1 water to figs. Boil over a low fire for an hour and allow to cool. Treat the mash like the blended product as per fresh figs, and repeat the entire procedure with wine yeast, water valve, etc.

3. **Medicinal uses of fig fruit wine**
   a. Tumors of trachea or lungs, chronic cough, chest pains: take a small glassful once or twice a day, or as prescribed by a physician.
   b. As a gargle for sore throat.
   c. A small glassful or two a day, or as prescribed by a physician, for chronic diarrhea.
   d. A small glassful or two a day as a diuretic.
   e. Similarly, for increasing lactation.
   f. Similarly, as an emmenagogue.
   g. Similarly, as a laxative.

Fig Fruit Wine *a la* Dioscorides

Place figs and water in a vessel closed with linen for 10 days. Water can be replaced with same amount of juice from pressing fresh figs or grapes, which improves the wine. May be repeated every 10 days up to five times, the last time used for vinegar. Salt or seawater can be added to prevent spoilage. Thyme and fennel may also be placed on the bottom of the vessel, then figs, and finally thyme and fennel, until full (Lansky et al. 2008).

domains) and fibrinogen-like domains for the ficolins. When the offending other is recognized, the complement activation cycle is initiated by the MBL and the ficolins via the lectin pathway of complement activation. Ficolins and MBL rely on attached serine proteases (MASPs), SP-A, and SP-D on direct opsonization, neutralization, and agglutination (Holmskov et al. 2003). Ficolins may also exert an important antibiotic effect against pathogenic bacteria, including *Streptococcus* Group B and *Salmonella* (Takahashi and Aoyagi 2008).

FIGURE 3.11 **(See color insert after page 128.)** *Ficus carica*, sagittal section (watercolor by Zipora Lansky).

Collectins and ficolins bind to oligosaccharide structures on the surface of microorganisms, leading to the killing of bound microbes through complement activation and phagocytosis. There is no major sequence homology between the two groups other than that the collagen-like sequences are over the N-terminal halves of the oligomers, and the C-terminal halves contain the C-type (CRDs) in collectins and the fibrinogen domains in ficolins. The molecules evolved to recognize surface sugar codes of microbes and their binding, and to prepare the microbes as targets for elimination by phagocytic cells (Lu et al. 2002). Also, Hong et al. (2009) demonstrated that a complement factor induces apoptosis in prostate cancer cells and that the factor's downregulation resulted in an increase in prostate hyperplasia and cancerous transformation.

Ficolins are proteins that contain both a collagen-like and fibrinogen-like domain; they are lectins with a binding affinity for *N*-acetylglucosamine (GlcNAc). They activate complement in conjunction with MBL (mannose-binding lectin)-associated serine protease (MASP). They can activate the complement pathway and thus mediate innate immunity. Two types of human serum ficolins, ficolin/P35 and Hakata antigen, are linked to MASPs and sMAP, a truncated protein of MASP 2 (Matsushita and Fujita 2001).

Evolutionary studies at the molecular level have revealed the presence of complement innate immunity system in invertebrates, ascidians, our closest invertebrate relatives (Fujita 2002), cnidarians (Wood-Charlson and Weis 2009), and shrimp (Zhang et al. 2009). In fact, lectin-mediated innate immune systems are common both to plants and animals (Zanker 2008). A lectin is a key factor in the anticancer effects of mistletoe (Lyu and Park 2009).

Lectins in figs represent an extremely promising avenue for immunopotentizing research. Because lectins can discriminate between different types of carbohydrates, there may also be a role for specific *Ficus* lectins in the context of personalized medicine, whereby medicines are prescribed on an at least partially individualized basis. The lectins in fig further represent a living link in the coevolution of our own

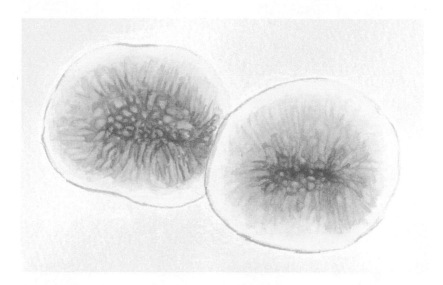

COLOR FIGURE 1.3 *Ficus carica* fruit, sagittal section (watercolor by Zipora Lansky).

COLOR FIGURE 1.4 *Ficus carica* fruit (27.7.2009, Hebrew University Botanical Garden, Mount Scopus, Jerusalem, Israel: by Krina Brandt-Doekes).

COLOR FIGURE 2.2 Figs from *Ficus carica*, including a coronal section (by Kaarina Paavilainen).

COLOR FIGURE 2.8 *Ficus deltoidea* branchlet (19.7.2009, University of Turku, Botanical Garden, Ruissalo, Turku, Finland, by Kaarina Paavilainen).

COLOR FIGURE 2.9 A small *Ficus deltoidea* (19.7.2009, University of Turku, Botanical Garden, Ruissalo, Turku, Finland, by Kaarina Paavilainen).

MISSION   CHECHIK   BROWN TURKEY

BURSA   BRUNSWICK   KADOTA

COLOR FIGURE 3.7 *Ficus carica* varieties examined for anthocyanin content and antioxidant activity in the study by Solomon et al. 2006 (Solomon et al. 2006. Antioxidant activities and anthocyanin content of fresh fruits of common fig (*Ficus carica* L.). *J Agric Food Chem.* 54: 7717–23. Reprinted with permission). The darkest varieties possess the highest amounts of anthocyanins.

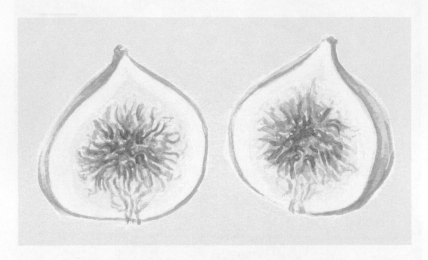

COLOR FIGURE 3.11   *Ficus carica*, sagittal section (watercolor by Zipora Lansky).

COLOR FIGURE 4.1   Three *Ficus carica* leaves with immature fruit (watercolor by Zipora Lansky).

COLOR FIGURE 6.5 Ficus microcarpa, stem (18.8.2009, Haifa, Israel, watercolor by Zipora Lansky)

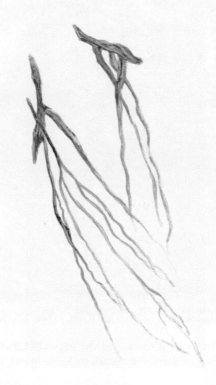

COLOR FIGURE 7.2 Aerial roots of *Ficus microcarpa* (18.8.2009, Haifa, Israel, watercolor by Zipora Lansky).

COLOR FIGURE 7.3 *Ficus carica* roots over the edge (29.7.2009, Hebrew University Botanical Garden, Mount Scopus, Jerusalem, Israel, by Krina Brandt-Doekes).

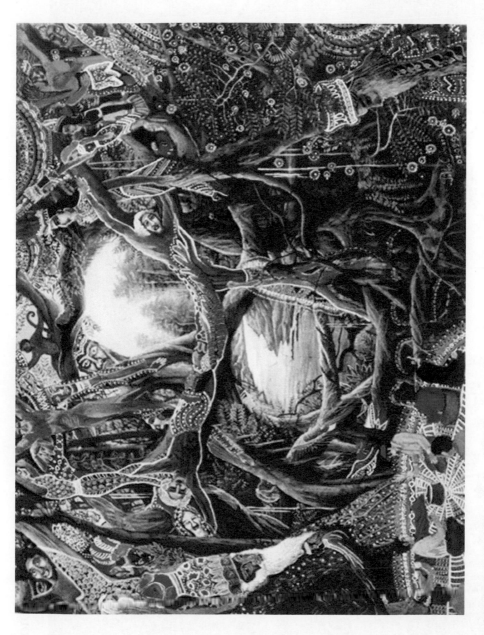

COLOR FIGURE 10.2 *Espíritu del Renaco,* by Pablo Amaringo. Gouache on paper. Copyright © by Pablo Amaringo, 1990.

COLOR FIGURE 10.3 *Renacal al Amanecer*, by Anderson Debernardi. Copyright © by Anderson Debernardi.

TABLE 3.3
Historical Uses of *Ficus* spp. Fruit: Formulas

How to Make Other Medicinal Products from Fresh or Dried Figs

Product	Method	Uses
Dried or fresh fruit aqueous decoction	Clean figs, and cull those that look bad. Cover with water in a stainless steel pot or glass vessel. Bring to a boil, cover, and allow to simmer for 1 h. Strain.	Gargle for sore, inflamed throat, tonsils, trachea, and hoarseness. As a drink for chronic cough, respiratory problems, and constipation.
Dried fruit aqueous decoction with hyssop	Decoct dried figs and dried hyssop as per preceding method. Ratio of figs to hyssop is dependent on individual taste or a physician's prescription.	Taken orally as an expectorant, for ordinary and chronic cough, lung problems/pains, chest diseases, difficulty of breathing, quinsy, pulmonary tuberculosis, hydropsy, and epilepsy as prescribed by a physician.
Dried fig decocted in wine	Decoct dried figs substituting red or white dry grape wine as described earlier.	Taken internally for swellings of jaw or soft flesh; as a suppository for stomach pain.
Dried fig decocted in honeyed wine	Decoct figs as described earlier, adding one cup of honey per liter of wine.	Taken internally or as a gargle for swollen tonsils, parotid glands, and throat.
Dried fig decocted with hyssop and savory in sweet wine	Decoct dried figs and dried herbs as described earlier. Ratio of figs to hyssop and savory is dependent on individual taste or a physician's prescription.	Taken internally on an empty stomach for chronic cough and pulmonary problems.
Figs soaked in distilled spirits	Slit the figs and soak them over night in distilled wine.	Two to three figs are taken orally in the morning for shortness of breath, for inflamed throat, and as an expectorant.
Dried fig, pomegranate peel aqueous decoction	Decoct dried figs and organically grown pomegranate peels in water for 1 to 2 h. Ratio of pomegranate peels to figs 1:3 or as directed by a physician.	Taken as a gargle for swellings/ inflammation of parotid glands.
Dried fig aqueous decoction with fenugreek flour or fenugreek flour and honey	Decoct dried figs in water with fenugreek flour for 1 h. Add honey according to taste.	Taken internally for pleurisy and peripneumony, asthma, and dry cough; as an expectorant.
Chopped dried fruit and leaf with sugar	Chop dried fruit with a few fresh leaves of fig tree and add sugar to taste.	Eat in small amounts for chronic diarrhea as directed by a physician.
Dried fig aqueous decoction with sugar	Decoct dried figs in water with sugar until the mixture thickens. Ratio of sugar to figs 1:3.	As a poultice for paronychia.

(continued on next page)

TABLE 3.3 (continued)

Historical Uses of *Ficus* spp. Fruit: Formulas

Product	Method	Uses
Fresh fig antidote	Mash fresh fruit and mix with salt, finely chopped rue, and groundnuts, or walnuts and rue.	Taken daily internally as an antidote and for preservation against poisons and epidemical diseases. Famous recipe by king Mithridates.
Fresh figs mixed with rue	Mash fresh fruit and mix with finely chopped rue.	Taken internally as an antidote against poisons.
Decoction of figs with rue	Decoct fresh figs in water or wine with finely chopped rue for 30 min. Sieve the decoction.	The fluid used as a clyster for stomach pain.
Figs mixed with nuts and almonds	Mix chopped figs with ground nuts and almonds. Ratio of figs to nuts to almonds 2:1:1.	Taken internally for loss of appetite. Good nutritional qualities. As an antidote against poisons.
Fresh figs mixed with nuts	Mix chopped sweet, ripe, fresh figs with ground nuts. Ratio of figs to nuts 3:1.	Taken internally as a laxative, against flatulence, for tendency to colic, for pains of back and joints. For urinary incontinence. As an antidote against poisons.
Dried figs stuffed with almonds	Slit the dried figs and insert an almond into each.	Taken internally against poisons. A delicacy.
Figs with aniseed	Figs spiced with aniseed are eaten for breakfast for 40 days.	Taken internally, helps to gain weight when necessary.
Figs with pepper	Fresh and dry figs are sliced and spiced with pepper.	One to two figs taken internally in the morning for removal of kidney stones.
Figs with nuts and pistachio nuts	Mix chopped figs with ground nuts and pistachios. Ratio of figs to nuts to pistachios 2:1:1.	Helps in emaciation; increases intelligence.
Figs with natron	Mash fresh figs with a little natron.	Taken internally as a laxative.
Figs with safflower and natron	Mash fresh or dry figs with safflower and a little natron.	Taken internally before meal for colic, paralysis, as a laxative.
Unripe figs with mustard	Sharp mustard is ground and mixed with water, and finely chopped unripe figs are added to the mixture.	Used as ear drops/inserted into ears as a paste for tinnitus and itching of ears.
Decoction of figs with mustard	Fresh figs are decocted in water till they reach a honey-like consistency, and froth from mustard seeds boiled in water is added to the mixture.	Used as ear drops for tinnitus. Externally on itch.
Figs with herbs	Fresh or dry figs taken with thyme, pepper, ginger, pennyroyal, savory, calamint, or hyssop.	Eaten 1½ h before a meal as a remedy for hard tumors, hepatomegaly, and/or splenomegaly.

innate immunity with that of plants in general and *Ficus* in particular. The intriguing hypothesis that physical contact with or ingestion of *Ficus* seed lectins may boost or otherwise modulate innate immunity in human systems could be tested in vitro, ex vivo, and in situ, and is certainly worthy of further investigations. Ficolins are employed by humans for biological defense (Fujita et al. 2000). Did the human mind coin this *Ficus*-like name (ficolin) unconsciously to point further investigation into the relationship between *Ficus* seed lectins and innate immunity, and the transfer of chemical evolution in the fig for its defense to be brought to the service of our human defense (Endo et al. 2006)?

The seeds (achenes) of the "jelly fig," *Ficus awkeotsang* Makino, yield a gel-forming polygalacturonide hydrocolloid with a pectinesterase (EC 3.1.1.11), responsible for the spontaneous gel formation of the awkeotsang polygalacturonide—this was isolated from the red tepals attached to the pedicels of seeds of it (Komae et al. 1989). The gel contains a pectimethylesterase for metabolizing methane and methanol that lowers methanol content in carambola wine production (Wu et al. 2005). The fat content of the seeds is 63% α-linolenic acid (Chua et al. 2008a), analogous to the conjugated linolenic acids found in pomegranate seeds. The same pectimethylesterase that lowers the methanol in carambola wine also induces apoptosis and cell cycle arrest in leukemia cells—90% in a recent examination of its potential as a leukemia treatment conducted by Chang et al. in 2005. The DNA fragment of *F. awkeotsang* that encodes for this compound was successfully introduced into a methylotrophic yeast, *Pichia pastoris* (Peng et al. 2005), and considerable work exploring the further genomic exploitation of this capability (Chua et al. 2007, 2008b) is taking place. The gel, which may be obtained by boiling the figs, is used to prepare traditional drinks and desserts in southern China and Southeast Asia, and is also being studied as a novel controlled release drug delivery system (Miyakazi et al. 2004).

There are additional benefits from the use of this possibly unique jelly matrix from *F. awkeotsang*. For example, a chitinase, which is active against fungi, was isolated from the jelly and found to be stable during prolonged storage at 4°C. Its antifungal potency was confirmed by inhibition of spore formation in *Colletotrichum gloeosporioides* (Li et al. 2003). This is important information, if not for direct use as an antifungal agent, then at least for the preservative quality of the medicinal material. Wang and Tzen (2005) offer their view that numerous enzymes with antibacterial and antifungal properties occur in the jelly, having evolved to protect the fig from rapid decay following ripening. So, putatively, the jelly matrix is well protected from decomposition by virtue of its antibiotic and antifungal arsenal of attack proteins that target the structural integrity of the bacteria and fungi.

Misaki et al. (2004) emphasize that a simple cold water extract is sufficient to release the jelly in the seeds. If the aforementioned enzymes are inactivated, it is possible also to extract and isolate a polysaccharide complex. These polysaccharides act as valuable dietary fibers, including a soluble highly methoxylated galacturonan with powerful cholesterol-lowering effect in hypercholesterolemic rat serum and liver.

Because of its anticancer properties, the jelly fig gel could serve as an ideal matrix for a complex anticancer drug, at least for its hydrophilic components. Ongoing work continues to elucidate the capability of such a matrix to achieve a sustained release

FIGURE 3.12 *Ficus awkeotsang* Marino (jelly fig) gel in a Vietnamese dessert *súóng sa bông cò* (http://en.wikipedia.org/wiki/File:Aiyu_jelly_by_abon_in_Taiwan.jpg).

FIGURE 3.13 *Ficus awkeotsang* Marino, the "jelly fig," Vietnamese *vùa chín* (http://i246. photobucket.com/albums/gg116/hoalaco/fruitaiyu.jpg).

active of drug components from within a hydrophilic matrix tablet (Miyazaki and Yakou 2006).

Dried figs are an extremely popular and ancient means of usage and storage of *Ficus* fruits. Recently, a number of studies have highlighted the susceptibility of dried figs to infestation by molds; that is, fungi that grow in multicellular filaments known as hyphae, which can produce noxious compounds that may specifically induce cancer or interfere with mammalian host resistance to cancer. Aflatoxin is among the best known of these toxins, which can promote and induce hepatic cancer and represent a major hepatocarcinoma risk factor in developing countries, especially Africa and the Middle East (Groopman et al. 2008). Because of this ancient proclivity to such infestation, figs may have evolved some of their anticancer chemistry as part of an intricate relationship with and defense against molds and their toxins. Certainly, the composition of sugars and amino acids in figs, which is significantly altered during such infestations, constitutes in the dried fig an almost ideal medium for fungal growth (Senyuva et al. 2008). Stress from drought has the effect of rendering these conditions even more favorable for the fungi, with higher levels of aflatoxins noted in the figs during drought years (Bircan et al. 2008). To some degree, the problem is being attenuated by treating the dried figs with ozone (Zorlugenc et al. 2008), though new methods of inspection and quantification of the risk are clearly indicated.

In addition to aflatoxins produced by fungi of the genus *Aspergillus*, other important mycotoxins have also been recently identified. These include HT-2 toxin, a metabolite of the tricothecene toxin, T-2 (Dohnal et al. 2008); patulin, a genotoxic and antibacterial; zearalenone, a potent estrogenic compound; and fumonisin B2 (Senyuva and Gilbert 2008). Some of these compounds have been earmarked as biological weapons for modern human warfare.

Fumonisin B1 is a known ceramide synthase inhibitor and has recently been added to the list of mycotoxins found in dried figs (Karbancioglu-Güler and Heperkan 2009). Ceramide synthase is the enzyme responsible for catalyzing the synthesis of ceramide, a phospholipid found in cell membranes. Ceramide is both a cell membrane structural component and a key signaling molecule helping to regulate actions outside and inside of cells. Especially, it is known to modulate the cancer process, facilitating differentiation and apoptosis, and inhibiting proliferation in malignant cells. Cross talk between these functions, as well as response to fungi, is at least partially dependent on phospholipids. Phospholipids figure prominently in the response of a model grass, *Brachypodium distachyon*, to pathogenic fungus attack and, significantly, also in amplifying and balancing antifungal treatment in humans (Lansky and Nevo 2009). These interactions further augment the possibility that anticancer compounds within the fig fruits may have evolved in part to thwart attack by fungal opportunists looking to take advantage of the fruits' naturally fungus-friendly internal milieus.

An aqueous ethanolic extract of ripe fruits from *F. carica* was studied for antispasmodic and antiplatelet actions in isolated rabbit jejunum and in an ex vivo model of human platelets, respectively, and shown to be active in both settings (Gilani et al. 2008). Since the neurotransmitter serotonin (5-hydroxytryptamine) is an important modulator of both smooth muscle contraction and platelet aggregation, it is likely that these figs possessed some compound active on serotonin. In fact, Singh and Goel

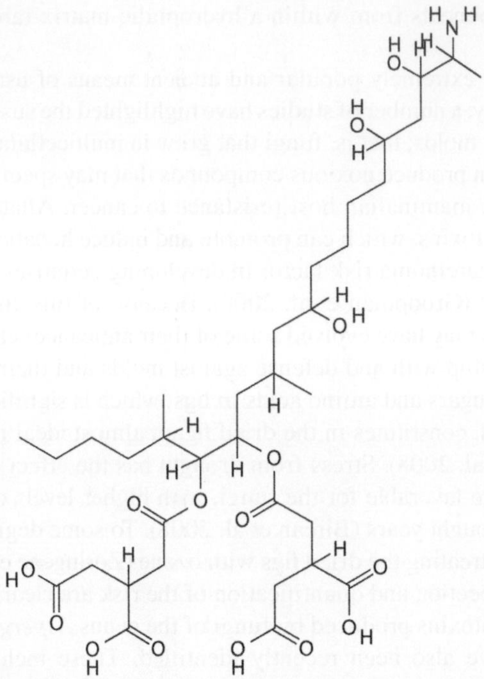

FIGURE 3.14 Fumonisin B1. Hepatotoxic, nephrotoxic, and genotoxic mycotoxin that may occur in dried figs. The toxin is produced by molds of the *Fusarium* genus and is probably also toxic to figs. The anticancer chemistry of figs may have evolved in part to overcome toxicity of fungi producing such toxins. (Karbancioglu-Güler, F., and D. Heperkan. 2009. Natural occurrence of fumonisin B1 in dried figs as an unexpected hazard. *Food Chem Toxicol*. 47: 289–92; figure from *PubChem*. The PubChem Project. 2001. [Online Database]. National Center for Biotechnology Information. U.S. National Library of Medicine. 8600 Rockville Pike, Bethesda, MD 20894. URL: http://pubchem.ncbi.nlm.nih.gov/ [accessed August 16, 2009].)

(2009) working with the figs from the Bo tree, *F. religiosa*, showed a methanolic extract of the fruits suspended in dimethylsulfoxide (DMSO) to have a significantly protective effect in male Swiss albino mice (*Mus musculus*) against seizures induced by either maximum electroshock (MES) or picrotoxin in a dose-dependent manner from 25, 50, or 100 mg/kg upon intraperitoneal (i.p.) injection with zero neurotoxicity. Seizures induced by pentylenetetrazol, however, were not protected against. The protective effects observed were interrupted by cyproheptadine, a nonselective (5-HT [1/2]) serotonin antagonist (4 mg/kg, i.p), confirming serotonergic involvement in the protective mechanism. Further, the extract produced a similar dose-dependent potentiation of pentobarbitone-induced sleep. The protection observed was comparable to that of phenytoin and diazepam, conventional antiepileptic agents.

Singh and Goel (2009) inform us that *F. religiosa* is also known in Ayurveda as possessing anticonvulsant properties, and that in a much earlier study the species was found to have higher amounts of serotonin than other *Ficus* species (Bliebtrau 1968). Serotonergic neural transmission is known to modulate a wide variety of experimentally induced seizures and is involved in seizure protection in various animal models of epilepsy by its altering effect on various GABAergic and glutamatergic functions (Bagdy et al. 2007). Singh and Goel further speculate that the high serotonin content in the parts of *F. religiosa* might transfer into clinical effects in humans whereby serotonin metabolism could be favorably influenced, perhaps even for religious or visionary effects.

Similarly, extracts of fruits from *F. capensis* showed good activity against sickling of red blood cells in sickle cell anemia (Mpiana et al. 2008), a genetic disorder of hemoglobin in which the abnormal shapes the erythrocytes assume may lead to vaso-occlusive crises and shortened life span. The disease is a major health concern, especially in Africa in regions where malaria is also prominent. The vein is related to the antispasmodic effects since serotonin also acts as an important mediator in platelets and may also play a role in the antisickling activity, even though preliminary investigations on serotonin, platelets, and sickle cells failed to show a specific correlation (Buchanan and Holtkamp 1983). The possibility that modulation of serotonin action may indirectly influence sickling has not been ruled out. Gastrointestinal function is also governed to a very large, previously unappreciated degree by serotonin fluctuations, in such a way that high plasma serotonin levels are associated with diarrhea and low serotonin levels with constipation, and much activity in the pharmaceutical industry is currently under way to find new and more specific serotonin inhibitors for this purpose (Camilleri 2009).

Specifically, serotonin may also play an important role in ulcerogenesis through modulation of the histamine system (Lazebnik et al. 2008); the latter is already a common pharmaceutical target in the treatment of gastric and duodenal ulcers. Thus, the findings of Rao et al. (2008), which showed a 50% ethanol–water extraction of the fruits of *F. glomerulosa* administered orally to rats in doses of 50, 100, and 200 mg/kg body weight to dose dependently inhibit ulcer index in pylorus ligation, ethanol and cold restraint stress-induced ulcers, might have a basis in a serotonergic effect. The authors point out that the extract also prevented the oxidative damage of gastric mucosa by blocking lipid peroxidation, downregulating superoxide dismutase and H^+K^+ ATPase, while increasing catalase activity, but even these changes could be linked to serotonergic mechanisms—for example, serotonin production is upregulated in rice leaves undergoing senescence (Kang et al. 2009), and inhibiting serotonin reuptake in neurons remains the mainstay of current antidepressant therapy for the elderly (Mukai and Tampi 2009). If fig fruits do indeed contain compounds that can modulate serotonin metabolism, as they seem to, then the full potential for psychoactive drugs opens from their gentle and safe flesh. In addition, because of the wide distribution of serotonin and its receptors in the periphery, other metabolic functions in mammals may be modulated, including cardiovascular and gastrointestinal functions, bone growth pathways, and glucose uptake (Jonnakuty and Gragnoli 2008).

The fruit of another *Ficus* species that has been extensively employed medicinally is *F. microcarpa*. The fruits were noted to contain terpenes, flavonoids, aliphatic compounds, and steroids, and to have been used traditionally in China for the relief of coughs and as an expectorant, platelet activating factor inhibitor, and an anti-inflammatory agent (Liu et al. 2008).

Work with *F. platyphylla* has focused on its anthelmintic effects. El-Sayyad et al. (1986) studied the air-dried powdered leaves and bark as well as the fresh fruit extract, containing tartrate, ascorbate, citrate, malate, kojate, maleate, and malonate, and the latex. The fresh fruit extract exerted anthelmintic effects in chickens against the parasite *Ascardia galli*.

In a large screening trial of extracts from plant parts used in Yemeni ethnomedicine, 90 crude extracts, including dichloromethane, methanol, and aqueous extracts from 30 medicinal plants used to treat common infections, were screened in vitro for antimicrobial activities against three Gram-positive bacteria and two Gram-negative bacteria, *Candida maltosa*, and five opportunistic human fungal pathogens (two yeasts, three hyphomycetes). Extracts from *Tamarindus indica* flowers and *F. vasta* fruits were the most active (Al-Fatimi et al. 2007).

Fruits of *F. glomerata*, for which Rao et al. (2008) showed gastric ulcer-protective effects in rats, also yields an 80% ethanol-in-water extract that was consistently active in lowering blood glucose when fed at the same time as a glucose load to normal or type 1 or type 2 model diabetic mice (Jahan et al. 2009). A butanol extract of the fruits showed antioxidant activity in silico, and yielded a pure compound, caffeoyl quinate, which also possessed significant antioxidant activity.

The action of the fig extract may be related to its anthocyanin content. The extract loses over half of its antisickling activity after 6 hours' exposure to ultraviolet light. Anthocyanins are particularly labile to light.

Traditional uses of figs in medicine have been documented over thousands of years and are reflected in medieval and ancient writings by a number of prominent physicians from within this period. Table 3.7, and the other similar tables in subsequent chapters, employ a scoring system based on the number of times a reference discusses the use of figs for the purpose indicated. If the number of citations

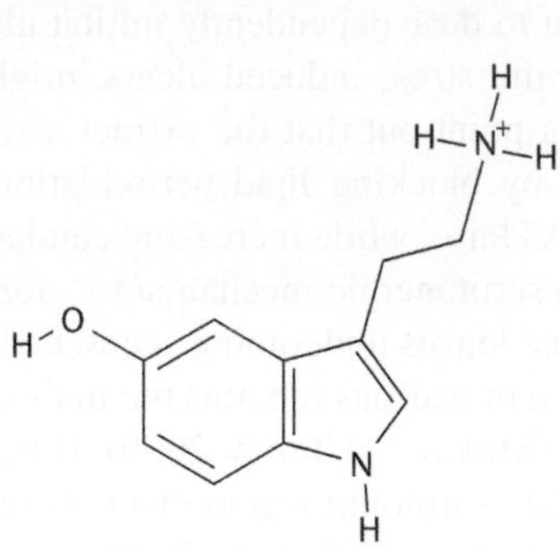

FIGURE 3.15 Serotonin (5-hydroxytryptamine) is a common denominator in the action of platelets and smooth muscle contraction as well as higher mental processes. Since *Ficus carica* fruit extracts inhibit both platelet aggregation and intestinal spasticity, involvement of this indoleamine in these actions is suggested (i.e., an antiserotonergic effect) that is worthy of specific investigation. (Singh, D., and R.K. Goel. 2009. Anticonvulsant effect of *Ficus religiosa*: role of serotonergic pathways. *J Ethnopharmacol.* 123: 330–4; figure from *PubChem*. The PubChem Project. 2001. [Online Database]. National Center for Biotechnology Information. U.S. National Library of Medicine. 8600 Rockville Pike, Bethesda, MD 20894. URL: http://pubchem.ncbi.nlm.nih.gov/ [accessed August 16, 2009].)

TABLE 3.4
Pharmacological Studies of *Ficus* spp. Fruit (including seeds and fruit peel)

Ficus Species	Pharmacological Action	Details	Reference
awkeotsang	Pectinesterase inhibitor (PEI) inhibits intrinsic carambola PE activity; reduces methanol content in carambola wine	Crude pectinesterase (PE) inhibitor (PEI) extracted from jelly-fig achenes (JFA) (*F. awakeosang* Makino) was added to carambola (*Averrhoa carambola* L.) puree to determine the change in methanol production during fermentation. Addition of pectin or microbial pectic enzyme to puree increased dose dependently the methanol content in fermented products. Decreasing ratio (from 1:0 to 1:19, v:v) of pectic enzyme to diluted crude PEI solution in the puree–enzyme mixture decreased the PE activity remarkably.	Wu et al. 2005
awkeotsang	Cold water extract of seeds: forms spontaneously a pudding-like gel Dietary fiber effect: remarkable cholesterol-lowering effect on hypercholesterolemic rat serum and also liver	Cold water extract of seeds of *F. awkeotsang* Makino forms spontaneously a pudding-like gel. In the purification process, an acidic polysaccharide was obtained from the water extract of heat-treated, enzyme-inactivated seeds. Fractionation of the sol. polysaccharide mainly containing galacturonic acid (85%) yielded an essentially linear, highly methoxylated α-1,4-linked galacturonan (DS of methoxyl, 69.3%, MW 3.4 × 105). Comparison of gel formation from the native and the heat-treated seeds strongly suggested that release of the methoxyl groups by endogeneous Me esterase must be responsible for the spontaneous gelation process.	Misaki et al. 2004
benghalensis	Galactose-specific lectin = *F. benghalensis* agglutinin (FBA)		Singha et al. 2007
benjamina	Causes anaphylactic reactions; causes oral allergy syndrome (OAS) with respiratory symptoms	Two cases of the oral allergy syndrome (OAS) to *F. benjamina* in patients whose main allergic manifestations were related to sensitization to grass and birch pollens. The patients were characterized by clinical history, skin prick tests (SPT) with commercial and inhouse extracts, prick-by-prick test, specific IgE measurements and challenge tests.	Antico et al. 2003

(continued on next page)

TABLE 3.4 (continued)
Pharmacological Studies of *Ficus* spp. Fruit (including seeds and fruit peel)

Ficus Species	Pharmacological Action	Details	Reference
benjamina (continued)		PBS-soluble and insoluble extracts of both fig skin and pulp were examined for the presence of potential allergens by IgE immunoblotting. Both patients showed OAS followed by respiratory symptoms when challenged with fig. They were negative in both specific IgE detection and SPT with fig extracts and many other plant materials, including *F. benjamina* and *Hevea brasiliensis*, while grass and birch pollens gave positive results. Prick-by-prick tests and SPT with in-house extracts indicated that the fig skin had a much higher allergenicity than the pulp. Despite negative IgE detection by the CAP assay, immunoblotting experiments showed that potential fig allergens were PBS soluble and present only in the skin of the fruit.	
carica	Antioxidant	Six commercial fig (*F. carica*) varieties differing in color (black, red, yellow, and green) were analyzed for total polyphenols, total flavonoids, antioxidant capacity, and amount and profile of anthocyanins.	Solomon et al. 2006
carica	Antispasmodic; antiplatelet effect	Aqueous-ethanolic extract of *Ficus carica* was studied for antispasmodic effect on the isolated rabbit jejunum preparations and for antiplatelet effect using ex vivo model of human platelets. In isolated rabbit jejunum, it (0.1–3.0 mg/mL) produced relaxation of spontaneous and low K(+) (25 mM)-induced contractions with negligible effect on high K(+) (80 mM) similar to that caused by cromakalim. Pretreatment of the tissue with glibenclamide caused rightward shift in the curves of low K(+)-induced contractions. Similarly, cromakalim inhibited the contractions induced by low K(+), but not of high K(+), while verapamil equally inhibited the contractions of K(+) at both concentrations. The extract (0.6 and 0.12 mg/mL) inhibited the adenosine 5′-diphosphate and adrenaline-induced human platelet aggregation.	Gilani et al. 2008

TABLE 3.4 (continued)

Pharmacological Studies of *Ficus* spp. Fruit (including seeds and fruit peel)

Ficus Species	Pharmacological Action	Details	Reference
cunia	Seed lectin binding to the bacteria *E. coli*, *Pseudomonas aeruginosa*, *Klebsiella pneumoniae*, *Bacillus subtilis*, and *Staphylococcus aureus*		Adhya et al. 2006
glomerata	Gastroprotective activity; antiulcer; preventing the oxidative damage of gastric mucosa by blocking lipid peroxidation and by significant decrease in superoxide dismutase, H+K+ATPase and increase in catalase activity	50% ethanolic extract of *F. glomerata* fruit (50, 100 and 200 mg/kg body wt.) administered orally, twice daily for 5 days to prevent pylorus ligation, ethanol (EtOH) and cold restraint stress (CRS)-induced ulcers. Estimation of H+K+ATPase activity and gastric wall mucus were performed in EtOH-induced ulcer and antioxidant enzyme activities in supernatant mitochondrial fraction of CRS-induced ulcers. The extract showed dose-dependent inhibition of ulcer index in pylorus ligation, ethanol, and cold restraint stress-induced ulcers.	Rao et al. 2008
glomerata	Inhibits glucose-6-phosphatase from rat liver	*F. glomerata* Roxb fruit extracted with water and organic solvents.	Rahman et al. 1994
platyphylla	Prepared latex of the fruit was proteolytic in egg albumin. The fresh fruit extract and the latex had anthelmintic activity against *Ascardia galli* in chickens.		El-Sayyad et al. 1986
racemosa	Lectins specifically agglutinate leukemia cells but not normal blood cells	Semipurified saline extracts of seeds from *F. racemosa* were tested for leukoagglutinating activity against whole leukocytes and mononuclear cells from patients with chronic myeloid leukemia (34), acute myeloblastic leukemia (5), acute lymphoblastic leukemia (7), chronic lymphocytic leukemia (3), various lymphoproliferative/hematologic disorders (54), and normal healthy subjects (50).	Agrawal and Agarwal 1990

TABLE 3.5
Chemicals Identified or Isolated from Fruit of Various *Ficus* spp.

Ficus Species	Compounds	Details	Reference
awkeotsang	Cold water extract of seeds forms gel, with acidic polysaccharide containing galacturonic acid (85%) and linear, highly methoxylated α-1,4-linked galacturonan (DS of methoxyl, 69.3%, MW 3.4 × 105); dietary fiber	In the purification process, an acidic polysaccharide was obtained from the water extract of heat-treated, enzyme-inactivated seeds. The soluble, highly methoxylated galacturonan exhibited a remarkable cholesterol-lowering effect on hypercholesterolemic rat serum and also liver.	Misaki et al. 2004
benghalensis	Galactose specific lectin = *F. benghalensis* agglutinin (FBA)	Lectin was purified by affinity repulsion chromatography on fetuin-agarose and was a monomer of molecular mass 33 kDa. Carbohydrate-binding activity of FBA was independent of any divalent cation. FBA did not bind with simple saccharides; however, sugar ligands with aromatic aglycons showed pronounced binding.	Singha et al. 2007
benjamina	Bergapten	Petroleum ether extracts of powdered air-dried fruits of *F. benjamina* chromatographed on alumina and eluted with 1:5 petroleum ether-benzene gave bergapten	Ahmad et al. 1971
carica	Anthocyanins; cyanidin-3-*O*-rhamnoglucoside (cyanidin-3-*O*-rutinoside); cyanidin; polyphenols; flavonoids	Six commercial fig varieties (*F. carica*) differing in color (black, red, yellow, and green) were analyzed for total polyphenols, total flavonoids, antioxidant capacity, and amount and profile of anthocyanins. Hydrolysis revealed cyanidin as the major aglycon. Proton and carbon NMR confirmed cyanidin-3-*O*-rhamnoglucoside (cyanidin-3-*O*-rutinoside; C3R) as the main anthocyanin in all.	Solomon et al. 2006
carica	Alkaloids; flavonoids; coumarins; saponins; sterols; terpenes		Gilani et al. 2008

TABLE 3.5 (continued)
Chemicals Identified or Isolated from Fruit of Various *Ficus* spp.

Ficus Species	Compounds	Details	Reference
carica "Mission"	Anthocyanin; chlorophylls a and b; β-carotene; lutein; violaxanthin; neoxanthin	Following application of Ethephon [16672-87-0] to "Mission" fig fruits (*F. carica*) during late period II of their development.	Puech et al. 1976
glomerata	Gallic acid; ellagic acid	High-performance thin layer chromatography (HPTLC) showed the presence of 0.57% and 0.36% wt./wt. of gallic acid and ellagic acid in FGE.	Rao et al. 2008
platyphylla	Fruits: tartrate, ascorbate, citrate, malate, kojate, maleate, and malonate Prepared latex of the fruit: aspartate, hydroxyproline, tryptophan, tyrosine, arginine, lysine, alanine, histidine, and two unidentified acids		El-Sayyad et al. 1986
pumila	Three new sesquiterpenoid glucosides: pumilasides A, B, and C ([1S,4S,5R,6R,7S,10S]-1,4,6-trihydroxyeudesmane 6-*O*-β-d-glucopyranoside, [1S,4S,5S,6R,7R,10S]-1,4-dihydroxymaaliane 1-*O*-β-d-glucopyranoside and 10α, 11-dihydroxycadin-4-ene 11-*O*-β-d-glucopyranoside); benzyl β-d-glucopyranoside; (E)-2-methyl-2-butenyl β-d-glucopyranoside; rutin	The structures of new compounds were characterized by spectral and chemical methods.	Kitajima et al. 2000

FIGURE 3.16 l-Tryptophan. (*NIST Chemistry WebBook*. 1996. [Online Database]. NIST Standard Reference Database Number 69. The National Institute of Standards and Technology (NIST). URL: http://webbook.nist.gov/chemistry/contact.html [accessed August 14, 2009].).

is regular, the reference is given in regular type and scored as "1." If the number of citations is strong, the reference is given in *italic* and scored as "2." If the number of citations is strongest, the reference is given in **bold** and scored as "3." The final score in the last column is the sum of these scores for the individual references. The particular parts of the fruit used are starred, but this information does not enter into the calculation of the total score for the function.

The indications are shown in the order of importance based on their score values.

Any medication and/or food could, according to the medieval understanding, have also negative effects, but they could be corrected by other herbs or foods. In the

TABLE 3.6
The Authors Quoted in Tables 3.7, 3.10, 4.4, 5.2, 6.6, and 7.4

Medieval/Ancient Authors

Author	Works	Biography
1. Pliny	*The Natural History*	23–79, Italy
2. Dioscorides	*The Materials of Medicine*	c. 40–90, Anazarbos (Turkey)
3. Galen	*Complete Works*	2nd century, Pergamum/Rome
4. Al-Tabari	*The Paradise of Wisdom*	800–875, Persia
5. Ishaq Israeli	*Complete Works*	855–955, Kairouan (Tunis)
6. Al-Razi	*The Comprehensive Book on Medicine*	c. 865–925, Persia
7. Ibn Sina	*The Canon of Medicine*	980–1037, Persia
8. Ibn Butlan	*Tables of Health*	d. 1066, Baghdad/Cairo/Antioch
9. Anonymous	*Flower of Medicine of the School of Salerno*	11th century, Italy
10. Albertus Magnus	*Book on Growing Things*	1206–1280, Germany
11. Ibn al-Baytar	*Comprehensive Book of Simple Drugs and Foods*	d. 1248, Spain/Egypt/Syria
12. Rufinus de Rizardo	*Herbal*	13th century, Italy
13. Hieronymus Bock	*Herbal on the Differences, Names and Properties of Herbs*	1498–1554, Germany
14. Leonhart Fuchs	*Notable Commentaries on the History of Plants*	1501–1566, Germany
15. Jacob Theodor Tabernaemontanus	*New Complete Herbal*	1522–1590, Germany
16. Gregorio López	*Treasure of Medicines*	1542–1596, Mexico
17. Al-Antaki	*Memorandum Book for Hearts and Comprehensive Book of Wonderful Marvels*	d. 1599, Syria/Cairo
18. John Gerard	*Herbal*	1545–1611/1612, England
19. Fr. Blas de la Madre de Dios	*Book of Domestic Medicines*	d. 1626, Philippines
20. John Parkinson	*The Botanical Theater*	1567–1650, England

TABLE 3.7
Use of the Fruit of Fig (including peel, seeds) in Ancient and Medieval Sources

Indication	Fruit	Dry Fruit	Fresh Fruit	Fruit Wine	Juice	Ash of Fruit	Immature Fruit	References	Score
Nutritive	*	*	*					1, 2, 5, 6, 7, 8, 9, 10, 11, 12, 13, 14, 15, 17, 18, 20	34
Warming	*	*	*		*		*	1, 2, 3, 4, 5, 6, 7, 8, 10, 11, 12, 13, 14, 15, 16, 17, 18, 20	32
Laxative	*	*	*	*			*	1, 2, 4, 6, 7, 8, 9, 11, 12, 13, 14, 15, 17, 20	24
Against all hard swellings/ tumors, hardening	*	*						1, 2, 3, 6, 7, 10, 11, 13, 14, 15, 16, 17, 20	19
Bad for stomach	*	*	*	*			*	1, 2, 6, 7, 8, 11, 12, 13, 14, 15, 17, 18, 20	19
Analgesic	*	*	*	*				1, 2, 5, 6, 7, 11, 12, 13, 14, 15, 16, 17, 18, 20	19
Sweet taste	*	*	*					1, 2, 6, 7, 8, 10, 11, 14, 15, 18, 20	18
Against materia congesting chest, lungs	*	*						1, 2, 5, 8, 11, 12, 13, 14, 15, 16, 17, 18	18
Against all hot swellings, inflammations	*	*						1, 2, 5, 6, 7, 11, 12, 14, 15, 16, 17, 18	17
Moisturizing	*	*	*				*	3, 5, 6, 7, 8, 10, 11, 13, 14, 15, 17, 18, 20	17
Generates lice	*	*	*					5, 6, 7, 8, 9, 10, 11, 12, 13, 14, 15, 16, 17, 18, 20	17
Against cough	*	*		*				1, 2, 5, 7, 11, 12, 13, 14, 15, 16, 17, 18, 20	16

(continued on next page)

TABLE 3.7 (continued)
Use of the Fruit of Fig (including peel, seeds) in Ancient and Medieval Sources

Indication	Fruit	Dry Fruit	Fresh Fruit	Fruit Wine	Juice	Ash of Fruit	Immature Fruit	References	Score
Causes flatulence	*	*		*			*	2, **5**, *6*, 7, 8, 11, **12**, 13, 14, 15, 17	16
Benefits chronic diseases	*	*		*				1, *2*, 5, 6, 7, *11*, 12, 13, *14, 15*, 18, 20	16
Harmful	*	*	*	*				1, 5, **6**, 7, 10, 12, 13, **14**, 15, 17, 20	15
Cleans chest	*	*						1, 2, 5, 8, 11, 12, *13, 14*, 15, 16, 17, 18	14
Against swelling, tumors in general	*	*					*	*1*, 2, 5, 7, 9, 10, 12, 13, *14, 15*, 16	14
Cooling, calming heat	*		*				*	1, 2, 5, 6, 7, 8, 10, 11, 12, 13, 14, 15, 18	13
Rarefying	*	*		*				2, 3, *6*, 7, *8*, 10, 11, *14*, 18	13
Dissolving	*	*					*	3, **6,** 7, 10, *11*, 15, 18	13
Against skin inflammations	*	*					*	1, 2, *6*, 7, 10, 11, *14*, 16, **18**	13
Creates bad blood	*	*		*				2, 5, 6, 7, 8, 10, 12, 13, *14*, 15, 18, 20	13
Against all swellings/ tumors of ears (mumps?)	*	*						*1*, 6, 7, 11, 12, *14*, 15, *18*, 20	12
Against tuberculosis, scrofula	*	*					*	*1*, 2, 5, 6, 9, 11, *12*, 14, 15, 20	12
Causes thirst		*						1, 2, 5, 6, 7, 11, 12, 13, *14*, 15, 20	12
Against dropsy	*	*						1, 2, 6, 7, 11, 13, *14*, 15, 16, 18, 20	12

TABLE 3.7 (continued)
Use of the Fruit of Fig (including peel, seeds) in Ancient and Medieval Sources

Indication	Fruit	Dry Fruit	Fresh Fruit	Fruit Wine	Juice	Ash of Fruit	Immature Fruit	References	Score
Against stomach pain	*	*						1, 2, 5, 6, 11, 12, 13, 14, 15, 16, 17	11
Benefits body	*	*	*				*	5, 6, 7, 10, 11, 12, 13, 14	11
Against warts	*				*		*	1, 2, 6, 7, 11, 14, 17	10
Against epilepsy	*	*	*					1, 2, 6, 7, 10, 11, 13, 14, 17, 20	10
Against lung diseases	*	*						2, 11, 12, 13, 14, 15, 16, 18, 20	10
Cleaning, absterging	*	*			*		*	3, 6, 7, 11, 12, 14, 15, 17	10
Benefits kidneys	*	*	*					2, 6, 7, 8, 11, 12, 13, 14, 17, 20	10
Sudorific	*	*						1, 2, 5, 6, 7, 8, 10, 11, 12, 14	10
Against dyspnea	*	*						1, 13, 14, 15, 17, 18, 20	9
Spicy, sharp	*	*					*	3, 5, 6, 7, 11, 14, 18	9
Cleans kidneys	*	*					*	4, 5, 6, 11, 12, 14, 18	9
Benefits stomach, abdomen	*	*		*				2, 7, 11, 12, 14, 15, 18, 20	9
Against chilblains	*	*				*		1, 2, 6, 11, 13, 14, 15, 16, 20	9
Against tinnitus	*	*			*		*	2, 5, 6, 7, 11, 12, 13, 14, 16	9
Benefits bladder	*	*	*					1, 2, 6, 7, 11, 12, 13, 14, 20	9
Alleviates thirst	*		*					1, 2, 6, 7, 11, 12, 13, 14, 15	9
Quickly nutritious	*	*	*					6, 7, 10, 11, 14	9
Cleans kidneys of materia but not stones	*	*					*	4, 5, 6, 11, 12, 14, 18	9

(continued on next page)

TABLE 3.7 (continued)
Use of the Fruit of Fig (including peel, seeds) in Ancient and Medieval Sources

Indication	Fruit	Dry Fruit	Fresh Fruit	Fruit Wine	Juice	Ash of Fruit	Immature Fruit	References	Score
Against kidney stones	*	*	*				*	6, 7, 8, 14, *15*, **18**	9
Drying, against moisture	*	*	*				*	*5, 8*, 10, 11, 14, 17	8
Ulcers in thighs	*	*						1, 2, 6, 11, *14*, 15, 20	8
Against paronychia	*	*						1, 2, 6, 7, 11, 13, 14, 15	8
Softening (in general)	*	*	*				*	1, 6, 7, 9, 11, 14, 15, 18	8
Improves color after an illness	*	*						2, 6, 7, 8, 11, 13, 14, 20	8
Benefits trachea	*	*						2, 6, 7, 11, 12,13, 14, 20	8
Benefits throat	*	*						1, 2, 6, 11, 12, 14, 18, 20	8
Against urinary problems	*	*	*	*				1, 2, 5, 7, 12, 14, 15, 17	8
Against dysuria, diuretic	*	*	*	*				1, 2, 5, 7, 12, 14, 15, 17	8
Against tonsillitis	*	*						2, 5, 11, 12, 14, 15, 16	7
Thick/coarse, thickens, coarsens	*	*					*	*5, 8*, 10, 11, 12	7
Exits body quickly	*	*	*				*	**6,** 7, 10, 11, 14	7
Against all swellings/ tumors of throat	*	*						6, 7, 8, 13, 15, 16, 20	7
Against all swellings/ tumors of trachea		*		*				2, 5, 6, 7, 11, 12, 14	7
Against all swellings/ tumors of spleen	*	*						4, 6, 7, 8, 11, *14*	7
Against poison	*	*						8, 13, 14, 15, 17, 18, 20	7
Against rabid dog bite	*			*			*	5, 6, 7, 10, 11, 12, 13	7

TABLE 3.7 (continued)
Use of the Fruit of Fig (including peel, seeds) in Ancient and Medieval Sources

Indication	Fruit	Dry Fruit	Fresh Fruit	Fruit Wine	Juice	Ash of Fruit	Immature Fruit	References	Score
Putrefies	*	*				*		5, 6, 8, 10, 12, 20	7
Fattening	*	*						7, 8, 9, 10, 11, 16, 17	7
Lacking nutrition	*							2, 6, 7, 11, 15, 18, 20	7
Cleans lungs, expectorant	*	*						5, 11, 12, 17, 18	6
Against asthma	*	*						1, 2, 6, 11, 14, 17	6
Malignant ulcers	*	*						2, 6, 11, 14, 15, 20	6
Against hot swellings/ tumors of trachea		*						2, 5, 6, 11, 12, 14	6
Maturating, ripening	*	*	*					1, 3, 7, 11, 14, 15	6
Diuretic	*	*		*				1, 2, 5, 7, 12, 14	6
Against phlegm	*	*						1, 6, 11, 14, 18	5
Cleans bladder	*	*						4, 5, 6, 11, 12	5
Cutting	*	*						6, 7, 10, 14	5
Ulcers	*							1, 13, 15	5
Against all swellings/ tumors of liver	*	*						4, 6, 7, 14	5
Against hard swelling, thickening of spleen	*	*						6, 7, 8, 11, 14	5
Against bubonic ulcers	*	*						2, 11, 13, 15, 18	5
Benefits chest	*	*						6, 7, 8, 12, 14	5
Cleans bladder	*	*						4, 5, 6, 11, 12	5
Nutritive that preserves health	*	*						1, 6, 12, 14	5
Strengthening		*						1, 13, 14, 15, 16	5
Quickly digested	*	*						5, 6, 11, 12	4

(continued on next page)

TABLE 3.7 (continued)
Use of the Fruit of Fig (including peel, seeds) in Ancient and Medieval Sources

Indication	Fruit	Dry Fruit	Fresh Fruit	Fruit Wine	Juice	Ash of Fruit	Immature Fruit	References	Score
Pushes materia to skin	*	*						7, 8, 10, 11	4
Good taste	*	*						8, 15, 18, 20	4
Discutient	*	*					*	1, 3, *14*	4
Cleans lungs	*	*						5, 11, 12, 17	4
Against putrid ulcers	*	*					*	1, 6, 11, 14	4
Against wrinkles	*	*						1, 10, 14, 16	4
Against sebaceous cysts	*	*					*	1, 6, 14, 16	4
Against ear itch		*					*	2, 6, 11, 14	4
Against liver obstruction	*	*						6, 7, 14, 15	4
Against dribbling	*							4, 6, 7, 11	4
Helps control urinating	*							4, 6, 7, 11	4
Against obstructions of spleen	*	*						6, 7, 14, 15	4
Benefits womb and women's diseases	*	*						1, 2, 6, 14	4
Against scorpion sting	*	*		*			*	1, 13, 14, 15	4
Food for elderly	*	*						1, 14, 16	3
Removes appetite			*					2, 6, 7	3
Nutrition unlike that of meat and grains	*							6, 7, 14	3
Generates soft flesh	*							14, 15, 20	3
Cholagogue		*						5, 12, 14	3
Less taste	*							2, 7, 11	3
Against flatulence	*	*					*	1, 11, 17	3
Against all swellings/ tumors of womb	*		*					7, 13, 20	3
Against phlegm in lungs, chest	*	*						1, 14, 18	3
Cleans stomach	*	*						6, 7, 11	3

TABLE 3.7 (continued)
Use of the Fruit of Fig (including peel, seeds) in Ancient and Medieval Sources

Indication	Fruit	Dry Fruit	Fresh Fruit	Fruit Wine	Juice	Ash of Fruit	Immature Fruit	References	Score
Against wounds	*						*	5, 10, 12	3
Against ulcers in the head							*	6, 7, 11	3
Against tumors, swellings in glands	*						*	5, 9, 12	3
Against pimples, pustules	*							1, 11, 12	3
Against tumors, swellings in glands	*						*	5, 9, 12	3
Against gout	*	*						14, 17, 18	3
Against roughness/ rawness of the throat	*	*	*					7, 13, 15	3
Causes itch, scabies	*	*						11, 14, 15	3
Causes pimples, pustules	*							1, 6, 14	3
Against marks on skin	*						*	6, 12, 17	3
Benefits lungs	*							6, 14, 18	3
Against chest diseases	*	*						15, 18, 20	3
Lactagogue				*				2, 6, 7	3
Harms swellings, tumors of liver	*	*						6, 7, 10	3
Against spleen disease	*	*						1, 14, 17	3
Harms swelling/ tumors of spleen		*	*					6, 7, 10	3
Against hardness, closure of womb	*		*					7, 13, 20	3
Against bubonic ulcers	*	*						13, 15, 20	3
Against bite of *mughali*	*						*	2, 6, 11	3
Causes impetigo	*		*					6, 7	2

(continued on next page)

TABLE 3.7 (continued)
Use of the Fruit of Fig (including peel, seeds) in Ancient and Medieval Sources

Indication	Fruit	Dry Fruit	Fresh Fruit	Fruit Wine	Juice	Ash of Fruit	Immature Fruit	References	Score
Bitter	*						*	12, 20	2
Against thick humors in kidneys		*						5, 12	2
Against thick humors in bladder		*						5, 12	2
Closing wounds, etc.	*							7, 10	2
Healing burns	*	*						1, 14	2
Against phlegm in stomach	*							6, 11	2
Against thick humors in kidneys		*						5, 12	2
Against thick humors in bladder		*						5, 12	2
Cleans bad humors		*						5, 12	2
Causes obstruction		*						8	2
Cleans spleen	*							6, 11	2
Cleans liver	*							6, 11	2
Cleans bad humors		*						5, 12	2
Against ordinary swellings/ tumors of trachea		*		*				7, 12	2
Against nonulcerating carcinoma		*						1, 14	2
Against all swelling, tumors in soft flesh		*						6, 14	2
Against all swellings/ tumors of jaws	*	*						1, 14	2
Against honeycomb ulcers							*	6, 11	2

TABLE 3.7 (continued)
Use of the Fruit of Fig (including peel, seeds) in Ancient and Medieval Sources

Indication	Fruit	Dry Fruit	Fresh Fruit	Fruit Wine	Juice	Ash of Fruit	Immature Fruit	References	Score
Against nonulcerating carcinoma		*						1, 14	2
Against hardness, hardenings of the womb	*		*					7, 20	2
Against pimples/ pustules having viscous humors	*						*	5, 12	2
Against allergic skin reactions							*	6, 11	2
Corrects discoloring							*	10, 14	2
Against scabies	*	*						7, 20	2
Against toothache	*							17, 20	2
Against madness	*	*						11, 17	2
Against joint pain	*	*						11, 17	2
Against back problems	*	*						8, 11	2
Causes elephantiasis				*				2, 6	2
Against roughness of eyelids							*	6, 11	2
Against pneumonia, pleurisy		*						1, 14	2
Against peripleumonia		*						1, 14	2
Against ordinary swellings/ tumors of trachea		*						7, 12	2
Eases breathing	*							13, 15	2
Softens chest	*							4, 9	2
Against chest fullness	*							13, 15	2

(continued on next page)

TABLE 3.7 (continued)
Use of the Fruit of Fig (including peel, seeds) in Ancient and Medieval Sources

Indication	Fruit	Dry Fruit	Fresh Fruit	Fruit Wine	Juice	Ash of Fruit	Immature Fruit	References	Score
Against chest pain	*			*				7, 17	2
Against diseases of the buttocks	*	*						1, 14	2
Against colic	*	*						11, 17	2
Benefits liver diseases		*						1, 14	2
Harms liver		*						8, 17	2
Harms spleen	*	*						8, 17	2
Against kidney disease		*						1, 14	2
Turns urine red	*							15, 16	2
Emmenagogue				*				2, 6	2
Eases birth	*	*						18, 19	2
Aphrodisiac	*							8, 9	2
Against astringent fever	*							1, 14	2
Creates good blood		*						5, 12	2
Nutritive for athletes		*						1, 14	2
Makes measles erupt quicker	*							13, 15	2
Makes smallpox break out quicker	*							13, 15	2
Causes fever	*							8, 20	2
Causes feverish shivering	*							1, 14	2
Difficult to digest		*						5, 12	2
Against bite of *ibn aras*							*	6, 7	2
Against bite of *scolopendrum*	*							2, 6	2
Quickly passing through	*	*						6, 14	2
Ripening phlegm	*							18	1
Against vapors in chest	*							13	1
Earthy substance							*	5	1
Ripening phlegm	*							18	1

TABLE 3.7 (continued)

Use of the Fruit of Fig (including peel, seeds) in Ancient and Medieval Sources

Indication	Fruit	Dry Fruit	Fresh Fruit	Fruit Wine	Juice	Ash of Fruit	Immature Fruit	References	Score
Antiphlegmatic	*							11	1
Milk taste					*			1	1
Sour taste	*							8	1
For sharp vinegar				*				2	1
Counteracts bad smell	*							8	1
Generates superfluities		*						6	1
Against thick humors	*							17	1
Burning, stinging	*	*						18	1
Causes ulcers	*	*						7	1
Viscous	*							8	1
Cleans ulcers	*							10	1
Against spoiled putrid materials		*						20	1
Opening and against obstruction	*							17	1
Creates thick, coarse blood	*							16	1
Against oozing, spreading ulcers		*						20	1
Against all swellings/tumors of lungs				*				7	1
Against stye in eyelid	*							6	1
Against liver inflammation, hot swelling, tumor	*	*						14	1
Dissolves thin, fine swellings, ulcers		*						6	1
Against gland problems	*						*	6	1
Against pimples, pustules in head	*						*	5	1
Against hemorrhoids	*							17	1

(continued on next page)

TABLE 3.7 (continued)
Use of the Fruit of Fig (including peel, seeds) in Ancient and Medieval Sources

Indication	Fruit	Dry Fruit	Fresh Fruit	Fruit Wine	Juice	Ash of Fruit	Immature Fruit	References	Score
Against skin problems	*							13	1
For coloring of hands	*							15	1
Against leprosy	*							13	1
Against morphea							*	10	1
Corrects color of warts							*	10	1
Against proud flesh	*							17	1
Against itch	*							20	1
Good for head	*							12	1
Harms head	*							8	1
Against tooth disease						*		1	1
Harms teeth	*							11	1
Increases intelligence	*							17	1
Against worry, obsession	*							17	1
Against sprains							*	6	1
Against hemiplegia	*							17	1
Causes sleepiness	*							14	1
Against hoarseness	*							20	1
Harms voice	*							1	1
Against roughness of trachea	*							17	1
Against lung obstructions	*							15	1
Against catarrh from head to chest to lungs	*							11	1
Against pains of sides	*							15	1
Against palpitations of heart	*							17	1

TABLE 3.7 (continued)
Use of the Fruit of Fig (including peel, seeds) in Ancient and Medieval Sources

Indication	Fruit	Dry Fruit	Fresh Fruit	Fruit Wine	Juice	Ash of Fruit	Immature Fruit	References	Score
Warms stomach		*						20	1
Against results of too much hotness of stomach	*							8	1
Against diarrhea	*							17	1
Constipating							*	8	1
Putrefies in stomach	*							20	1
Causes stomach pain	*							8	1
Benefits liver		*						6	1
Strengthens liver	*							17	1
Harms hot liver	*							15	1
Thickens, coarsens liver		*						8	1
Causes liver obstruction		*						8	1
Benefits spleen		*						6	1
Harms hot spleen	*							15	1
Thickens, coarsens spleen		*						8	1
Causes splenic obstruction		*						8	1
Against kidney leanness	*							17	1
Warms kidney		*						11	1
Opens kidney channels	*						*	18	1
Against kidney pain			*					15	1
Against bladder pain		*						18	1
Removes dead fetus	*							19	1
Against yaws	*						*	6	1
Against poisonous animal bites	*							18	1

(continued on next page)

TABLE 3.7 (continued)
Use of the Fruit of Fig (including peel, seeds) in Ancient and Medieval Sources

Indication	Dry Fruit	Fresh Fruit	Fruit Wine	Juice	Ash of Fruit	Immature Fruit	References	Score
Against coagulated milk in stomach (poisoning)						*	1	1
Against consumption of white lead						*	1	1
Against consumption of bulls' blood						*	1	1
Against corruption of air	*						18	1
Animal feed	*						10	1
Resistant to spoilage	*						8	1
Against lack of appetite	*						10	1

TABLE 3.8
Side Effects of *Ficus* spp. Fruit: Medieval

The Possible Negative Effects of Figs According to Historical Sources, and How to Correct Them

Variations	Notes
Side effect: Heavy for stomach	Corrective: Aniseed and oxymel with water taken internally; drinking cold water after eating figs.
Side effect: Causes bloating	Corrective: Aniseed and oxymel with water taken internally; oxymel and salty sauce.
Side effect: Causes obstructions in the intestines	Corrective: Nuts and sweet almonds taken internally.
Side effect: Stomachache	Corrective: Oxymel taken internally.
Side effect: Nutritionally bad	Corrective: Nuts and almonds taken internally.
Side effect: Causes an increase in lice	Corrective: Nuts, Syrian hyssop, or aniseed taken internally; rubbing the body with borax and chick-pea flour.
Side effect: Harms weak liver	Corrective: Nuts, Syrian hyssop, or aniseed.
Side effect: Harms spleen	Corrective: Nuts, Syrian hyssop, or aniseed.
For undefined side effects	Corrective: Oxymel; calamint, and hyssop, or ginger; almonds; pepper; vinegar.

TABLE 3.9
Historical Uses of Ashes and Lyes from *Ficus* spp.: Formulas

How to Prepare Ashes and Lyes from Different Parts of *Ficus*

Ashes: Made by burning the fruit, basal shoots, branches, or wood of the cultivated or wild fig trees.
Lye: Made of ashes of burnt branches or wood of wild and cultivated fig trees. Ashes must be steeped
in water long and often. Used by moistening a sponge in it often and applying it immediately, or
internally or externally after mixing it with other ingredients (Dioscorides 1902, Al-Razi 1955–1970).

Variations	Notes
Ashes mixed with wax and rose oil	Externally for healing burns and preventing scarring
Ashes mixed with water and oil	For bruises; ruptures; convulsions
Ashes of leaves	For gangrenes; fleshy excrescences
Lye made of ashes of branches	Applied with wool or sponge on gangrenes and warts
	Dropped into hollow, putrid ulcers for closing them
Ashes of young shoots burnt twice and mixed with white lead	Divided into cakes, for the cure of ulcerations of the eyes
Ashes of wood	Externally for cicatrizing wounds
Lye made of ashes of the wood	Externally on putrid ulcers; internally and externally for the bite of spider
	Internally for dysentery, diarrhea
Ashes of the wood mixed with olive oil	Used externally for cleaning ulcers and marks on the skin; for nerve pain; for cramps; for paralysis; for weakness of eyesight; as a dentifrice, whitens the teeth; makes the hair black; sudorific.
	As a suppository for dysentery, diarrhea
	Internally, with water, for bruises; ruptures; cramps; for dissolving coagulated blood
Ashes of wood mixed with honey	For cold sores
Burned, pulverized fruit mixed with wax	Used externally for chilblains and burns

following table, there are some examples of potential side effects of fig products and their correctives.

An important way of using the different parts of the *Ficus* spp. was the preparation of ashes and lyes. The following tables give an overview of their historical medicinal uses. The following two tables are introduced at this point in the text, since ashes of fruits are sometimes employed. Nonetheless, the tables apply also to ashes prepared from leaves, and especially from the stems and woody parts of the tree.

For references used in Table 3.10, see Table 3.6. The scoring system is based on the number of times a reference discusses the use of ashes or lye prepared from *Ficus* for the purpose indicated. If the number of citations is regular, the reference is given in regular type and scored as "1." If the number of citations is strong, the reference is given in italic and scored as "2." If the number of citations is strongest, the reference is given in **bold** and scored as "3." The final score in the last column is the sum of these scores for the individual references. The particular types of lye and ashes used

TABLE 3.10

Historical Uses of Ashes and Lye Made of *Ficus* sp.

Indication	Ashes	Lye of Ashes	Ashes of Fruit	Ashes of Leaves	Ashes of Branches/Twigs	Lye of Ashes of Branches	Ashes of Tree	Lye of Ashes of Tree	Ashes of Wood	Lye of Ashes of Wood	References	Score
Closing (wounds, etc.)	*	*			*	*	*				1, **2**, 6, *14*, **20**	13
Against injuries due to falls from a height, particularly internal bleeding	*	*			*	*	*		*		1, 2, 6, **13**, *14*, 15, 20	12
Against wounds	*	*			*	*	*				1, **2**, 6, *14*	10
Against spoiled putrid materials	*	*			*	*	*				2, 6, 7, **15**, 17, 20	8
Dissolving coagulated blood					*	*	*				1, 2, 6, *13*, 14, 15, 20	8
Against abdominal, intestinal ulcers	*	*			*	*	*				1, 2, 6, 7, 14	7
Against cramps, convulsions	*	*			*	*	*				1, 2, *14*, 20	7
Against proud flesh				*	*		*				1, 2, 6, 14	6
Against dysentery	*				*	*	*		*		1, 6, 7, 14	5
Against gangrene				*	*		*				1, 14, 17, 20	5
Analgesic	*				*	*	*		*		1, 2, 6, 7, 14	5
Against malignant ulcers	*	*			*		*				6, 7, 13, 15	4
Against ruptures	*	*			*		*				1, 2, 14, 20	4
Helpful in chronic diseases					*		*				2, 6, 20	4
Cleaning, absterging	*	*	*		*		*				1, 2, 6, 14	4
Cleans ulcers			*							*	7, *17*, 20	4
Against chilblains	*				*					*	1, 20	3
Against fluxes					*		*		*		2, 6, 20	3

									References	Count
Against nerve pain					*		*	*	2, 6, 7	3
Against oozing, spreading ulcers					*				20	3
Against putrid ulcers				*	*			*	2, 6, 15	3
Caustic				*	*		*	*	2, 15, 20	3
Hemostatic				*					2, 17, 20	3
Against bruises					*				1, 14	2
Against coagulated milk in stomach (poisoning)		*					*		7, 10	2
Against colic			*		*		*		1, 14	2
Against diarrhea					*			*	6, 7	2
Against ergotism			*			*	*		15, 17	2
Against fistulas				*			*		2, 6	2
Against shrew mouse bite				*			*	*	2, 7	2
Against skin problems	*a								1, 14	2
Against tetanus			*						1, 14	2
Against tooth disease			*						1, 14	2
Against ulcers in eyes	*b								1, 14	2
Against weakness of sight			*		*				1, 14	2
Astringent			*						2, 14	2
Causes ulcers				*		*			6, 15	2
Healing burns				*	*			*	1, 14	2
Spicy, sharp				*					1, 14	2
Sudorific						*		*	2, 6	2
Warming				*					1, 14	2
Against diseases of the buttocks							*		17	1
Antidote for consumption of gypsum									2	1

(continued on next page)

TABLE 3.10 (continued)

Historical Uses of Ashes and Lye Made of *Ficus* sp.

Indication	Ashes	lye of Ashes	Ashes of Fruit	Ashes of Leaves	Ashes of Branches/Twigs	Lye of Ashes of Branches	Ashes of Tree	Lye of Ashes of Tree	Ashes of Wood	Lye of Ashes of Wood	References	Score
Against hemiplegia									*		6	1
Against hemorrhoids									*		6	1
Against looseness of uvula					*						1	1
Against marks on skin	*										17	1
Against paronychia										*	7	1
Against sebaceous cysts						*					20	1
Against spreading ulcers									*		17	1
Against stomach illness						*					2	1
Against ulcers in thighs						*				*	7	1
Against warts	*										20	1
Benefits gums	*	*									17	1
Blackens hair		*									17	1
Putrefies			*								6	1
Whitens the teeth	*										17	1

[a] Ashes of stalks springing from roots.

[b] Ashes of stalks springing from roots.

are starred, but this information does not enter into the calculation of the total score for the function. The indications are shown in the order of importance based on their score values.

SUMMARY

Fig fruits are possibly the oldest intentionally cultivated food of humankind. Rich in easily digestible natural sugars, they also contain rich amounts of anthocyanins, antioxidant flavonoids that contribute the figs' colors and have the ability to regulate signaling pathways that guide cellular metabolism. Fig seeds are packed with lectins that are linked to human immunity. The anthocyanins and lectins together may explain the ability of fig fruits to produce healing effects such as reducing tumors, whether of cancerous or infectious origin. The jelly in figs, particularly in the species *F. awkeotsang* (the "jelly fig"), offers potential for a unique drug delivery matrix with intrinsic antifungal properties. Although figs are an extremely safe food and medicine, care should be exercised to cull dried fruits that may be infested with fungi that produce toxins.

REFERENCES

Adhya, M., B. Singha, and B.P. Chatterjee. 2006. *Ficus cunia* agglutinin for recognition of bacteria. *Indian J Biochem Biophys*. 43: 94–7.

Agrawal, S., and S.S. Agarwal. 1990. Preliminary observations on leukaemia specific agglutinins from seeds. *Indian J Med Res*. 92: 38–42.

Ahmad, P.I., S.A. Ahmad, and A. Zaman. 1971. Chemical examination of *Millettia pendula* and *Ficus benjamina* (fruits). *J Indian Chem Soc*. 48: 979.

Al-Antaki, D.b.A. 1356 (A.H.). *Tadhkirah ula li-lbab wa-l-jami' li-l-'ajab al-'ujab.* [Memorandum Book for Hearts and Comprehensive Book of Wonderful Marvels]. Misr: Al-Matba'ah al-'Utmaniyyah.

Albertus Magnus. 1867. *De vegetabilibus libri VII: Historiae naturalis pars XVIII*. [Book on Growing Things]. Ed. E.H.F. Meyer and K. Jessen. Berlin: G. Reimeri.

Al-Fatimi, M., M. Wurster, G. Schröder, and U. Lindequist. 2007. Antioxidant, antimicrobial and cytotoxic activities of selected medicinal plants from Yemen. *J Ethnopharmacol*. 111: 657–66.

Al-Razi, M.b.Z. 1955–1970. *Kitab al-hawi fi al-tibb*. [The Comprehensive Book on Medicine]. Vol. XXI: 1, pp. 197–207, 256–259. Hyderabad: Matba'at Majlis Dairat al-Ma'arif al-'Utmaniyyah.

Al-Tabari, A.b.R. 1928. *Firdaws al-hikmah fi al-tibb* [The Paradise of Wisdom], ed. M.Z. Al-Siddiqi. Berlin: Aftab.

Antico, A., G. Zoccatelli, C. Marcotulli, and A. Curioni. 2003. Oral Allergy Syndrome to Fig. *Int Arch Allergy Immunol*. 131: 138–42.

Bagdy, G., V. Kecskemeti, P. Riba, and R. Jakus. 2007. Serotonin and epilepsy. *J Neurochem*. 100: 857–73.

Bekatorou, A., A. Sarellas, N.G. Ternan et al. 2002. Low-temperature brewing using yeast immobilized on dried figs. *J Agric Food Chem*. 50: 7249–57.

Bircan, C., S.A. Barringer, U. Ulken, and R. Pehlivan. 2008. Increased aflatoxin contamination of dried figs in a drought year. *Food Addit Contam*. 2: 1–9.

Bjerre, M., T.K. Hansen, and A. Flyvbjerg. 2008. Complement activation and cardiovascular disease. *Horm Metab Res*. 40: 626–34.

Blas, Fr. 1984. *El libro de medicinas caseras de Fr. Blas de la Madre de Dios: Manila, 1611.* [Book of Home Remedies], ed. and comm. F. Guerra. Madrid: Ediciones cultura hispánica.

Bliebtrau, J.N. 1968. *The Parable of the Beast.* New York: Macmillan Company (p. 74, via Singh and Goel, 2009).

Bock, H. 1964. *Kreütterbuch darin underscheidt Nammen und Würckung der Kreütter, standen.* [Herbal in which are the Different Names and Properties of Herbs]. Josiam Rihel, Strassburg, 1577. Reprint Konrad Kölbl, München.

Buchanan, G.R., and C.A. Holtkamp. 1983. Evidence against enhanced platelet activity in sickle cell anaemia. *Br J Haematol.* 54: 595–603.

Camilleri, M. 2009. Serotonin in the gastrointestinal tract. *Curr Opin Endocrinol Diabetes Obes.* 16: 53–9.

Chang, J.H., Y.T. Wang, and H.M. Chang. 2005. Pectinesterase inhibitor from jelly fig (*Ficus awkeotsang* Makino) achene induces apoptosis of human leukemic U937 cells. *Ann N Y Acad Sci.* 1042: 506–15.

Chua, A.C., W.M. Chou, C.L. Chyan, and J.T. Tzen. 2007. Purification, cloning, and identification of two thaumatin-like protein isoforms in jelly fig (*Ficus awkeotsang*) achenes. *J Agric Food Chem.* 55: 7602–8.

Chua, A.C., E.S. Hsiao, Y.C. Yang, L.J. Lin, W.M. Chou, and J.T. Tzen. 2008b. Gene families encoding 11S globulin and 2S albumin isoforms of jelly fig (*Ficus awkeotsang*) achenes. *Biosci Biotechnol Biochem.* 72: 506–13.

Chua, A.C., P.L. Jiang, L.S. Shi, W.M. Chou, and J.T. Tzen. 2008a. Characterization of oil bodies in jelly fig achenes. *Plant Physiol Biochem.* 46: 525–32.

Dioscorides, P. 1902. *Des Pedanios Dioscurides aus Anazarbos Arzneimittellehre in fünf Büchern,* trans. and comm. J. Berendes. Stuttgart: Ferdinand Enke.

Dohnal, V., A. Jezkova, D. Jun, and K. Kuca. 2008. Metabolic pathways of T-2 toxin. *Curr Drug Metab.* 9: 77–82.

El-Sayyad, S.M., H.M. Sayed, and S.A. Mousa. 1986. Chemical constituents and preliminary anthelmintic activity of *Ficus platyphylla* (Del). *Bull Pharm Sci Assiut Univ.* 9: 164–77.

Endo, Y., Y. Liu, and T. Fujita. 2006. Structure and function of ficolins. *Adv Exp Med Biol.* 586: 265–79.

Fang, C. 2007. [Grape wine prepared from grape and fig]. *Faming Zhuanli Shenqing Gongkai Shuomingshu.* Chinese Patent Application: CN 2006-10052538 20060719.

Flos medicinae scholae Salerni. 1852–59. In *Collectio Salernitana: Ossia documenti inediti, e trattati di medicina appartenenti alla scuola medica Salernitana,* Vol. 5, ed. G.E.T. Henschel, C. Daremberg, and S. de Renzi. Napoli: Filiatre-Sebezio.

Fuchs, L. 1549. *De historia stirpium commentarii insignes.* [Notable Commentaries on the History of Plants]. Lyon: B. Arnolletum. Electronic edition 1995 by Bibliothèque nationale de France.

Fujita, T. 2002. Evolution of the lectin-complement pathway and its role in innate immunity. *Nat Rev Immunol.* 2: 346–53. Review.

Fujita, T., Y. Endo, and M. Matsushita. 2000 [Activation of complement lectin pathway by human ficolin]. *Seitai no Kagaku* 51: 244–9.

Galen. Galenus, C.G. 1964–1965. *Opera omnia* [Complete Works], ed. C.G. Kühn [Facsimile reprint of Leipzig: C. Cnobloch, 1821–33 edition]. Hildesheim: Olms.

Gerard, J. 1633. *The Herball or Generall Historie of Plantes.* London: Adam Islip Ioice Norton and Richard Whitakers.

Gilani, A.H., M.H. Mehmood, K.H. Janbaz, A.U. Khan, and S.A. Saeed. 2008. Ethnopharmacological studies on antispasmodic and antiplatelet activities of *Ficus carica*. *J Ethnopharmacol.* 119: 1–5.

Groopman, J.D., T.W. Kensler, and C.P. Wild. 2008. Protective interventions to prevent aflatoxin-induced carcinogenesis in developing countries. *Annu Rev Public Health* 29: 187–203.

Holmskov, U., S. Thiel, and J.C. Jensenius. 2003. Collectins and ficolins: Humoral lectins of the innate immune defense. *Annu Rev Immunol.* 21: 547–78.

Hong, Q., C.I. Sze, S.R. Lin et al. 2009. Complement C1q activates tumor suppressor WWOX to induce apoptosis in prostate cancer cells. *PloS One* 4: e5755.

Ibn al-Baytar, a.M. 1992/1412. *Kitab al-jami' li-mufradat al-adwyah w-al-aghdhyah.* [Comprehensive Book of Simple Drugs and Foods]. Beirut: Dar al-kutub al-'ilmiyah.

Ibn Butlan, M.b.H. 1531. *Tacuini sanitatis Elluchasem Elimithar: de sex rebus non naturalibus, earum naturis, operationibus, and rectificationibus, publico omnium usui, conservandae sanitatis, recens exarati.* [Tables of Health]. Argentorati: Apud Ioannem Schottum.

Ibn Sina, H.a.A. 1877/1294. *Kitab al-qanun fi-l-tibb.* [The Canon of Medicine]. Beirut: Dar Sadir.

Israeli, I. 1515. *Omnia opera Ysaac in hoc volumine contenta: cumquibusdam alijs opusculis.* [Complete Works]. [Lyons: Jean de La Plate for Barthelemy Trot]. Electronic edition 1995 by Bibliothèque nationale de France (BnF).

Jahan, I.A., N. Nahar, M. Mosihuzzaman et al. 2009. Hypoglycaemic and antioxidant activities of *Ficus racemosa* Linn. fruits. *Nat Prod Res.* 23: 399–408.

Jones, J.D., and J.L. Dangl. 2006. The plant immune system. *Nature* 444: 323–9.

Jonnakuty, C., and C. Gragnoli. 2008. What do we know about serotonin? *J Cell Physiol.* 217: 301–6.

Kang, K., Y.S. Kim, S. Park, and K. Back. 2009. Senescence induced serotonin biosynthesis and its role in delaying senescence in rice leaves. *Plant Physiol.* May 13, 2009 [Epub ahead of print].

Karbancioglu-Güler, F., and D. Heperkan. 2009. Natural occurrence of fumonisin B1 in dried figs as an unexpected hazard. *Food Chem Toxicol.* 47: 289–92.

Kitajima, J., K. Kimizuka, and Y. Tanaka. 2000. Three new sesquiterpenoid glucosides of *Ficus pumila* fruit. *Chem Pharm Bull. (Tokyo)* 48: 77–80.

Komae, K., Y. Sone, M. Kakuta, and A. Misaki. 1989. Isolation of pectinesterase from *Ficus awkeotsang* seeds and its implication in gel-formation of the awkeotsang polygalacturonide. *Agric Biol Chem.* 53: 1247–54.

Lansky, E.P., and E. Nevo. 2009. Plant immunity may benefit human medicine. *Open Syst Biol J.* 2: 18–19.

Lazebnik, L.B., G.N. Sokolova, A.E. Lychkova, and I.E. Trubitsyna. 2008. [The pathogenetic role of the serotonin-histamine-cyclic nucleotide system in ulcerogenesis]. *Vestn Ross Akad Med Nauk* 1: 9–12.

Li, Y.C., C.T. Chang, E.S. Hsiao, J.S. Hsu, J.W. Huang, and J.T. Tzen. 2003. Purification and characterization of an antifungal chitinase in jelly fig (*Ficus awkeotsang*) achenes. *Plant Cell Physiol.* 44: 1162–7.

Liu, L., L. Wang, T. Wang, X. Liu, and X. Yang. 2008. [Phytochemical and pharmacological research progress in *Ficus microcarpa* L. f.]. *Shizhen Guoyi Guoyao* 19: 390–2.

López, G. 1982. *El tesoro de medicinas de Gregorio López 1542–1596.* [Treasure of Medicines], ed. and comm. F. Guerra. Madrid: Instituto de cooperación iberoamericana, Ediciones cultura hispánica.

Lu, J., C. Teh, U. Kishore, and K.B. Reid. 2002. Collectins and ficolins: Sugar pattern recognition molecules of the mammalian innate immune system. *Biochim Biophys Acta Gen Subj.* 1572: 387–400.

Lyu, S.Y., and W.B. Park. 2009. Mistletoe lectin modulates intestinal epithelial cell-derived cytokines and B cell IgA secretion. *Arch Pharm Res.* 32: 443–51.

Matsushita, M., and T. Fujita. 2001. Ficolins and the lectin complement pathway. *Immunol Rev.* 180: 78–85.

Misaki, A., K. Komae, and M. Kakuta. 2004. [Gel-forming polysaccharide of Aw-keo (*Ficus awkeotsang*): structure, and enzyme-induced gel-formation and food utilization]. *Foods Food Ingredients J. Japan* 209: 277–284.

Miyazaki, Y., and S. Yakou. 2006. [The gelation mechanism of jelly-fig and its pharmaceutical applications]. *Pharm Technol Japan* 22: 137–141.

Miyazaki, Y., S. Yakou, and K. Takayama. 2004. [Study on jelly fig extract as a potential hydrophilic matrix for controlled drug delivery]. *Int. J. Pharm.* 287: 39–46.

Mpiana, P.T., V. Mudogo, D.S. Tshibangu et al. 2008. Antisickling activity of anthocyanins from *Bombax pentadrum*, *Ficus capensis* and *Ziziphus mucronata*: Photodegradation effect. *J Ethnopharmacol.* 120: 413–8.

Mukai, Y., and R.R. Tampi. 2009. Treatment of depression in the elderly: A review of the recent literature on the efficacy of single- versus dual-action antidepressants. *Clin Ther.* 31: 945–61.

NIST Chemistry WebBook. 1996. [Online Database]. NIST Standard Reference Database Number 69. The National Institute of Standards and Technology (NIST). URL: http://webbook.nist.gov/chemistry/contact.html [accessed August 14, 2009].

Parkinson, J. 1640. *The Theater of Plants, Or, An Herball of a Large Extent.* London: T. Cotes. Available through Early English Books Online, http://eebo.chadwyck.com.

Peng, C.C., E.S. Hsiao, J.L. Ding, and J.T. Tzen. 2005. Functional expression in pichia pastoris of an acidic pectin methylesterase from jelly fig (*Ficus awkeotsang*). *J Agric Food Chem.* 53: 5612–6.

Pliny (*the Elder*). Pliniux, C. 1967–1970. *C. Plini Secundi naturalis historiae, libri XXXVII.* [The Natural History], ed. K.F.T. Mayhoff. Stuttgart: B.C. Teubner.

PubChem. The PubChem Project. 2001. [Online Database]. National Center for Biotechnology Information. U.S. National Library of Medicine. 8600 Rockville Pike, Bethesda, MD 20894. URL: http://pubchem.ncbi.nlm.nih.gov/ [accessed August 16, 2009].

Puech, A.A., C.A. Rebeiz, and J.C. Crane. 1976. Pigment changes associated with application of ethephon ((2-chloroethyl)phosphonic acid) to fig (*Ficus carica* L.) fruits. *Plant Physiol.* 57: 504–9.

Rahman, N.N., M. Khan, and R. Hasan. 1994. Bioactive components from *Ficus glomerata*. *Pure Appl Chem.* 66: 2287–90.

Rao, C.V., A.R. Verma, M. Vijayakumar, and S. Rastogi. 2008. Gastroprotective effect of standardized extract *Ficus glomerata* fruit on experimental gastric ulcers in rats. *J Ethnopharmacol.* 115: 323–326.

Ray, S., H. Ahmed, S. Basu, and B.P. Chatterjee. 1993. Purification, characterisation, and carbohydrate specificity of the lectin of *Ficus cunia*. *Carbohydr Res.* 242: 247–63.

Rufinus de Rizardo. 1946. *Herbal of Rufinus.* Ed. L. Thorndike. Chicago: The University of Chicago Press.

Senyuva, H.Z., J. Gilbert, S. Oztürkoglu, S. Ozcan, and N. Gürel. 2008. Changes in free amino acid and sugar levels of dried figs during aflatoxin B1 production by *Aspergillus flavus* and *Aspergillus parasiticus*. *J Agric Food Chem.* 56: 9661–6.

Senyuva, H.Z., and J. Gilbert. 2008. Identification of fumonisin B2, HT-2 toxin, patulin, and zearalenone in dried figs by liquid chromatography-time-of-flight mass spectrometry and liquid chromatography-mass spectrometry. *J Food Prot.* 71: 1500–4.

Singh, D., and R.K. Goel. 2009. Anticonvulsant effect of *Ficus religiosa*: Role of serotonergic pathways. *J Ethnopharmacol.* 123: 330–4.

Singha, B., M. Adhya, and B.P. Chatterjee. 2007. Multivalent II [beta-d-Galp-(1–>4)-beta-D-GlcpNAc] and Talpha [beta-D-Galp-(1–>3)-alpha-D-GalpNAc] specific Moraceae family plant lectin from the seeds of *Ficus bengalensis* fruits. *Carbohydr Res.* 342: 1034–43.

Solomon, A., S. Golubowicz, Z. Yablowicz et al. 2006. Antioxidant activities and anthocyanin content of fresh fruits of common fig (*Ficus carica* L.). *J Agric Food Chem.* 54: 7717–23.

Sun, L. 2008. [Method for producing high-selenium and high-germanium beverage]. *Faming Zhuanli Shenqing Gongkai Shuomingshu.* Chinese Patent Application CN 2007-10014893 20070914, 2008.

Tabernaemontanus, I.T. 1970. *Neu vollkommen Kräuter-Buch, darinnen uber 3000. Kräuter, mit schönen und kunstlichen Figuren, auch deren Underscheid und Würckung, samt ihren Namen in mancherley Sprachen beschrieben.* [New Complete Herbal], ed. C. Bauhinus and H. Bauhinus, 4th ed. Grünwald bei München: K. Kölbl. Reprint of Basel: J.L. König, 1731 edition.

Takahashi, S., and Y. Aoyagi. 2008. [Role of ficolin in preventing bacterial infections]. *Rinsho Kensa* 52: 887–892.

Toi, M., H. Bando, C. Ramachandran et al. 2003. Preliminary studies on the anti-angiogenic potential of pomegranate fractions in vitro and in vivo. *Angiogenesis* 6: 121–8.

Van der Walt, J.P., and I.T. Tscheuschner. 1956. *Saccharomyces delphensis* nov. spec.; a new yeast from South African dried figs. *Antonie Van Leeuwenhoek* 22: 162–6.

Wang, M.M.C., and J.T.C. Tzen. 2005. Achene proteins in jelly fig (*Ficus awkeotsang*) and their potential biotechnological application. *Adv Plant Physiol.* 8: 191–200.

Wang, L., and T. Yue. 2005. [Study on brewing technology of fermentation fig wine]. *Zhongguo Niangzao* 10: 59–62.

Wood-Charlson, E.M., and V.M. Weis. 2009. The diversity of C-type lectins in the genome of a basal metazoan, *Nematostella vectensis. Dev Comp Immunol.* 33: 881–9.

Wu, F. 2005. [Wine of *Ficus carica* fruit and its preparation]. *Faming Zhuanli Shenqing. Gongkai Shuomingshu.* Chinese Patent Application CN 2004-10014512 20040329, 2005.

Wu, J.S., M.C. Wu, C.M. Jiang, Y.P. Hwang, S.C. Shen, and H.M. Chang. 2005. Pectinesterase inhibitor from jelly-fig (*Ficus awkeotsang* Makino) achenes reduces methanol content in carambola wine. *J Agric Food Chem.* 53: 9506–11.

Zänker, K.S. 2008. General introduction to innate immunity: Dr. Jekyl/Mr. Hyde quality of the innate immune system. *Contrib Microbiol.* 15: 12–20.

Zhang, X.W., W.T. Xu, X.W. Wang et al. 2009. A novel C-type lectin with two CRD domains from Chinese shrimp *Fenneropenaeus chinensis* functions as a pattern recognition protein. *Mol Immunol.* 46: 1626–37.

Zorlugenç, B., F. Kiroğlu Zorlugenç, S. Oztekin, and I.B. Evliya. 2008. The influence of gaseous ozone and ozonated water on microbial flora and degradation of aflatoxin B(1) in dried figs. *Food Chem Toxicol.* 46: 3593–7.

4 Leaves

If fig fruits were our first food and first medicinal food, fig leaves may have been our first real medicine. In general, the leaves of fruit trees contain stronger antioxidants than the fruits. Fruit tree leaves are less in the edible class than are fruits, and are closer to medicines. The higher degree of medicality is associated with higher levels of toxicity.

FIGURE 4.1 **(See color insert after page 128.)** Three *Ficus carica* leaves with immature fruit (watercolor by Zipora Lansky).

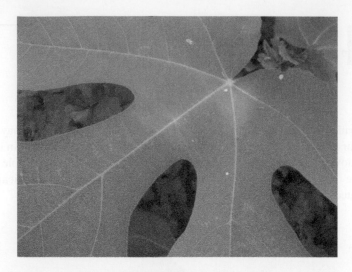

FIGURE 4.2   *Ficus carica*, leaf and its veins (29.7.2009, Hebrew University Botanical Garden, Mount Scopus, Jerusalem, Israel, by Krina Brandt-Doekes).

FIGURE 4.3   *Ficus carica*, leaf (29.7.2009, Hebrew University Botanical Garden, Mount Scopus, Jerusalem, Israel, by Krina Brandt-Doekes).

ALKALOIDS

One class of compounds often associated with both toxicity and potent physiological or therapeutic activity is the alkaloids. Alkaloids are generally complex carbon- and nitrogen-containing compounds. Fig leaves (and twigs, bark, and roots) are a significant source of the subgroup of alkaloids known as *phenanthroindolizidine alkaloids*, known to occur in the leaves of two *Ficus* species: *F. septica* and *F. hispida* (Damu et al. 2005; Peraza-Sánchez et al. 2002).

FIGURE 4.4 *Ficus retusa*, leaf (19.7.2009, University of Turku, Botanical Garden, Ruissalo, Turku, Finland, by Kaarina Paavilainen).

FIGURE 4.5 *Ficus sarmentosa* var. *nipponica*, leaves (19.7.2009, University of Turku, Botanical Garden, Ruissalo, Turku, Finland, by Kaarina Paavilainen).

FIGURE 4.6   *Ficus sycomorus*, leaves (19.7.2009, University of Turku, Botanical Garden, Ruissalo, Turku, Finland, by Kaarina Paavilainen).

These alkaloids are found in lupines (*Lupinus* spp., a genus of legumes) and other plant species, and also in amphibians, especially in the skins of Madagascan frogs of the genus *Mantella*, which likely obtain these alkaloids from dietary sources, namely, ants, beetles, and possibly mites. These compounds, similar to the indolizidine alkaloids in general, have medicinal effects, including anticancer. The parent alkaloid in this class is indolizidine 203A, of which over 80 variants, including stereoisomers, have been identified.

Traditional wisdom has usually held that the toxic alkaloids in frog skin evolved as a defense against predators, though recent investigation of one indolizidine alkaloid (235B) in frog skin exhibited significant antibiotic activity against the Gram-positive bacillus *Bacillus subtilis*, suggesting a use by the frog to prevent bacterial infestation of its skin (Michael 2007, 2008), and potential for these alkaloids, including those in fig leaf, for antibiotic function in medical practice (Mandal et al. 2000a). A specific antibacterial effect may be part of the medicinally prepared fig leaf extract's action against diarrhea (Mandal and Ashok Kumar 2002).

A more complex phenanthroindilizidine alkaloid in this series, tylophorine, is named for the genus of the plant from which it was originally isolated, namely, *Tylophora atrofolliculata*. Naturally occurring variants of this compound have been found in leaves, stems, and twigs of two fig species, *F. septica* and *F. hispida*.

In a study of 12 related phenanthroindolizidine alkaloids of which 5 were prepared synthetically and 7 were isolated from *T. atrofolliculata*, two, DCB-3503 and tylophorinidine, a tylophorine derivative, were found to be the most potent in the series as potential antitumor agents (Gao et al. 2007).

FIGURE 4.7 *Ficus triangularis*, leaves (19.7.2009, University of Turku, Botanical Garden, Ruissalo, Turku, Finland, by Kaarina Paavilainen).

The potency of these alkaloids for tumor growth suppression, as measured by the molarity needed to suppress growth in several different human cancer cell lines, was found to be comparable to that of most current chemotherapeutic agents. Second, tylophorine analogs induced albumin expression and decreased α-fetoprotein expression in HepG2 liver cancer cells, suggesting that tylophorine analogs could induce HepG2 differentiation (Gao et al. 2004). Differentiation is the process by which tumor cells, which are undifferentiated versions of their normal cellular counterparts, may revert to their normal, noncancerous state by regaining functions that were lost during the dedifferentiation, carcinogenic or cancer-promoting process (Kawaii and Lansky 2004).

Gao et al. (2004, 2007) also showed that the tylophorine analogs exerted inhibitory effects on cyclic AMP response elements, activator protein-1 sites, or nuclear factor-κ B (NF-κ-B) binding site-mediated transcriptions. These effects, which relate to decreasing cellular inflammation in the cell, were independent of the cytotoxic effect of these compounds, and were selective for cancer cells over their normal

FIGURE 4.8 *Ficus lingua*, leaves (19.7.2009, University of Turku, Botanical Garden, Ruissalo, Turku, Finland, by Kaarina Paavilainen).

counterparts. Elsewhere, the compounds inhibited growth in multidrug resistant cell lines as well as in the drug-sensitive ones. For these reasons, tylophorine analogs in particular and phenanthroindolizidine alkaloids in general represent an extremely promising class of drugs for relatively nontoxic cancer therapy.

Other work with phenanthroindolizidine alkaloids tylophorine and ficuseptine-A, the latter also found in leaves of *F. septica*, revealed these compounds' anti-inflammatory actions in an in vitro model of acute inflammation that was separate from, though most likely related to, their antitumor properties. The action was related to suppression of nitric oxide production and did not include cytotoxicity to the model RAW264.7 cells, though the compounds were toxic to cancer cells (Damu et al. 2005, 2009).

Structures of some of the "new" phenanthroindolizidine alkaloids recently discovered in *F. septica* stems, along with six alkaloids that were previously noted, are shown in Figures 4.12–4.15. The compounds are ficuseptines B-D (1-3), 10R,13aR-tylophorine *N*-oxide (4), 10R,13aR-tylocrebrine *N*-oxide (5), 10S,13aR-tylocrebrine *N*-oxide (6), 10S,13aR-isotylocrebrine *N*-oxide (7), and 10S,13aS-isotylocrebrine *N*-oxide (8).

Yang and coworkers (2006) studied tylophorine and ficuseptine A for their anti-inflammatory mechanisms. Both phenanthroindolizidine alkaloids potently suppressed nitric oxide production without cytotoxicity to the lipopolysaccharide stimulated, noncancerous RAW264.7 cells under study in contrast to their growth-suppressive effect on cancer cells. This anti-inflammatory effect of tylophorine was studied in more depth and was found to involve inhibition of the induced

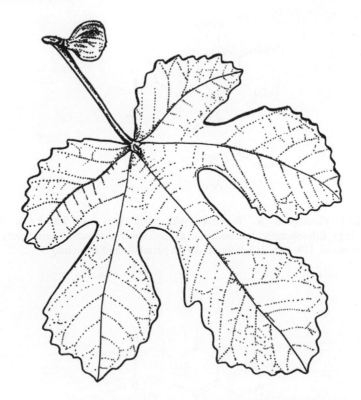

FIGURE 4.9 *Ficus carica* leaf and a small fig (pen and ink, by Zipora Lansky).

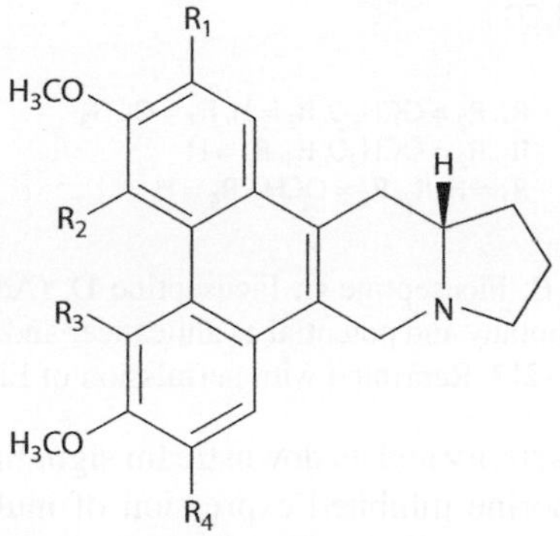

FIGURE 4.10 Tylophorine $R_1 = OCH_3$, $R_2 = H$, $R_3 = H$, $R_4 = OCH_3$. (Adapted from Lansky et al. 2008. *Ficus* spp. (fig): Ethnobotany and potential as anticancer and anti-inflammatory agents. *J Ethnopharmacol.* 119: 195–213. Reprinted with permission of Elsevier.)

protein levels of tumor necrosis factor-α (TNF-α), inducible nitric-oxide synthase (iNOS), and cyclooxygenase (COX)-II. It further inhibited activation of murine iNOS and COX-II promoter activity. Tylophorine also enhanced phosphorylation of Akt, decreasing expression and phosphorylation of c-Jun protein, thus inhibiting AP1 activity. Tylophorine also blocked mitogen-activated protein/extracellular

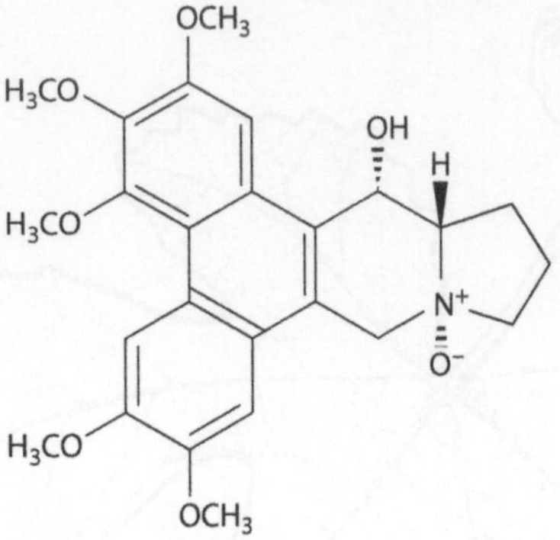

FIGURE 4.11 Ficuseptine A. (Reprinted with permission from Lansky et al. 2008. *Ficus* spp. (fig): Ethnobotany and potential as anticancer and anti-inflammatory agents. *J Ethnopharmacol.* 119: 195–213. Reprinted with permission of Elsevier.)

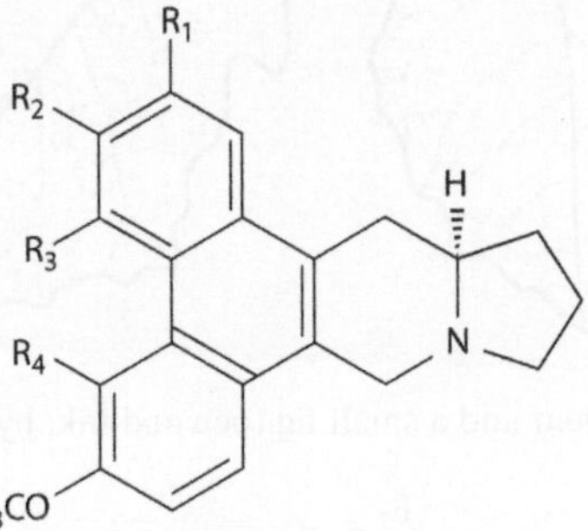

$R_1, R_2 = OCH_2O, R_3 = H, R_4 = OCH_3$
$R_1, R_2 = OCH_2O, R_3, R_4 = H$
$R_1 = H, R_2, R_3 = OCH_3, R_4 = H$

FIGURE 4.12 Ficuseptine B; Ficuseptine C; Ficuseptine D. (Adapted from Lansky et al. 2008. *Ficus* spp. (fig): Ethnobotany and potential as anticancer and anti-inflammatory agents. *J Ethnopharmacol.* 119: 195–213. Reprinted with permission of Elsevier.)

signal-regulated kinase 1 activity and its downstream signaling activation of NF-κ-B and AP1. In short, tylophorine inhibited expression of multiple pro-inflammatory factors and downhill signaling pathways leading to inflammatory processes. Since inflammation is also closely related to cancer development and progression, these actions are also important for better understanding the anticancer effects of these alkaloids.

COUMARINS

Similar to alkaloids, coumarins, that is, compounds derived from the parent compound coumarin, are relatively toxic and highly physiologically active compounds

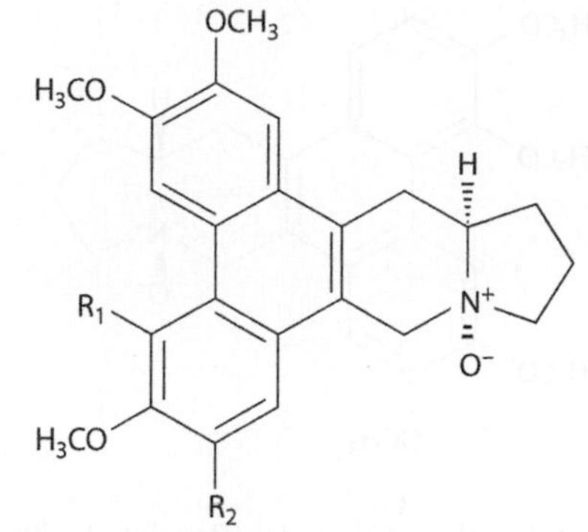
OCH3
H3CO
H
R1
N+
O−
H3CO
R2
R1 = OCH3, R2 = H
R1 = H, R2 = OCH3

FIGURE 4.13 10*R*,13*aR*-tylocrebrine *N*-oxide; 10*R*,13*aR*-tylophorine *N*-oxide. (Adapted from Lansky et al. 2008. *Ficus* spp. (fig): Ethnobotany and potential as anticancer and anti-inflammatory agents. *J Ethnopharmacol.* 119: 195–213. Reprinted with permission of Elsevier.)

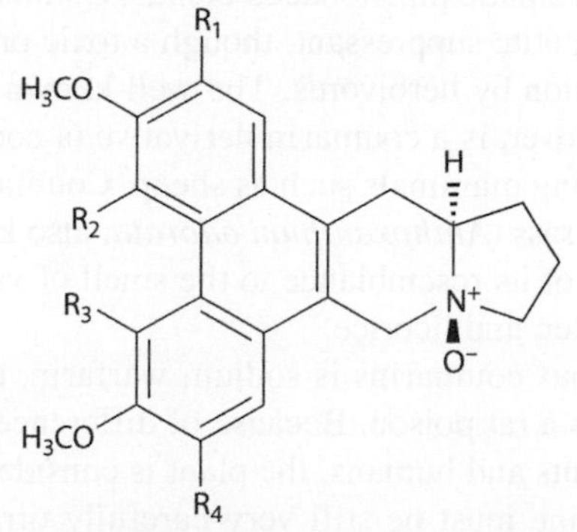
R1
H3CO
H
R2
R3
N+
O−
H3CO
R4
R1 = H, R2 = H
R1 = H, R2 = OCH3, R3 = H, R4 = OCH3
R1 = OCH3, R2 = H, R3 = OCH3, R4 = H
R1 = OCH3, R2 = H, R3 = H, R4 = OCH3

FIGURE 4.14 10*S*,13*aR*-antofine *N*-oxide; 10*S*,13*aR*-isotylocrebrine *N*-oxide; 10*S*,13*aR*-tylocrebrine *N*-oxide; 10*S*,13*aR*-tylophorine *N*-oxide. (Adapted from Lansky et al. 2008. *Ficus* spp. (fig): Ethnobotany and potential as anticancer and anti-inflammatory agents. *J Ethnopharmacol.* 119: 195–213. Reprinted with permission of Elsevier.)

that plants most likely utilize for their defense. Coumarins may be employed by plants in their leaves or bark, as well as in their seeds. They occur in cinnamon (*Cassia* sp.) bark and also in mullein (*Verbascum* sp.) of the figwort family (Scrophulariaceae), and provide an aromatic and attractive component to plants, as well as a poisonous one. Perhaps this aspect is used by the plants to help trick predators into eating the toxic elements. Humans have seized upon this aromatic quality by combining

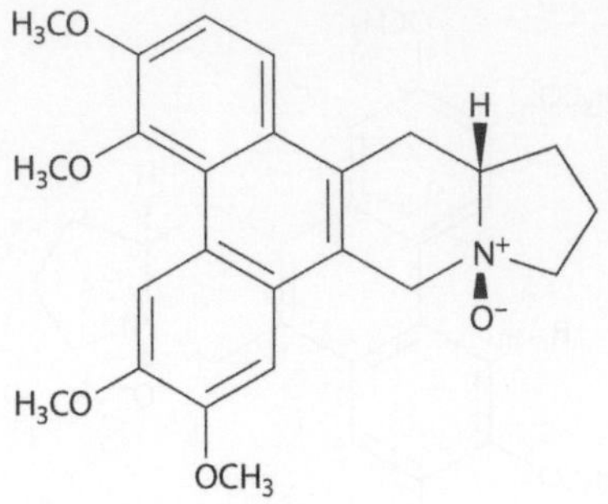

FIGURE 4.15   10*S*,13*a*S-isotylocrebrine *N*-oxide. (Reprinted with permission from Lansky et al. 2008. *Ficus* spp. (fig): Ethnobotany and potential as anticancer and anti-inflammatory agents. *J Ethnopharmacol.* 119: 195–213.)

coumarin-containing plant parts, such as the tonka bean, *Dipleteryx odorata*, a tree of the Fabaceae family, with other plant material. Tonka beans have been notoriously employed in the perfume industry, and most especially as a tobacco additive. Though regulations have pretty much eliminated their employment in cigarettes, they are sometimes still used in aromatic pipe tobacco blends. Coumarins are also employed by grasses and are an appetite suppressant, though a toxic one, helping these plants modulate their consumption by herbivores. The well-known phytoestrogen coumestrol, found in soy and clover, is a coumarin derivative (a coumestan) that may also induce infertility in grazing mammals such as sheep. Coumarin also lends its sweet toxicity to sweet vernal grass (*Anthoxanthum odorata*; also known as bison grass or as *vanilla grass* because of its resemblance to the smell of vanilla), sweet woodruff (*Galium odorata*), lavender, and licorice.

One of the most famous coumarins is sodium warfarin, used in medicine as an anticoagulant and also as a rat poison. Because of differences in metabolism of the compound between rodents and humans, the plant is considerably more toxic to the former. Its use in medicine must be still very carefully titrated though to prevent unwanted toxic or even fatal effects.

The class of coumarins found in fig leaves is most prominently the furocoumarins, such as psoralen and bergapten, which are more copious in *F. carica* than coumarins such as umbelliferone, 4′,5′-dihydropsoralen and marmesin by 2.5–7x, depending upon the season, according to Innocenti et al. (1982). Psoralen and bergapten were isolated from the water extract of the leaves of *F. carica* and found to exert potent antitumor action (Meng et al. 1996). A screen of 11 different coumarin compounds against human bladder carcinoma cell line E-J proliferation in vitro by Yang et al. in 2007 showed potent inhibitory effects of coumarins, including those found in fig leaf. Some preliminary pharmacokinetic analysis of the disposition of *F. hirta* Vahl leaf psoralen in rats has recently been completed, and it highlighted the potential beneficial interactions accruing from the combining of this material with plant biomass from other sources and species (Li et al. 2009).

FIGURE 4.16 *Ficus carica*, leaves (27.7.2009, Seven Species Garden, Binyamina, Israel, by Helena Paavilainen).

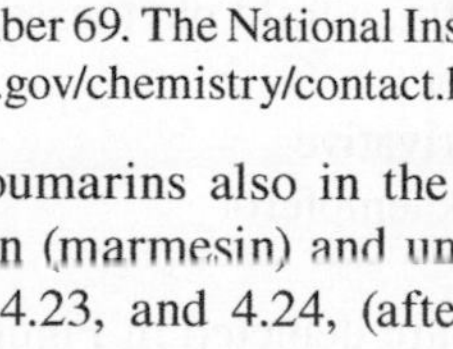

FIGURE 4.17 Coumarin. (*NIST Chemistry WebBook.* 1996. [Online Database]. NIST Standard Reference Database Number 69. The National Institute of Standards and Technology (NIST). URL: http://webbook.nist.gov/chemistry/contact.html [accessed August 14, 2009].)

Earlier studies identified coumarins also in the leaf of *F. carica*, inclusive of bergapten, 4′,5′ dihydropsoralen (marmesin) and umbelliferone. These compounds are depicted in Figures 4.22, 4.23, and 4.24, (after Damjanić and Akacić, 1974; Innocenti et al., 1982).

Furocoumarin glucosides isolated from the methanolic extract of the leaves of *F. ruf-icaulis* Merr. var. *antaoensis* included 5-O-β-d-glucopyranosyl-6-hydroxyangelicin,

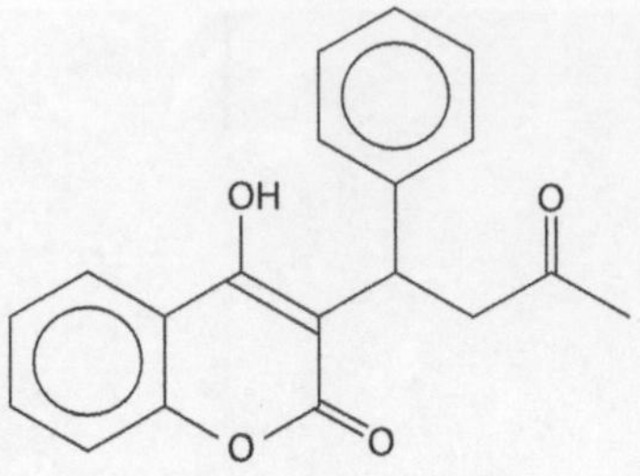

FIGURE 4.18 Warfarin. (*NIST Chemistry WebBook*. 1996. [Online Database]. NIST Standard Reference Database Number 69. The National Institute of Standards and Technology (NIST). URL: http://webbook.nist.gov/chemistry/contact.html [accessed August 14, 2009].)

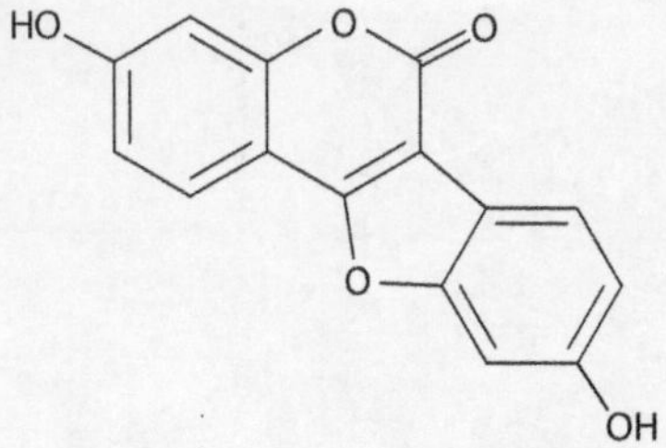

FIGURE 4.19 Coumestrol. (Reprinted with permission from Lansky and Newman 2007. Punica granatum (pomegranate) and its potential for prevention and treatment of inflammation and cancer. *J Ethnopharmacol*. 109: 177–206. Reprinted with permission of Elsevier.)

6-*O*-β-d-glucopyranosyl-5-hydroxyangelicin, 5,6-*O*-β-d-diglucopyranosylangelicin, 8-*O*-β-d-glucopyranosyl-5-hydroxypsoralen, 5-*O*-β-d-glucopyranosyl-8-hydroxypsoralen, 3,4,5-trihydroxydehydro-α-ionol-9-*O*-β-d-glucopyranoside (Chang et al. 2005). Rutin and isoquercitrin also occurred.

FLAVONOIDS

At least four different flavonoid glycosides are found in *Ficus* leaves. These are

1. Genistin, the glycoside of genistein, an isoflavone and classic "phytoestrogen"
2. Rutin, glycoside of quercetin, a mild phytoestrogen, good antioxidant, and signal transduction modifier
3. Isoquercitrin, quercetin derivative
4. Kaempferitrin, relative of kaempferol

These four flavonoid glycosides are depicted in Figure 4.29 and Figures 4.31–4.33. All flavonoids are antioxidant with signal transduction modifying capability. In general, the sugar moieties included in these compounds may reduce or eliminate their potency unless the sugars are first cleaved by digestion or some other process such as

FIGURE 4.20 *Ficus carica* leaves (27.7.2009, Seven Species Garden, Binyamina, Israel, by Helena Paavilainen).

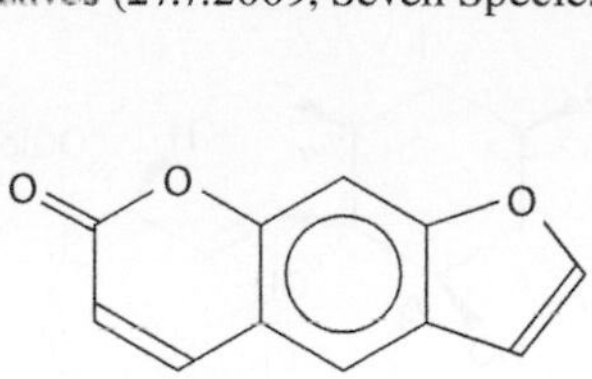

FIGURE 4.21 Psoralen. (*NIST Chemistry WebBook*. 1996. [Online Database]. NIST Standard Reference Database Number 69. The National Institute of Standards and Technology (NIST). URL: http://webbook.nist.gov/chemistry/contact.html [accessed August 14, 2009].)

fermentation. This may be an underlying reason for the traditional fermentation of herbal decoctions in early medicine and plant alchemy.

Genistein is the *aglycone* of genistin, the compound formed when the sugar moiety has been removed. Genistein is, as previously suggested, the compound most people usually think of as an example of a phytoestrogen. It is a physiologically active compound that plants most likely utilize in their long-term defense strategy,

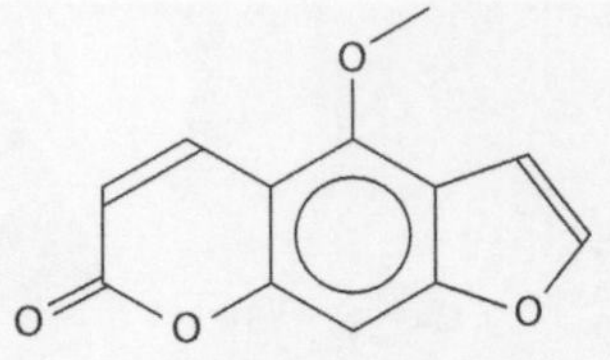

FIGURE 4.22 Bergapten. (*NIST Chemistry WebBook*. 1996. [Online Database]. NIST Standard Reference Database Number 69. The National Institute of Standards and Technology (NIST). URL: http://webbook.nist.gov/chemistry/contact.html [accessed August 14, 2009].)

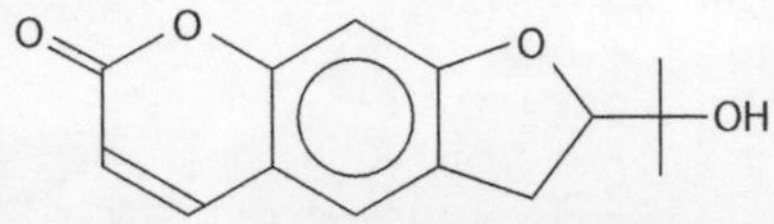

FIGURE 4.23 4′, 5′ Dihydroxypsoralen (marmesin). (*NIST Chemistry WebBook*. 1996. [Online Database]. NIST Standard Reference Database Number 69. The National Institute of Standards and Technology (NIST). URL: http://webbook.nist.gov/chemistry/contact.html [accessed August 14, 2009].)

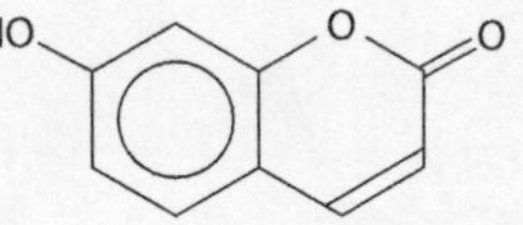

FIGURE 4.24 Umbelliferone. (*NIST Chemistry WebBook*. 1996. [Online Database]. NIST Standard Reference Database Number 69. The National Institute of Standards and Technology (NIST). URL: http://webbook.nist.gov/chemistry/contact.html [accessed August 14, 2009].)

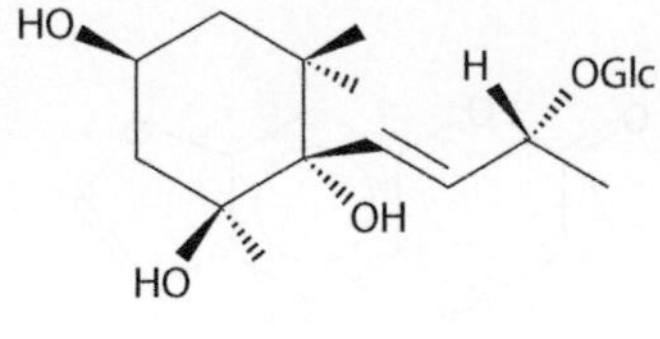

FIGURE 4.25 3,4,5-Trihydroxydehydro-α-ionol-9-*O*-β-d-glucopyranoside. (Reprinted with permission from Lansky et al. 2008. *Ficus* spp. (fig): Ethnobotany and potential as anticancer and anti-inflammatory agents. *J Ethnopharmacol*. 119: 195–213. Reprinted with permission of Elsevier.)

such as inducing sterility over time in their predators. It may also be a potent inhibitor of platelet aggregation, as discovered in a study by Kim and Yun-Choi (2008), working with a methanolic extract of *Sophora japonica* that contained both genistin and genistein. In this setting, the aglycone was considerably more active than the glycoside. Genistein was further found, in human glioma multiforme cells, to inhibit topoisomerase II, an enzyme needed for separating the daughter strands of DNA

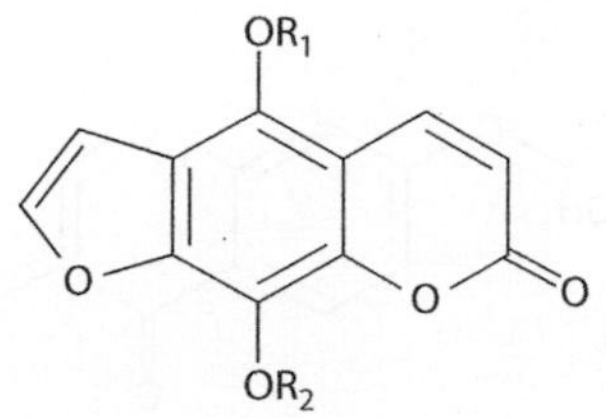

R₁ = Glc, R₂ = H
R₁ = H, R₂ = Glc

FIGURE 4.26 5-*O*-β-d-glucopyranosyl-8-hydroxypsoralen; 8-*O*-β-d-glucopyranosyl-5-hydroxypsoralen. (Adapted from Lansky et al. 2008. *Ficus* spp. (fig): Ethnobotany and potential as anticancer and anti-inflammatory agents. *J Ethnopharmacol.* 119: 195–213. Reprinted with permission of Elsevier.)

R₁ = Glc, R₂ = Glc
R₁ = Glc, R₂ = H
R₁ = H, R₂ = Glc

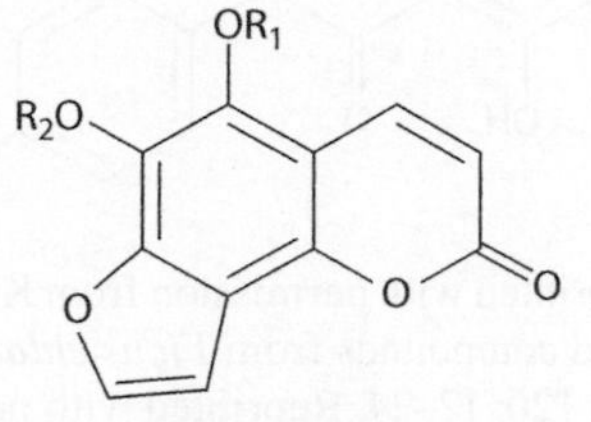

FIGURE 4.27 5,6-*O*-β-d-diglucopyranosylangelicin; 5-*O*-β-d-glucopyranosyl-6-hydroxy-angelicin; 6-*O*-β-d-glucopyranosyl-5-hydroxyangelicin. (Adapted from Lansky et al. 2008. *Ficus* spp. (fig): Ethnobotany and potential as anticancer and anti-inflammatory agents. *J Ethnopharmacol.* 119: 195–213. Reprinted with permission of Elsevier.)

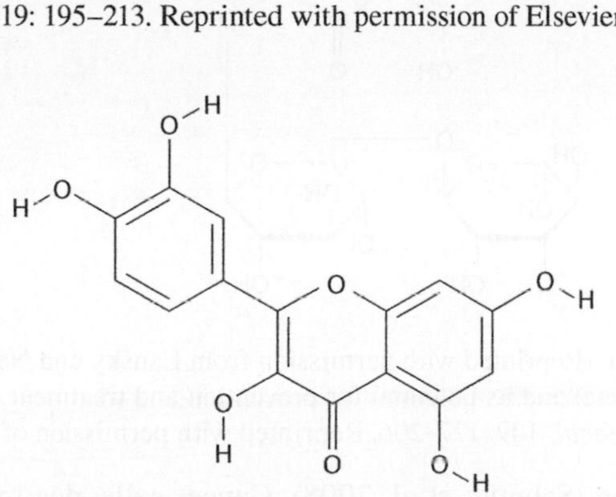

FIGURE 4.28 Quercetin. (*PubChem*. The PubChem Project. 2001. [Online Database]. National Center for Biotechnology Information. U.S. National Library of Medicine. 8600 Rockville Pike, Bethesda, MD 20894. URL: http://pubchem.ncbi.nlm.nih.gov/ [accessed August 16, 2009].)

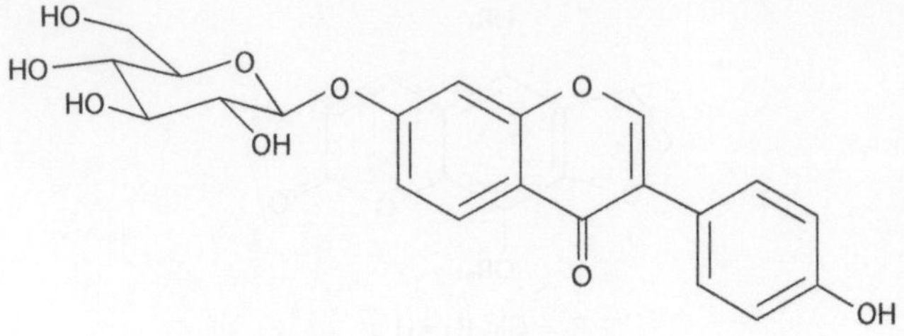

FIGURE 4.29 Genistin. (Reprinted with permission from Lansky et al. 2008. *Ficus* spp. (fig): Ethnobotany and potential as anticancer and anti-inflammatory agents. *J Ethnopharmacol.* 119: 195–213. Reprinted with permission of Elsevier.)

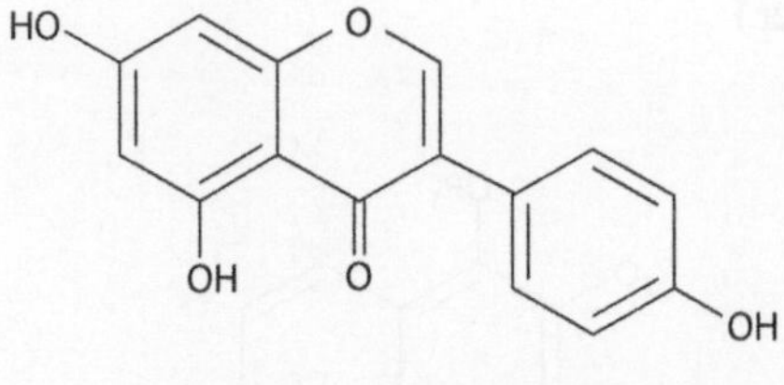

FIGURE 4.30 Genistein. (Reprinted with permission from Kuete et al. 2008. Antimicrobial activity of the crude extracts and compounds from *Ficus chlamydocarpa* and *Ficus cordata* (Moraceae). *J Ethnopharmacol.* 120: 17–24. Reprinted with permission of Elsevier.)

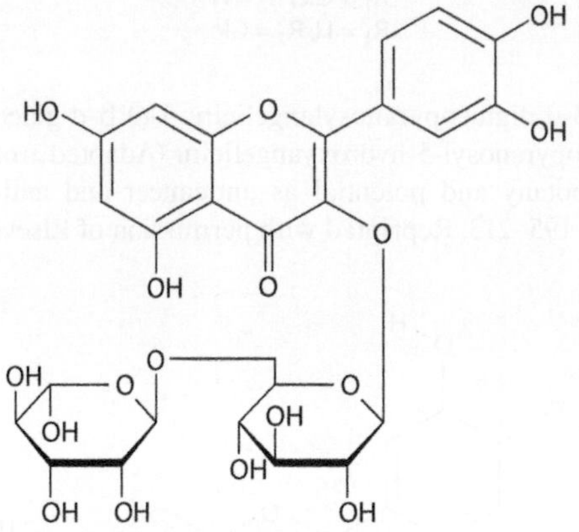

FIGURE 4.31 Rutin. (Reprinted with permission from Lansky and Newman 2007. *Punica granatum* (pomegranate) and its potential for prevention and treatment of inflammation and cancer. *J Ethnopharmacol.* 109: 177–206. Reprinted with permission of Elsevier.)

following replication (Schmidt et al. 2008). Cancer cells, due to their rapid dividing, may overexpress topoisomerase, and accordingly, its inhibition by genistein results in cell cycle arrest and apoptosis (programmed self-destruction). Most notably, genistein synergizes well with other anticancer compounds, in general, and with proapoptotic compounds. These compounds include (1) antibodies against cell

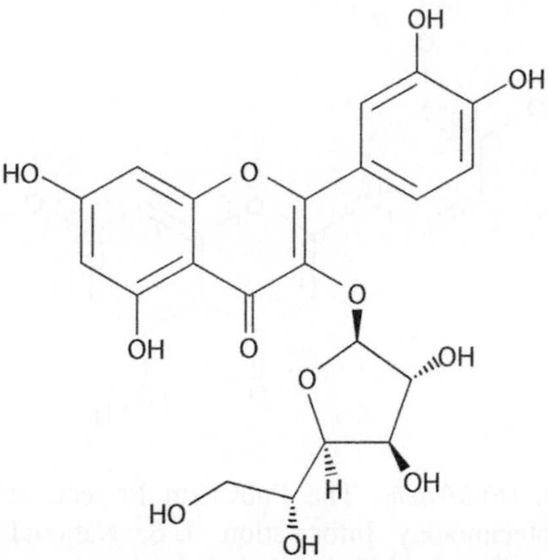

FIGURE 4.32 Isoquercitrin. (Reprinted with permission from Lansky et al. 2008. *Ficus* spp. (fig): Ethnobotany and potential as anticancer and anti-inflammatory agents. *J Ethnopharmacol*. 119: 195–213. Reprinted with permission of Elsevier.)

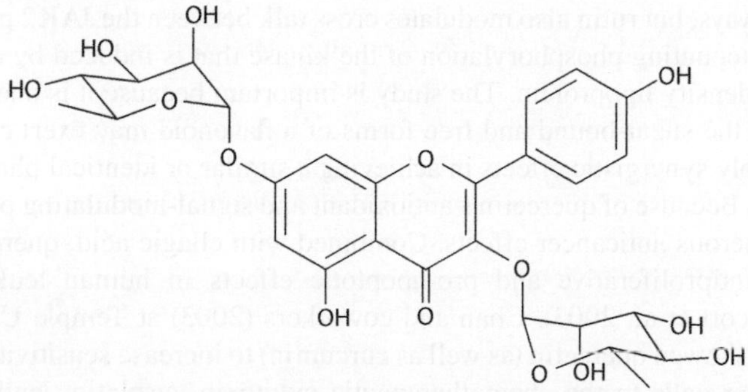

FIGURE 4.33 Kaempferitrin. (Reprinted with permission from Lansky et al. 2008. *Ficus* spp. (fig): Ethnobotany and potential as anticancer and anti-inflammatory agents. *J Ethnopharmacol*. 119: 195–213. Reprinted with permission of Elsevier.)

adhesion proteins in human ovarian cancer cells in vitro (Novak-Hofer et al. 2008), (2) β-lapachone, a plant-derived topoisomerase I inhibitor that induces apoptosis both by caspase-3 and minor death pathways in human prostate cancer cells (Kumi-Diaka et al. 2004), and (3) pomegranate extracts that, together with genistein, induce apoptosis in human estrogen receptor positive MCF-7 breast cancer cells (Jeune et al. 2005). It is thus likely that the genistin in *Ficus* leaves, after cleavage from its sugar to yield genistein, may yield a synergistic adjunct to compounds located elsewhere in either the leaves or some other part of the fig tree (such as the fruits) in inducing apoptosis in human cancer cells.

Rutin is a glycoside of quercetin, the most widely studied and probably the most widely occurring flavonol. A new study by J.S. Choi and associates (2009) at Kangwon National University in South Korea has revealed that both the aglycone (quercetin)

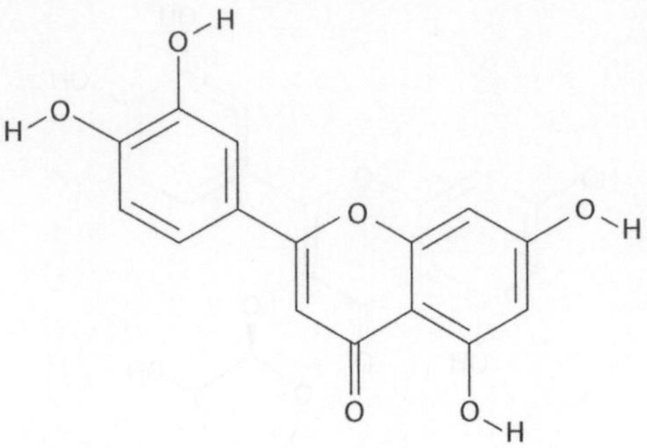

FIGURE 4.34 Luteolin. (*PubChem*. The PubChem Project. 2001. [Online Database]. National Center for Biotechnology Information. U.S. National Library of Medicine. 8600 Rockville Pike, Bethesda, MD 20894. URL: http://pubchem.ncbi.nlm.nih.gov/ [accessed August 16, 2009].)

and its glycoside (rutin) help to suppress apoptosis in noncancerous endothelial cells, a suppression that helps prevent atherosclerosis; however, that the two compounds do so in overlapping but complementary ways. Specifically, both modulate the JAK2-STAT3 pathways, but rutin also modulates cross talk between the JAK2 pathway and MAPK by attenuating phosphorylation of the kinase that is induced by an oxidized form of low-density lipoprotein. The study is important because it is one of the first to show how the sugar-bound and free forms of a flavonoid may exert complementary or possibly synergistic effects in achieving a similar or identical pharmacological objective. Because of quercetin's antioxidant and signal-modulating properties, it also has numerous anticancer effects. Combined with ellagic acid, quercetin exerts synergistic antiproliferative and pro-apoptotic effects in human leukemia cells (Mertens-Talcott et al. 2003). Chan and coworkers (2003) at Temple University in Philadelphia showed quercetin (as well as curcumin) to increase sensitivity of human ovarian cancer cells to the chemotherapeutic cytotoxin, cisplatin, lending further credence to the possibility that inclusion of flavonoids in cancer-therapeutic mixtures can create synergy through cross talk between signaling pathways and other poorly understood mechanisms for multitargeting and overcoming cancer resistance.

Independently, Vaya and Mahmood (2006) used high-performance liquid chromatography (HPLC) and mass spectrometry detectors to study *F. carica* leaves and found the major flavonoids present to be quercetin and luteolin with a total of 631 and 681 mg/kg extract, respectively. Luteolin is a particularly potent anticancer compound possessing 58% of the estrogenic agonist effect of genistein (Zand et al. 2000) and notable anti-invasive activity in human prostate cancer cells that is synergistically enhanced by the presence of the phenyl propanoid, caffeic acid, the gallic acid dimer, ellagic acid, and the octadecatrienoic fatty acid, punicic acid (Lansky et al. 2005). In a study of 39 different phenolic compounds found in the leaves of 14 *Ficus* species growing in the Budongo Forest of Uganda, inclusive of 14 flavonol-*O*-glycosides, 6 flavone-*O*-glycosides, and 15 flavone-*C*-glycosides, the patterns were found to be distinctive to the individual species and reliable means for species identification (Greenham et al. 2007). Novel flavonoid glycosides scutellarein

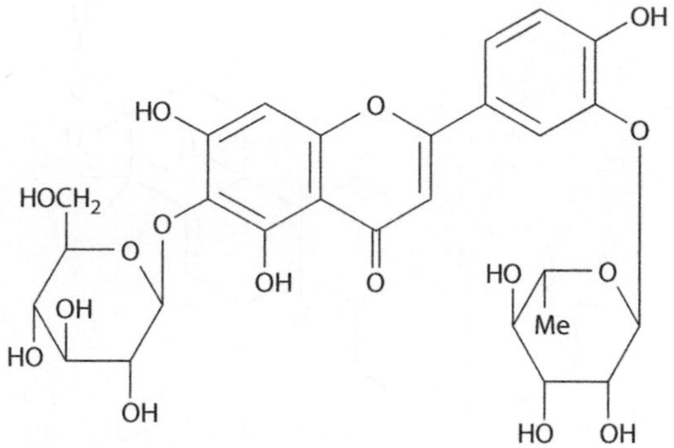

FIGURE 4.35 Infectoriin (luteolin 6-*O*-β-d-glucopyranoside 3'-*O*-α-l-rhamnoside). (Reprinted with permission from Jain et al. 1990. Isolation and characterization of luteolin 6-*O*-β-d-glucopyranoside 3'-*O*-α-l-rhamnoside from *Ficus infectoria. J Chem Res Synop.* 12: 396–7. Reproduced by permission of The Royal Society of Chemistry.)

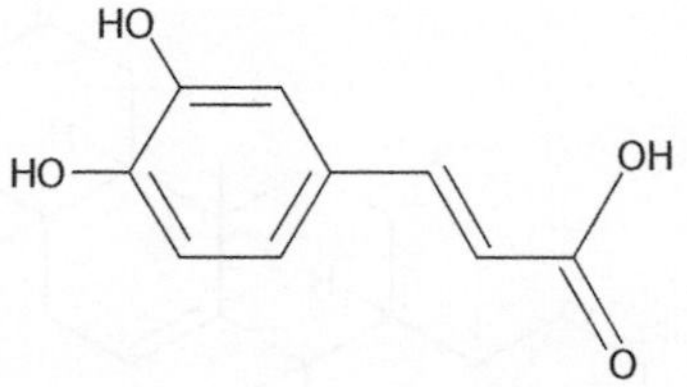

FIGURE 4.36 Caffeic acid. (Reprinted with permission from Lansky and Newman 2007. *Punica granatum* (pomegranate) and its potential for prevention and treatment of inflammation and cancer. *J Ethnopharmacol.* 109: 177–206. Reprinted with permission of Elsevier.)

6-*O*-α-l-rhamnopyranosyl(1→2)-β-d-galactopyranoside and luteolin 6-*O*-β-d-glucopyranoside 3'-*O*-α-l-rhamnoside (infectoriin) were isolated and characterized from the leaves of *F. infectoria* (Jain et al. 1990, 1991).

TERPENOIDS, STEROLS

Ten compounds, including the two acetophenones 4-hydroxy-3-methoxy-acetophenone and 3,4,5-trimethoxyacetophenone; the two triterpenoids α-amyrin acetate and β-amyrin acetate; the two sterols β-sitosterol and stigmasterol; and the three long-chain aliphatics palmitic acid, myristic acid, and 1-triacontanol, were isolated from the $CHCl_3$ fraction of the leaves of the Formosan *F. septica* with the phenanthroindolizidine alkaloid tylophorine, obtained from the basic fraction (Tsai et al. 2000). Triterpenoids rhoiptelenol, 3α-hydroxy-isohop-22(29)-en-24-oic acid, lupenyl acetate, α-amyrin acetate, β-amyrin acetate, lupeol, α-amyrin, β-amyrin, taraxerol, glutinol, ursolic acid, and betulinic acid were also isolated from leaves of *F. thunbergii* (Kitajima et al. 1994) and oleanolic acid from the leaves of *F. hispida* (Khan et al. 1990). *F. palmata* leaves contained β-sitosterol and bergapten (Ahmad et al. 1976).

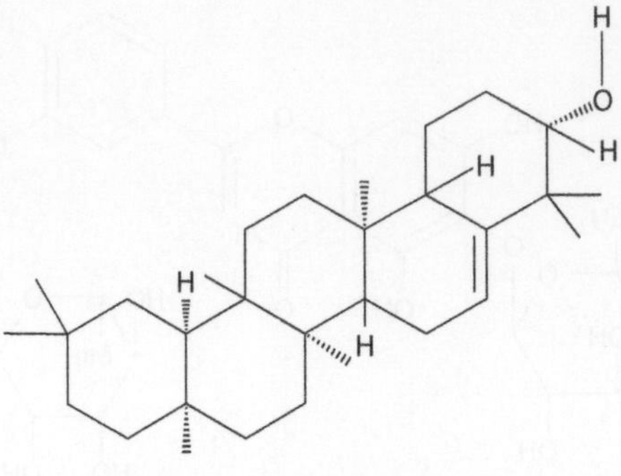

FIGURE 4.37 β-Glutinol. (*PubChem*. The PubChem Project. 2001. [Online Database]. National Center for Biotechnology Information. U.S. National Library of Medicine. 8600 Rockville Pike, Bethesda, MD 20894. URL: http://pubchem.ncbi.nlm.nih.gov/ [accessed August 16, 2009].)

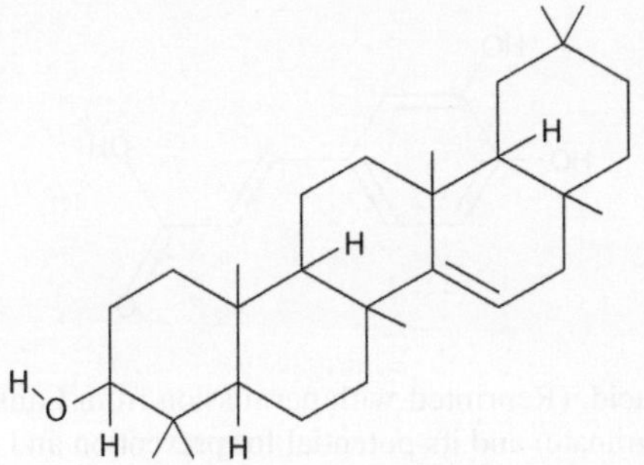

FIGURE 4.38 Taraxerol. (*PubChem*. The PubChem Project. 2001. [Online Database]. National Center for Biotechnology Information. U.S. National Library of Medicine. 8600 Rockville Pike, Bethesda, MD 20894. URL: http://pubchem.ncbi.nlm.nih.gov/ [accessed August 16, 2009].)

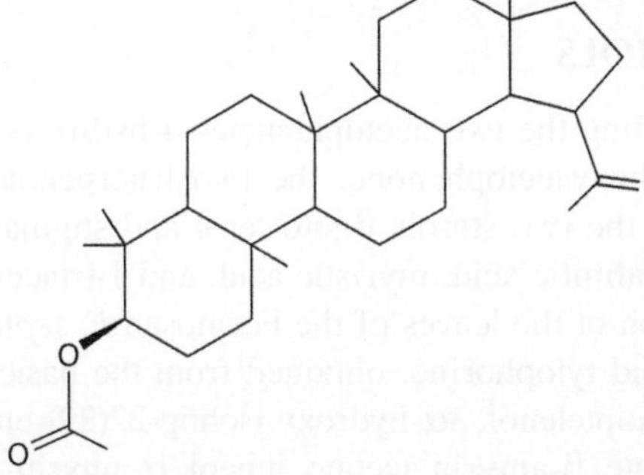

FIGURE 4.39 Lupenyl acetate. (*NIST Chemistry WebBook*. 1996. [Online Database]. NIST Standard Reference Database Number 69. The National Institute of Standards and Technology (NIST). URL: http://webbook.nist.gov/chemistry/contact.html [accessed August 14, 2009].)

FIGURE 4.40 Myristic acid. (*NIST Chemistry WebBook*. 1996. [Online Database]. NIST Standard Reference Database Number 69. The National Institute of Standards and Technology (NIST). URL: http://webbook.nist.gov/chemistry/contact.html [accessed August 14, 2009].)

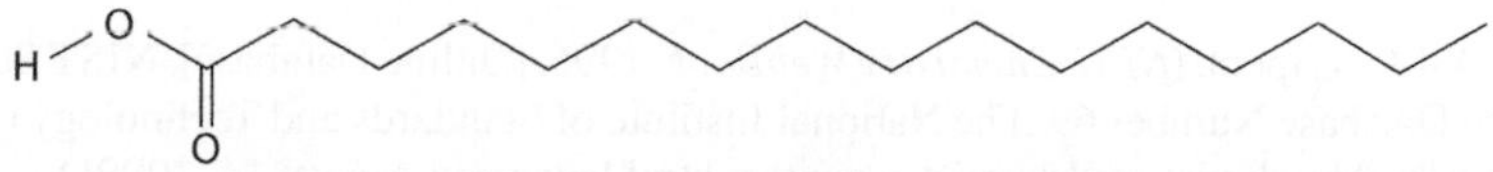

FIGURE 4.41 Palmitic acid. (*PubChem*. The PubChem Project. 2001. [Online Database]. National Center for Biotechnology Information. U.S. National Library of Medicine. 8600 Rockville Pike, Bethesda, MD 20894. URL: http://pubchem.ncbi.nlm.nih.gov/ [accessed August 16, 2009].)

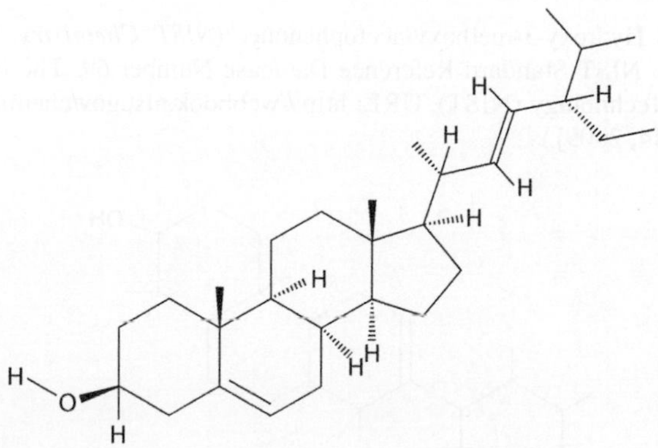

FIGURE 4.42 Stigmasterol. (*PubChem*. The PubChem Project. 2001. [Online Database]. National Center for Biotechnology Information. U.S. National Library of Medicine. 8600 Rockville Pike, Bethesda, MD 20894. URL: http://pubchem.ncbi.nlm.nih.gov/ [accessed August 16, 2009].)

PEPTIDES AND PROTEINS

Among proteins and peptides in *Ficus* leaves, a Cu Zn superoxide dismutase (SOD) was identified and partially characterized in *F. carica*. This SOD was found to be relatively stable and to possess approximately 12.78% of an alpha helix in its secondary structure (Zhang et al. 2008). Superoxide dismutase is a class of enzymes that catalyze the dismutation of superoxide (O_2^-) into oxygen and hydrogen peroxide and constitute a major antioxidant defense of organisms.

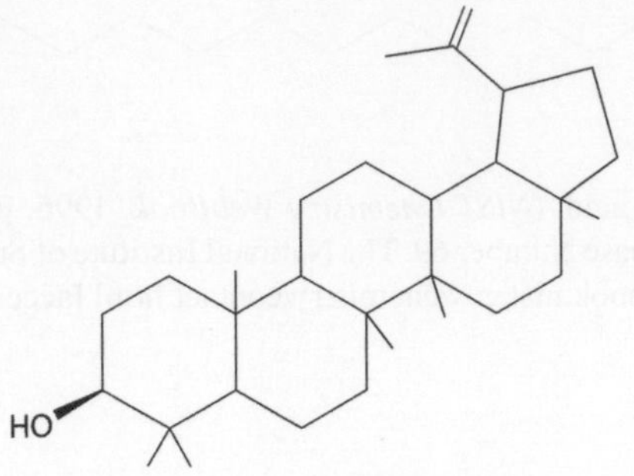

FIGURE 4.43 Lupeol. (*NIST Chemistry WebBook*. 1996. [Online Database]. NIST Standard Reference Database Number 69. The National Institute of Standards and Technology (NIST). URL: http://webbook.nist.gov/chemistry/contact.html [accessed August 14, 2009].)

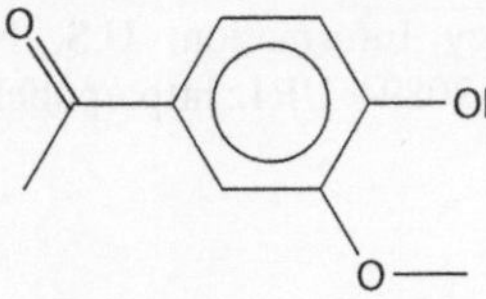

FIGURE 4.44 4-Hydroxy-3-methoxy-acetophenone. (*NIST Chemistry WebBook*. 1996. [Online Database]. NIST Standard Reference Database Number 69. The National Institute of Standards and Technology (NIST). URL: http://webbook.nist.gov/chemistry/contact.html [accessed August 14, 2009].)

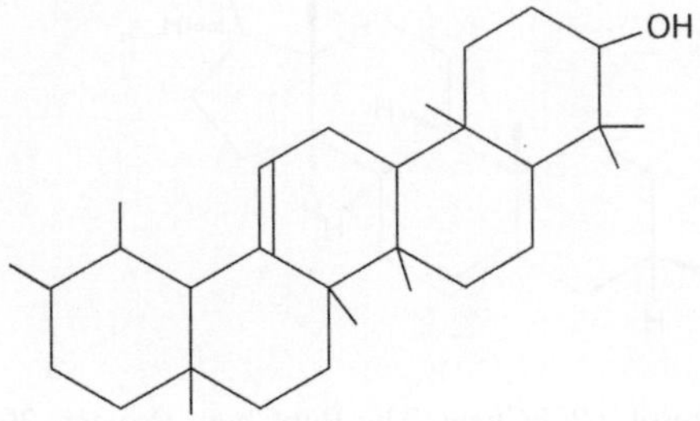

FIGURE 4.45 α-Amyrin. (*NIST Chemistry WebBook*. 1996. [Online Database]. NIST Standard Reference Database Number 69. The National Institute of Standards and Technology (NIST). URL: http://webbook.nist.gov/chemistry/contact.html [accessed August 14, 2009].)

NOCICEPTION

Extracts of *Ficus* leaves also exhibit antinociceptive effects in vivo. For example, an aqueous extract from the leaves of *F. deltoidea* was evaluated in three in vivo rodent models of nociception (pain perception), namely, acetic acid-induced abdominal writhing, formalin, and hot plate test. The results of the study showed that intraperitoneal administration of the *F. deltoidea* leaves aqueous extract at the

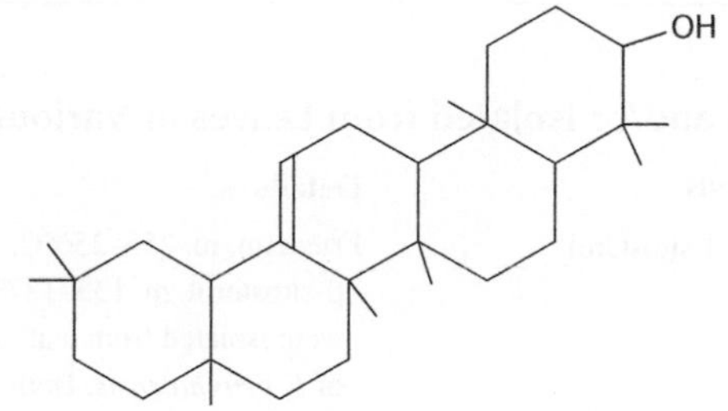

FIGURE 4.46 β-Amyrin. (*NIST Chemistry WebBook*. 1996. [Online Database]. NIST Standard Reference Database Number 69. The National Institute of Standards and Technology (NIST). URL: http://webbook.nist.gov/chemistry/contact.html [accessed August 14, 2009].)

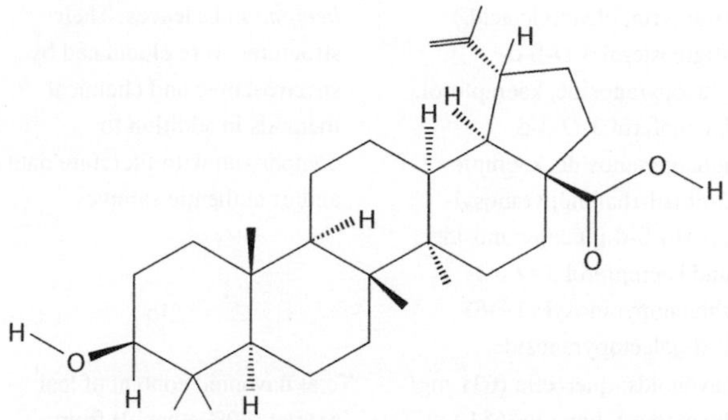

FIGURE 4.47 Betulinic acid. (*PubChem*. The PubChem Project. 2001. [Online Database]. National Center for Biotechnology Information. U.S. National Library of Medicine. 8600 Rockville Pike, Bethesda, MD 20894. URL: http://pubchem.ncbi.nlm.nih.gov/ [accessed August 16, 2009].)

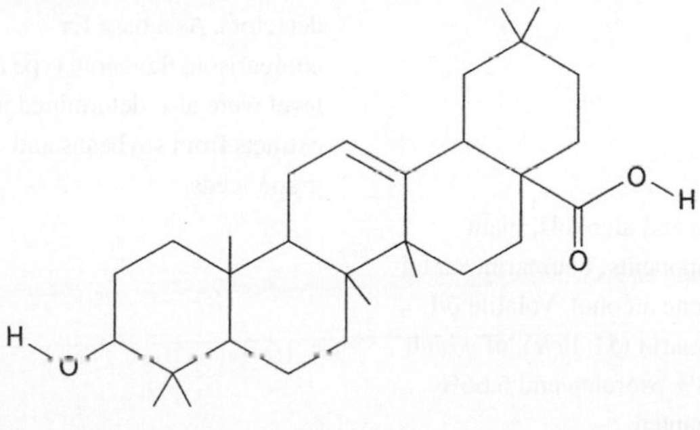

FIGURE 4.48 Oleanolic acid. (*PubChem*. The PubChem Project. 2001. [Online Database]. National Center for Biotechnology Information. U.S. National Library of Medicine. 8600 Rockville Pike, Bethesda, MD 20894. URL: http://pubchem.ncbi.nlm.nih.gov/ [accessed August 16, 2009].)

TABLE 4.1
Chemicals Identified and/or Isolated from Leaves of Various *Ficus* spp.

Ficus Species	Compounds	Details	Reference
benghalensis	Friedelin, β-sitosterol	Friedelin, m. 255–256°C, and β-sitosterol, m. 135–137°C, were isolated from leaf extracts of *F. benghalensis*. Both compounds were chromatographically identified by direct comparison with pure specimens.	Chatterjee and Chakraborty 1968
benjamina	β-Amyrin acetate, β-amyrone, 3β-friedelanol, taraxerol, stigmasterol, β-amyrin, α-amyrin, oleanolic acid, stigmasterol 3-*O*-β-d-glucopyranoside, kaempferol, kaempferol 3-*O*-β-d-glucopyranoside, kaempferol 3-*O*-α-l-rhamnopyranosyl-(1→6)-β-d-glucopyranoside, and kaempferol 3-*O*-α-l-rhamnopyranosyl-(1→6)-β-d-galactopyranoside.	The compounds were isolated for the first time from the methanolic extract of *F. benjamina* L. leaves. Their structures were elucidated by spectroscopic and chemical methods in addition to comparison with literature data and/or authentic samples.	Farag 2005
carica	Flavonoids: quercetin (631 mg/ kg extract), luteolin (681 mg/ kg extract)	Total flavonoid content of leaf extract (70% ethanol) from *F. carica* L. was determined by using reverse-phase high-performance liquid chromatography (HPLC) and analyzed by UV/VIS array and electrospray ionization (ESI)–mass spectrometry (MS) detectors. As a base for comparison, flavonoid type and level were also determined in extracts from soybeans and grape seeds.	Vaya and Mahmood 2006
carica	Esters and alcohols, main components: coumarin, santal terpene alcohol. Volatile oil: coumarin (51.46%), of which 44.8% psoralen and 6.66% bergapten.		Zhao et al. 2004

TABLE 4.1 (continued)
Chemicals Identified and/or Isolated from Leaves of Various *Ficus* spp.

Ficus Species	Compounds	Details	Reference
carica	Bergapten, psoralen, 59 compounds from the volatile fraction of GC/MS	Bergapten and psoralen were isolated and identified from the water extract of the leaf of *F. carica*, and 59 compounds were identified from the volatile fraction by GC/MS.	Meng et al. 1996
carica	Phenolic compounds: chlorogenic acid, rutin, psoralen	Sea sand disruption method (SSDM) and matrix solid-phase disruption (MSPD) were compared to solid–liquid extraction (SLE) for the extraction of phenolic compounds from the *F. carica* leaves. Recoveries above 85% were obtained for chlorogenic acid, rutin, and psoralen using the SSDM.	Teixeira et al. 2006
carica	Triterpenoids: calotropenyl acetate, methyl maslinate, lupeol acetate		Saeed and Sabir 2002
carica	Proteins; proteoses; peptone and amino acids containing tyrosine		Deleanu 1916
carica	Proteolytic enzymes	Proteolytic enzymes were extracted from H_2O extracts of leaves of *F. carica*, 1 g ground dried material with 40 mL H_2O at room temperature for 14–16 h. Proteolytic activities were determined on casein (I) and gelatin (II) solutions.	Gonashvili and Gonashvili 1968
exasperata	Tannins; flavonoids; saponins; no alkaloids; no anthraquinones		Ayinde et al. 2007
glomerata	New tetracyclic triterpene: gluanol acetate, 13α,14β,17βH,20αH-lanosta-8,22-diene 3β-acetate; β-amyrin and β-sitosterol,		Sen and Chowdhury 1971
hispida	Oleanolic acid		Khan et al. 1990

(continued on next page)

TABLE 4.1 (continued)
Chemicals Identified and/or Isolated from Leaves of Various *Ficus* spp.

Ficus Species	Compounds	Details	Reference
infectoria	New diglycoside: scutellarein 6-*O*-α-l-rhamnopyranosyl(1→2)-β-d-galactopyranoside	Structure determined mainly on the basis of chemical and spectral evidence.	Jain et al. 1991
infectoria	Novel flavone glycoside: infectoriin	Isolated from the leaves of *F. infectoria*, its structure determined by chemical and spectral evidence.	Jain et al. 1990
lyrata	New chromone: 5,6-dihydroxy-2-methylchromone; 7 known flavonoids: 5-hydroxy-7,3,3′,4′-tetramethoxyflavone, 5,4′-dihydroxy-6,7,8-trimethoxyflavone, 5,4′-dihydroxy-7,8-dimethoxyflavone, 4-methoxychalcone, 7,4′-dimethoxyapigenin, 5,7,4′-trihydroxy-2′,3′,6′-trimethoxyisoflavone, acacetin-7-*O*-glucoside, and acacetin-7-*O*-neohesperidoside; β-sitosterol-d-glucoside	The structures of the compounds were established on the basis of chemical and spectral evidence (IR, UV, 1H NMR, 13C NMR, and mass spectra).	Basudan et al. 2005
microcarpa	Two novel triterpenes: 29(20→19)abeolupane-3,20-dione and 19,20-secoursane-3,19,20-trione; (3β)-3-hydroxy-29(20→19) abeolupan-20-one, lupenone, and α-amyrone	Isolated from the leaves of *F. microcarpa* and characterized by spectroscopic means, including 2D-NMR techniques and chemical methods.	Kuo and Lin 2004
palmata	β-Sitosterol, bergapten		Ahmad et al. 1976
platyphylla	Three coumarin derivatives: 7-methoxycoumarin; α- and β-amyrin, ceryl alcohol, β-sitosterol, ursolic acid, an unknown flavonoid aglycon, two unknown flavonoids, and rutin.	Air-dried powdered leaves and bark of *F. platyphylla* were extracted with 70% EtOH and the concentrated extracts were successively fractionated with petroleum ether, CHCl₃, and EtOAc. Each fraction was examined by TCL, and the compounds were separated by preparative TLC and subsequent crystallization.	El-Sayyad et al. 1986

TABLE 4.1 (continued)
Chemicals Identified and/or Isolated from Leaves of Various *Ficus* spp.

Ficus Species	Compounds	Details	Reference
ruficaulis	5-*O*-β-d-glucopyranosyl-6-hydroxyangelicin, 6-*O*-β-d-glucopyranosyl-5-hydroxyangelicin, 5,6-*O*-β-d-diglucopyranosylangelicin, 8-*O*-β-d-glucopyranosyl-5-hydroxypsoralen, 5-*O*-β-d-glucopyranosyl-8-hydroxypsoralen, 3,4,5-trihydroxydehydro-α-ionol-9-*O*-β-d-glucopyranoside, rutin, and isoquercitrin.	The compounds were isolated from the methanolic extract of leaves of *F. ruficaulis* Merr. var. *antaoensis*. The structures of the first four compounds were elucidated by the analysis of their spectroscopic data.	Chang et al. 2005
semicordata	Flavonoids: (+)–catechin, quercetin, quercitrin	Flavonoids were isolated from dried leaves of *F. semicordata*. Their structure was elucidated by using spectral methods, especially MS, NMR techniques.	Nguyen et al. 2002
septica	Two acetophenones: 4-hydroxy-3-methoxy-acetophenone and 3,4,5-trimethoxyacetophenone; two triterpenoids: α-amyrin acetate and β-amyrin acetate; two steroids: β-sitosterol and stigmasterol; three long-chain aliphatic compounds: palmitic acid, myristic acid, and 1-triacontanol; one phenanthroindolizidine alkaloid: tylophorine	The compounds were isolated from the $CHCl_3$ sol. fraction of leaves of *F. septica*. Tylophorine was obtained from the basic fraction. The structures of these compounds were elucidated by spectroscopic methods.	Tsai et al. 2000
F. thonningii	Four stilbenes, including glycosides		Greenham et al. 2007
thunbergii	Two new triterpenoids: rhoiptelenol and 3α-hydroxy-isohop-22(29)-en-24-oic acid; known triterpenoids: lupenyl acetate, β-amyrin acetate, α-amyrin acetate, lupeol, β-amyrin, α-amyrin, taraxerol, glutinol, ursolic acid, and betulinic acid	Triterpenoids were isolated from the fresh leaves and stems of *F. thunbergii* Max., and their structures were determined by spectral and chemical methods.	Kitajima et al. 1994

(continued on next page)

TABLE 4.1 (continued)
Chemicals Identified and/or Isolated from Leaves of Various *Ficus* spp.

Ficus Species	Compounds	Details	Reference
Ficus spp.	Flavonoids: 14 flavonol *O*-glycosides, 6 flavone *O*-glycosides and 15 flavone *C*-glycosides; acylated flavonoid glycosides		Greenham et al. 2007

FIGURE 4.49 *Ficus deltoidea*, leaves (19.7.2009, University of Turku, Botanical Garden, Ruissalo, Turku, Finland, by Kaarina Paavilainen).

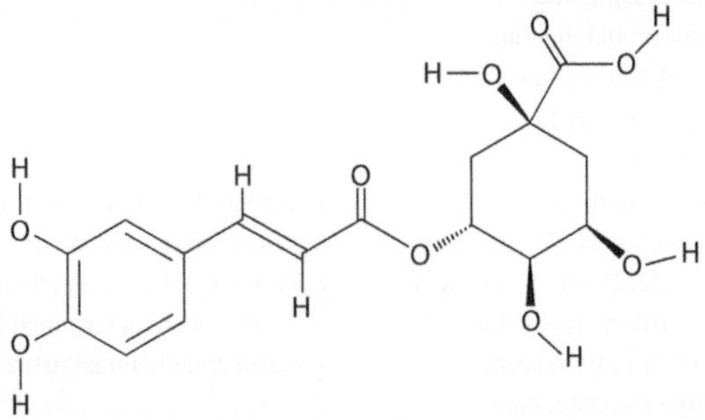

FIGURE 4.50 Chlorogenic acid. (*PubChem*. The PubChem Project. 2001. [Online Database]. National Center for Biotechnology Information. U.S. National Library of Medicine. 8600 Rockville Pike, Bethesda, MD 20894. URL: http://pubchem.ncbi.nlm.nih.gov/ [accessed August 16, 2009].)

dose of 1, 50, and 100 mg/kg 30 min prior to pain induction produced significant dose-dependent antinociception in all settings, indicating both central and peripherally mediated activities. Furthermore, the antinociceptive effect of the extract in the formalin and hot plate test was reversed by the nonselective opioid receptor antagonist naloxone, suggesting that the endogenous opioid system is involved in its analgesic action (Sulaiman et al. 2008). Also, ethanolic and aqueous extracts of the leaves of *F. benjamina* (var. *comosa*) exacted significant antinociceptive effects in an in situ analgesiometer test involving controlled, modulated delivery of electrical impulses to a rat tail, while the benzene fraction of the same leaves revealed a new triterpenoid, named serrat-3-one, and the known pentacontanyl decanoate, friedelin and β-sitosterol (Parveen et al. 2009). Similarly, a methanolic extract of *F. benjamina* leaves yielded a mix of both flavonoid glycosides and aglycones, triterpenoids, coumarins and sterols including β-amyrin acetate, β-amyrone, 3β-friedelanol, taraxerol, stigmasterol, β-amyrin, α-amyrin, oleanolic acid, stigmasterol 3-O-β-d-glucopyranoside, kaempferol, kaempferol 3-O-β-d-glucopyranoside, kaempferol 3-O-α-l-rhamnopyranosyl-(1→6)-β-d-glucopyranoside, and kaempferol 3-O-α-l-rhamnopyranosyl-(1→6)-β-d-galactopyranoside and exerted significant antipyretic, antinociceptive, and anti-inflammatory effects in vivo (Farag 2005).

Also, known flavonoids 5-hydroxy-7,3,3′,4′-tetramethoxyflavone, 5,4′-dihydroxy-6,7,8-trimethoxyflavone, 5,4′-dihydroxy-7,8-dimethoxyflavone, 4-methoxychalcone, 7,4′-dimethoxyapigenin, 5,7,4′-trihydroxy-2′,3′,6′-trimethoxyisoflavone (a rare flavonoid), acacetin-7-O-glucoside, and acacetin-7-O-neohesperidoside were isolated from the leaves of *F. lyrata* along with 5,6-dihydroxy-2-methylchromone and β-sitosterol-d-glucoside (Basudan et al. 2005). Novel triterpenoids 29(20→19) abeolupane-3,20-dione and 19,20-secoursane-3,19,20-trione, as well as (3β)-3-hydroxy-29(20→19)abeolupan-20-one, lupenone, and α-amyrone, were identified in the leaves of *F. microcarpa* (Kuo and Lin 2004). The condensed tannin (+)-catechin and flavonoids quercetin and quercitrin were found in leaves of *F. semichordata*

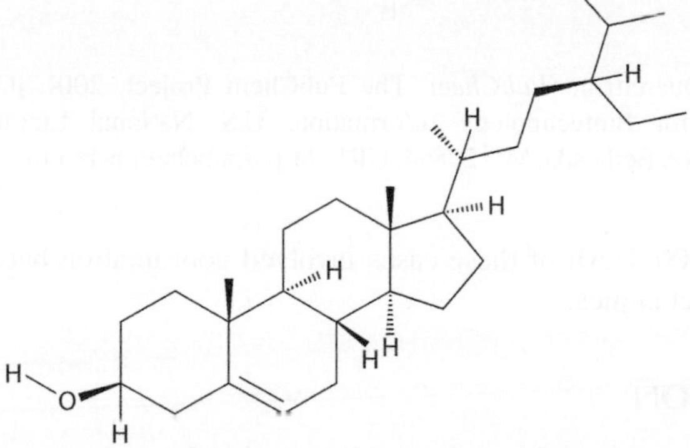

FIGURE 4.51 β-Sitosterol. (*PubChem. The PubChem Project. 2001. [Online Database]. National Center for Biotechnology Information. U.S. National Library of Medicine. 8600 Rockville Pike, Bethesda, MD 20894. URL: http://pubchem.ncbi.nlm.nih.gov/ [accessed August 16, 2009].)

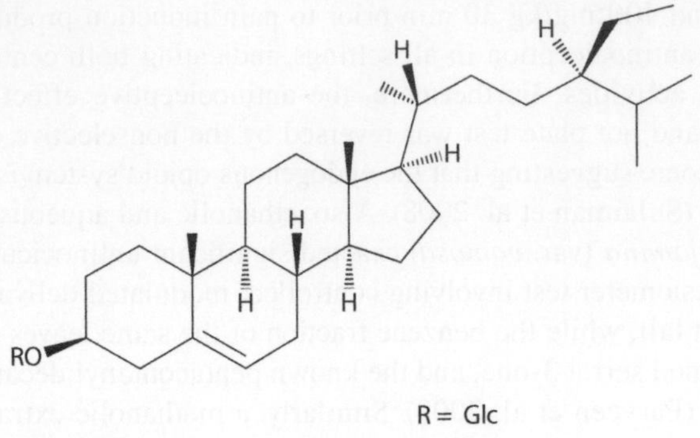

R = Glc

FIGURE 4.52 β-Sitosterol-β-d-glucoside. (Adapted from Lansky et al. 2008. *Ficus* spp. (fig): Ethnobotany and potential as anticancer and anti-inflammatory agents. *J Ethnopharmacol.* 119: 195–213. Reprinted with permission of Elsevier.)

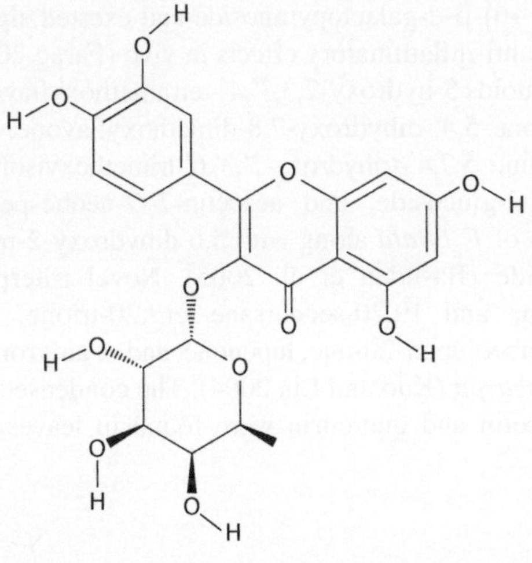

FIGURE 4.53 Quercitrin. (*PubChem*. The PubChem Project. 2001. [Online Database]. National Center for Biotechnology Information. U.S. National Library of Medicine. 8600 Rockville Pike, Bethesda, MD 20894. URL: http://pubchem.ncbi.nlm.nih.gov/ [accessed August 16, 2009].)

(Nguyen et al. 2002). All of these cases involved coordination between NMR and MS analytical techniques.

INFLAMMATION

Antioxidant defenses lend themselves well to anti-inflammatory applications, since oxidative tension is a promoting factor in many pro-inflammatory events. A methanol extract of *F. religiosa* leaf was tested in lipopolysaccharide (LPS)-induced production of NO and pro-inflammatory cytokines, such as tumor necrosis factor-α (TNF-α), interleukin-β (IL-1β), and IL-6 in BV-2 cells, a mouse microglial line,

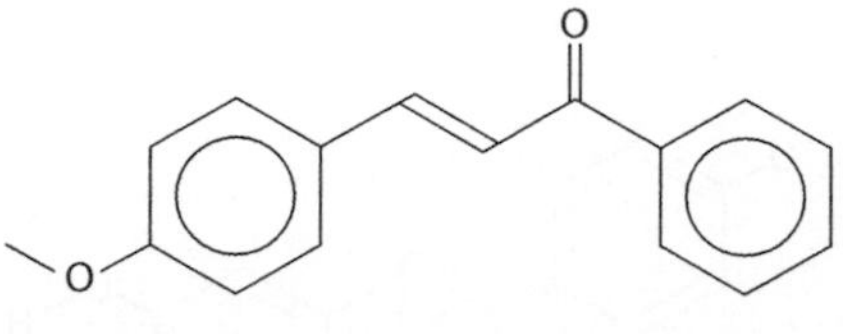

FIGURE 4.54 4-Methoxychalcone. (*NIST Chemistry WebBook*. 1996. [Online Database]. NIST Standard Reference Database Number 69. The National Institute of Standards and Technology (NIST). URL: http://webbook.nist.gov/chemistry/contact.html [accessed August 14, 2009].)

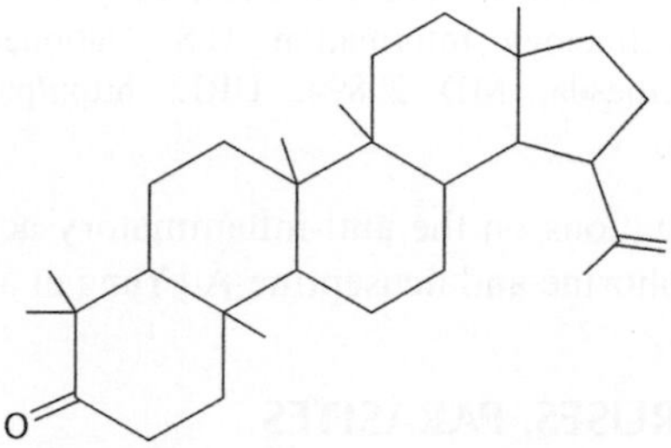

FIGURE 4.55 Lupenone. (*NIST Chemistry WebBook*. 1996. [Online Database]. NIST Standard Reference Database Number 69. The National Institute of Standards and Technology (NIST). URL: http://webbook.nist.gov/chemistry/contact.html [accessed August 14, 2009].)

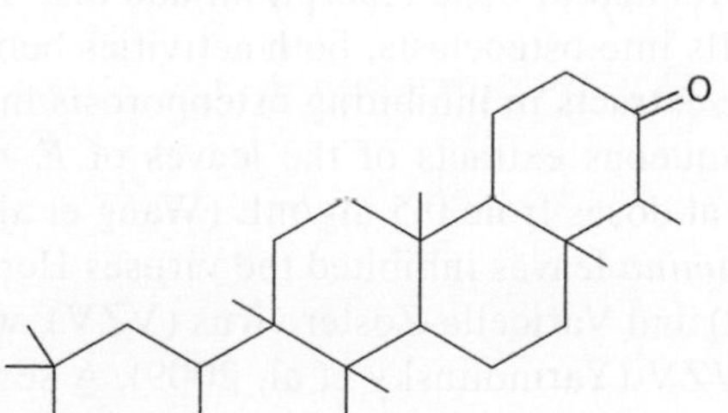

FIGURE 4.56 Friedelin. (*PubChem*. The PubChem Project. 2001. [Online Database]. National Center for Biotechnology Information. U.S. National Library of Medicine. 8600 Rockville Pike, Bethesda, MD 20894. URL: http://pubchem.ncbi.nlm.nih.gov/ [accessed August 16, 2009].)

and found to dose dependently inhibit LPS-induced production of NO and pro-inflammatory cytokines in a dose-dependent manner while simultaneously attenuating expression of m-RNA for pro-inflammatory cytokines and inducible nitric oxide synthase (iNOS). The study sheds further light on the ability of the *Ficus* leaf extract to disrupt inflammation at multiple control points extending across genomic, transcriptomic, and metabolomic levels within cells, with specific promise in microglial cells and neuroinflammatory disorders (Jung et al. 2008). These actions closely

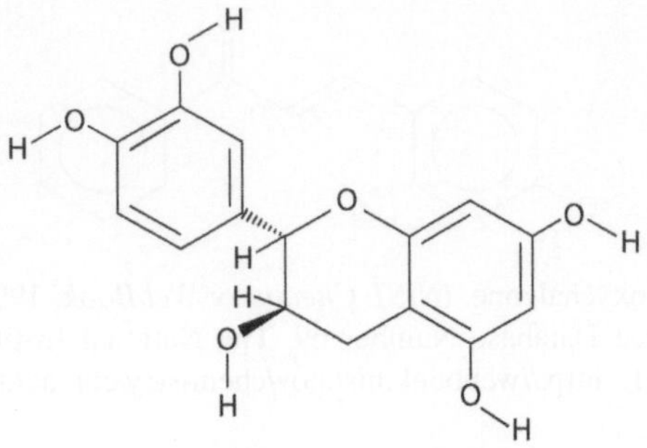

FIGURE 4.57 Catechin. (*PubChem*. The PubChem Project. 2001. [Online Database]. National Center for Biotechnology Information. U.S. National Library of Medicine. 8600 Rockville Pike, Bethesda, MD 20894. URL: http://pubchem.ncbi.nlm.nih.gov/ [accessed August 16, 2009].)

paralleled earlier investigations on the anti-inflammatory actions of phenanthroin-dolizidine alkaloids tylophorine and ficuseptine-A (Yang et al. 2006).

OSTEOPOROSIS, VIRUSES, PARASITES

Numerous other pharmacological actions have been described for leaves of various *Ficus* species. Water extract of leaves of *F. exasperata* Vahl effects significant decreases in mean arterial blood pressure (Ayinde et al. 2007). Hexane and ethyl acetate fractions of the leaves of *F. erecta* inhibited factors IL-1β, IL-6, and PGE2, believed to be related to bone resorption, and also inhibited differentiation of RAW 264.7 cells into osteoclasts, both activities believed to show potential for a role for these extracts in inhibiting osteoporosis in menopausal women (Yoon et al. 2007). Aqueous extracts of the leaves of *F. carica* killed human simplex virus in vitro at doses from 0.5 mg/mL (Wang et al. 2004), and ethanolic extracts of *F. benjamina* leaves inhibited the viruses Herpes Simplex Virus-1 and -2 (HSV-1, HSV-2) and Varicella-Zoster virus (VZV), while an extract of its fruits inhibited only VZV (Yarmolinsky et al. 2009). A sesquiterpenoid, verrucarin L acetate, was isolated from the leaves of *F. fistulosa* and found to inhibit the growth of *Plasmodium falciparum* at ID 50 doses below 1 ng/mL (Zhang et al. 2002).

TOXICOLOGY

An ethanolic extract of leaves from *F. exasperata* was tested along with ethanolic extracts of other selected medicinal plants traditionally employed in Nigeria in three different bench-top examinations for toxicity: (1) the brine shrimp lethality tests, (2) inhibition of telomerase activity, and (3) induction of chromosomal aberrations in vivo in rat lymphocytes. The *F. exasperata* leaf ethanolic extract was nontoxic in all three settings (Sowemimo et al. 2007).

TABLE 4.2
Pharmacological Studies of *Ficus* spp. Leaves

Ficus Species	Pharmacological Action	Details	Reference
benjamina	Significant anti-inflammatory, antinociceptive, and antipyretic activities in animal models	Methanolic extract of *F. benjamina* L. leaves exhibited significant anti-inflammatory, antinociceptive, and antipyretic activities in animal models.	Farag 2005
carica	Strong antitumor activity	Animal experiment showed that bergapten and psoralen, isolated from the water extract of the leaf of *F. carica*, had strong antitumor activity.	Meng et al. 1996
carica	Anti-HSV-1 effect; low toxicity; directly killing-virus effect on HSV-1	Water extract from the leaves of *F. carica* possessed distinct anti-HSV-1 effect. The MTC was 0.5 mg/mL, TDO was 15 mg/mL, and TI was 30.0 mg/mL. It possessed low toxicity and virucidal effect on HSV-1.	Wang et al. 2004
carica	Controls postprandial glycemia in IDDM	Decoction of *F. carica* leaves, as a supplement to breakfast, on diabetes control was given to insulin-dependent diabetes mellitus (IDDM) patients. Postprandial glycemia was significantly lower during supplementation with FC. Medium average capillary profiles were also lower in the two subgroups of patients during FC. Average insulin dose was 12% lower during FC in the total group.	Serraclara et al. 1998
carica	Antidiabetic; [in diabetes:] causes a decline in the levels of total cholesterol and a decrease in the total cholesterol/ HDL cholesterol ratio; reduction of the hyperglycemia	From the aqueous decoction of *F. carica* leaves, after treatment with HCl, centrifuging, treatment with sodium hydroxide (NaOH) and extraction with chloroform ($CHCl_3$). The administration of the organic phase rats with streptozotocin-induced diabetes led to a decline in the levels of total cholesterol and a decrease in the total cholesterol/ HDL cholesterol ratio (with respect to the control group), together with a reduction of the hyperglycemia.	Canal et al 2000

(continued on next page)

TABLE 4.2 (continued)
Pharmacological Studies of *Ficus* spp. Leaves

Ficus Species	Pharmacological Action	Details	Reference
carica	Hypotriglyceridemic effect	Hypertriglyceridemia was induced in rats with oral administration of 20% emulsion of long-chain triglycerides (LCT emulsion). After the intraperitoneal injection of serum saline (control group, $n = 10$) or *F. carica* leaf decoction (50 g dry wt/kg body wt; group A, $n = 10$), plasma triglyceride levels in the control group and group A were 5.9 ± 2.9 mmol/L and 5.5 ± 2.9 mmol/L just after the LCT emulsion protocol; 4.7 ± 2.7 mmol/L and 0.9 ± 0.4 mmol/L, $p < 0.005$, 60 min after the LCT protocol; and 3.6 ± 2.9 mmol/L and 1.0 ± 0.4 mmol/L, $p < 0.05$, 90 min after the LCT protocol.	Perez et al. 1999
deltoidea	Antinociceptive	Intraperitoneal administration of the *F. deltoidea* leaves aqueous extract at the dose of 1, 50, and 100 mg/kg, 30 min prior to pain induction produced significant dose-dependent antinociceptive effect in acetic acid-induced abdominal writhing, formalin and hot plate test, which indicating the presence of both central and peripherally mediated activities. The antinociceptive effect of the extract in the formalin and hot plate test was reversed by the nonselective opioid receptor antagonist naloxone, suggesting that the endogenous opioid system is involved in its analgesic mechanism of action.	Sulaiman et al. 2008
erecta	Antiosteoporotic activity	MG-63 cells were stimulated with IL-1β (10 ng/mL) to induce osteoporotic factors (IL-6, COX-2, and PGE2), and RAW 264.7 cells were stimulated with RANKL (100 ng/mL) to induce their differentiation into osteoclasts. *F. erecta* fractions decreased the mRNA expression of IL-6 and COX-2, and protein levels of COX-2 and PGE2 production. Among sequential solvent fractions, hexane and EtOAc fractions decreased differentiation into osteoclasts of RAW 264.7 cells.	Yoon et al. 2007

TABLE 4.2 (continued)
Pharmacological Studies of *Ficus* spp. Leaves

Ficus Species	Pharmacological Action	Details	Reference
exasperata	Antiulcer; hypotensive; showed a dose-related reduction in mean arterial blood pressure; probable stimulation of muscarinic receptors in the heart or release of histamine into the circulatory system	*F. exasperata* leaf water extract showed a dose-related reduction in mean arterial blood pressure. At 10 mg/kg, a reduction of 16.6 ± 1.1 mmHg was observed, whereas at 30 mg/kg, a fall in mean arterial pressure of 38.3 ± 0.6 mmHg was obtained. The hypotensive effect of the extract was significantly reduced with a prior administration of 2.5 mg of either atropine or chlorpheniramine. This suggests the probable stimulation of muscarinic receptors in the heart or release of histamine into the circulatory system, thereby causing the initial fall in blood pressure.	Ayinde et al. 2007
exasperata	Antiulcerogenic	*F. exasperata* leaf extract possesses significant antiulcerogenic properties in a dose-dependent way, protecting rats from aspirin-induced ulcerogenesis, delaying intestinal transit, increasing pH, and decreasing both the volume and acidity of gastric secretion.	Akah et al. 1998
fistulosa	Inhibits the growth of *Plasmodium falciparum*	Verrucarin L acetate, isolated in bioassay-directed fractionation of an extract prepared from the dried leaves and stem barks of *F. fistulosa* Reinw. ex Blume, was characterized as macrocyclic trichothecene sesquiterpenoid and found to inhibit the growth of *Plasmodium falciparum* with IC 50 values below 1 ng/mL.	Zhang et al. 2002
hispida	Inhibitory activity against castor oil-induced diarrhea and PGE(2)-induced enteropooling; causes reduction in gastrointestinal motility on charcoal meal test; antidiarrheal agent	Methanol extract of *F. hispida* L.	Mandal and Ashok Kumar 2002

(continued on next page)

TABLE 4.2 (continued)
Pharmacological Studies of *Ficus* spp. Leaves

Ficus Species	Pharmacological Action	Details	Reference
pumila	Antimicrobial activity against *Escherichia coli, Pseudomonas aeruginosa, Bacillus subtilis,* and *Candida albicans*	New neohopane from leaves of *F. pumila* showing antimicrobial activity against *E. coli, Ps. aeruginosa, B. subtilis,* and *C. albicans* with an average antimicrobial index of 0.5, 0.3, 0.3, and 0.7, respectively, at a concentration of 30 μg.	Ragasa et al. 1999
racemosa	Mosquito larvicidal	Crude hexane, ethyl acetate, petroleum ether, acetone, and methanol extracts of the leaf of *F. racemosa* showed moderate larvicidal effects against the early fourth-instar larvae of *Culex quinquefasciatus*. The larval mortality was observed after 24 h exposure.	Rahuman et al. 2008
racemosa	Anti-inflammatory	Anti-inflammatory activity of *F. racemosa* extract was evaluated on carrageenin-, serotonin-, histamine-, and dextran-induced rat hind paw edema models. The extract at doses of 200 and 400 mg/kg has been found to possess significant anti-inflammatory activity on the tested experimental models. The extract (400 mg/kg) exhibited maximum anti-inflammatory effect, that is, 30.4%, 32.2%, 33.9%, and 32.0% at the end of 3 h with carrageenin-, serotonin-, histamine-, and dextran-induced rat paw edema, respectively. In a chronic test, the extract (400 mg/kg) showed 41.5% reduction in granuloma weight. The effect produced by the extract was comparable to that of phenylbutazone, a prototype of a nonsteroidal anti-inflammatory agent.	Mandal et al. 2000b
racemosa	Antibacterial	Extracts of *F. racemosa* L. leaves showed significant antibacterial potential against *Escherichia coli* ATCC 10536, *Basillus pumilis* ATCC 14884, *Bacillus subtilis* ATCC 6633, *Pseudomonas aeruginosa* ATCC 25619, and *Staphylococcus aureus* ATCC 29737. The petroleum ether extract was the most effective against the tested organisms.	Mandal et al. 2000a

TABLE 4.2 (continued)
Pharmacological Studies of *Ficus* spp. Leaves

Ficus Species	Pharmacological Action	Details	Reference
religiosa	Delaying of neurodegeneration in human brain diseases; anti-inflammatory; inhibition of microglial activation; therapeutic potential for various neurodegenerative diseases	Inhibition of lipopolysaccharide (LPS)-induced production of NO and pro-inflammatory cytokines, such as tumor necrosis factor-α (TNF-α), interleukin-β (IL-1β), and IL-6 in BV-2 cells, a mouse microglial line; attenuating of the expression of mRNA and proteins of inducible nitric oxide synthase (iNOS) and pro-inflammatory cytokines, suggesting the blockage of transcription levels, respectively; downregulation of the extracellular signal-regulated kinase (ERK), c-Jun N-terminal kinase (JNK), and p38 mitogen-activated protein kinase (MAPK) signaling pathway; suppressing of the nuclear factor κB (NF-κB) activation; anti-inflammatory properties in LPS-induced activation of BV2 microglial cells	Jung et al. 2008

In a second study of the *aqueous* extract of *F. exasperata* leaves, the LD (50) (lethal dose to 50% of the animals) was determined for intraperitoneal injection to be 0.54 g/kg. Toxicity by oral administration was indeterminable. Acute toxicity over 24 h and following daily dosing for a 14-day period was assessed, but oral administration of 2.5, 5, 10, and 20 g/kg of the extract produced neither mortality nor changes in behavior or any other physiological activity in mice, leaving body weights and temperatures, and WBC count, platelets, and hemoglobin without significant alteration. However, the 14-day daily dose study did reveal a significant increase in body temperature ($p < 0.05$) and a significant decrease in the red blood cell count, hemoglobin count, and hematocrit values ($p < 0.05$). Thus, although the acute toxicity of the drug appeared very low, long-term studies are required, especially to further illuminate the potential erythropenic effect of chronic dosing (Bafor and Igbinuwen 2009).

HISTORICAL USES

Tables 4.3 and 4.4 give an overview of the historical medicinal uses of the leaves of different *Ficus* spp.

For references used in Table 4.4, see Chapter 3, Table 3.6. The scoring system is based on the number of times a reference discusses the use of *Ficus* leaves for the

TABLE 4.3
Historical Uses of *Ficus* spp. Leaf: Formulas

How to Make Fig Leaf Remedies for Internal and External Applications

Variations	Notes
Poultice made of (fresh) leaves	For scrofulous sores; removes warts; against the bite of a rabid dog
Poultice made of leaves and unripe fruit	For struma
Poultice made of fig leaves and branches	For rubbing coarse and itching eyelids; as a poultice for vitiligo
Poultice made of fresh leaves, used with the green figs or shoots, the milky juice of fig tree and wine	For scorpion stings: Shoots, or the green figs, are taken internally in wine, the milky juice is poured into the wound, and the leaves are applied to it
Mixed with wine	Externally for gangrenous ulcers
Mixed with vinegar	Externally for ulcers; for dog bites; for bites of rabid dogs
Mixed with honey	Externally for dog bites and favus
Mixed with leaves of poppy	Externally for extracting splinters of broken bones
Mixed with chick-pea flour and wine	Against poison of marine animals
Boiled leaves mixed with barley flour and natron	Externally for removal of warts
Boiled until soft with marshmallow (*Althaea*) roots	As a poultice for scrofula and tumors
Boiled with the root of black spurge laurel	For scabies in animals

purpose indicated. If the number of citations is regular, the reference is given in regular type and scored as "1." If the number of citations is strong, the reference is given in *italic* and scored as "2." If the number of citations is strongest, the reference is given in **bold** and scored as "3." The final score in the last column is the sum of these scores for the individual references. The particular ways of using the leaf are starred, but this information does not enter into the calculation of the total score for the function. The indications are shown in the order of importance based on their score values.

For the historical uses of ashes and lye prepared from the leaves of *Ficus* spp., see Chapter 3, Tables 3.9 and 3.10.

SUMMARY

Leaves from fig trees are rich in alkaloids, coumarins, triterpenoids, and flavonoids. They provide potent anti-inflammatory medicaments with strong anticancer actions. The leaves are used medically for antinociceptive as well as poison-clearing effects, and numerous other medical actions.

TABLE 4.4
Historical Uses of *Ficus* spp. Leaf

Indication	Leaf	Juice of Leaf	Ash of Leaf	References	Score
Against warts	*	*		2, 3, 6, 7, 11, 14, 15, 20	13
Opens the mouths of the veins (causing bleeding)	*	*		3, 6, 7, 10, 11, 13, 14, *15*, 20	11
Cleaning, absterging	*	*		3, 6, 7, 10, 11, 14, 20	9
Against putrid ulcers	*			1, 2, 7, 13, 15	7
Against swelling, tumors in general	*			**1**, 14, **18**	7
Opening hemorrhoids	*	*		6, 7, 10, 11, 13, 15, 20	7
Warming	*	*		3, 6, 7, 10, 11, 20	7
Against tuberculosis, scrofula	*			1, 14, 15, 18	6
Burning, stinging	*	*		3, 6, 11, 14, 15, 20	6
Causes ulcers	*	*		6, 7, 10, 11, 14, 20	6
Against roughness of eyelids	*			2, 6, 7	5
Against ulcers on the head	*			2, 6, 13, *15*	5
Softening (in general)	*	*		1, 6, 7, 10, 14	5
Against ordinary dog bite	*			1, 2, 6	4
Against scabies	*			2, 7, 13, 15	4
Against vitiligo	*			2, 6, 7, 11	4
Against allergic skin reactions	*			1, 2, 7	3
Against rabid dog bite	*			7, 10, 14	3
Against tetter	*			1, 7, 14	3
Animal feed	*			15, 18, 20	3
Cooling, calming heat	*	*		7, 10	3
Laxative	*	*		6, 11, 20	3
Against alopecia	*			1, 14	2
Against gangrene			*	1, 14	2
Against honeycomb ulcers	*			1, 2	2
Against proud flesh			*	1, 14	2
Discutient	*			1, 14	2
Pushes materia to skin	*	*		7, 10	2
Removes splinters of bones after an accident	*			1, 7	2
Spicy, sharp	*	*		18, 20	2
Sudorific	*			7, 10	2
Abortifacient	*			17	1

(continued on next page)

TABLE 4.4 (continued)
Historical Uses of *Ficus* sp. Leaf

Indication	Leaf	Juice of Leaf	Ash of Leaf	References	Score
Against all hard swellings/tumors, hardening	*			15	1
Against chilblains		*		7	1
Against dandruff	*			6	1
Against diarrhea	*			17	1
Against epilepsy	*			17	1
Against gland problems	*			6	1
Against leprosy	*			6	1
Against madness	*			17	1
Against marks on skin	*			6	1
Against morphea	*			10	1
Against poison of marine animals	*			1	1
Against scabies/eruptions in the eyelids	*			7	1
Against scars		*		7	1
Against scorpion sting	*			1	1
Against shrew mouse bite	*			1	1
Against skin problems	*			1	1
Against spreading ulcers	*			1	1
Against thick moisture in ulcers	*			7	1
Against worry, obsession	*			17	1
Against yaws	*			6	1
Helpful in body	*			10	1
Benefits chronic diseases	*			17	1
Causes ulcers in the intestines		*		20	1
Caustic	*			1	1
Corrects color of warts	*			10	1
Corrects discoloring	*			10	1
Corroding		*		15	1
Dissolving	*			6	1
Emmenagogue	*			17	1
Opening; against obstruction	*			18	1
Rarefying		*		3	1
Against itch	*			13	1

REFERENCES

Ahmad, S.A., S.A. Siddiqui, and A. Zaman. 1976. Chemical examination of *Callicarpa macrophylla*, *Lagerstromoea lanceolata*, *Ficus palmata* and *Taxodium mucronatum*. *J Indian Chem Soc*. 53: 1165–6.

Akah, P.A., O.E. Orisakwe, K.S. Gamaniel, and A. Shittu. 1998. Evaluation of Nigerian traditional medicines. II. Effects of some Nigerian folk remedies on peptic ulcer. *J Ethnopharmacol*. 62: 123–7.

Ayinde, B.A., E.K. Omogbai, and F.C. Amaechina. 2007. Pharmacognosy and hypotensive evaluation of *Ficus exasperata* Vahl (Moraceae) leaf. *Acta Pol Pharm*. 64: 543–6.

Bafor, E.E., and O. Igbinuwen. 2009. Acute toxicity studies of the leaf extract of *Ficus exasperata* on haematological parameters, body weight and body temperature. *J Ethnopharmacol*. 123: 302–7.

Basudan, O.A., M. Ilyas, M. Parveen, H.M. Muhisen, and R. Kumar. 2005. A new chromone from *Ficus lyrata*. *J Asian Nat Prod Res*. 7: 81–5.

Canal, J.R., M.D. Torres, A. Romero, and C. Perez. 2000. A chloroform extract obtained from a decoction of *Ficus carica* leaves improves the cholesterolaemic status of rats with streptozotocin-induced diabetes. *Acta Physiol Hung*. 87: 71–6.

Chan, M.M., D. Fong, K.J. Soprano, W.F. Holmes, and H. Heverling. 2003. Inhibition of growth and sensitization to cisplatin-mediated killing of ovarian cancer cells by polyphenolic chemopreventive agents. *J Cell Physiol*. 194: 63–70.

Chang, M.S., Y.C. Yang, Y.C. Kuo et al. 2005. Furocoumarin glycosides from the leaves of *Ficus ruficaulis* Merr. var. *antaoensis*. *J Nat Prod*. 68: 11–13, 634.

Chatterjee, D., and D.P. Chakraborty. 1968. Chemical examination of *Ficus bengalensis*. *J Indian Chem Soc*. 45: 285.

Choi, J.S., S.W. Kang, J. Li et al. 2009. Blockade of oxidized LDL-triggered endothelial apoptosis by quercetin and rutin through differential signaling pathways involving JAK2. *J Agric Food Chem*. [Epub ahead of print].

Damjanić, A., and B. Akacić. 1974. Furocoumarins in *Ficus carica*. *Planta Med*. 26: 119–23.

Damu, A.G., P.C. Kuo, L.S. Shi, C.Y. Li, C.R. Su, and T.S. Wu. 2009. Cytotoxic phenanthroindolizidine alkaloids from the roots of *Ficus septica*. *Planta Med*. March 18, 2009. [Epub ahead of print].

Damu, A.G., P.C. Kuo, L.S. Shi et al. 2005. Phenanthroindolizidine alkaloids from the stems of *Ficus septica*. *J Nat Prod*. 68: 1071–5.

Deleanu, N.T. 1916. The peptolytic enzyme of *Ficus carica*. *Bull Sect Sci Acad Roum*. 4: 345–54.

El-Sayyad, S.M., H.M. Sayed, and S.A. Mousa. 1986. Chemical constituents and preliminary anthelmintic activity of *Ficus platyphylla* (Del). *Bull Pharm Sci Assiut Univ*. 9: 164–77.

Farag, S.F. 2005. Phytochemical and pharmacological studies of *Ficus benjamina* L. leaves. *Mansoura J Pharm Sci*. 21: 19–36.

Gao, W., S. Bussom, S.P. Grill et al. 2007. Structure-activity studies of phenanthroindolizidine alkaloids as potential antitumor agents. *Bioorg Med Chem Lett*. 17: 4338–42.

Gao, W., W. Lam, S. Zhong, C. Kaczmarek, D.C. Baker, and Y.C. Cheng. 2004. Novel mode of action of tylophorine analogs as antitumor compounds. *Cancer Res*. 64: 678–88.

Gonashvili, S.G., and M.S. Gonashvili. 1968. Proteolytic enzymes of some Georgian plants. *Rastitel'nye Resursy* 4: 356–65.

Greenham, J.R., R.J. Grayer, J.B. Harborne, and V. Reynolds. 2007. Intra- and interspecific variations in vacuolar flavonoids among *Ficus* species from the Budongo Forest, Uganda. *Biochem Syst Ecol*. 35: 81–90.

Innocenti, G., A. Bettero, and G. Caporale. 1982. Determination of the coumarinic constituents of *Ficus carica* leaves by HPLC. *Farmaco Sci*. 37: 475–85.

Jain, N., M. Ahmad, M. Kamil, and M. Ilyas. 1990. Isolation and characterization of luteo-lin 6-*O*-β-d-glucopyranoside 3'-*O*-α -l-rhamnoside from *Ficus infectoria*. *J Chem Res Synop*. 12: 396–7.

Jain, N., M. Ahmad, M. Kamil, and M. Ilyas. 1991. Scutellarein 6-*O*-α-l-rhamno-pyranosyl(1→2)-→ -d-galactopyranoside: A new flavone diglycoside from *Ficus infectoria*. *J Chem Res Synop*. 8: 218–19.

Jeune, M.A., J. Kumi-Diaka, and J. Brown. 2005. Anticancer activities of pomegranate extracts and genistein in human breast cancer cells. *J Med Food* 8: 469–75.

Jung, H.W., H.Y. Son, C.V. Minh, Y.H. Kim, and Y.K. Park. 2008. Methanol extract of *Ficus* leaf inhibits the production of nitric oxide and pro-inflammatory cytokines in LPS-stimulated microglia via the MAPK pathway. *Phytother Res*. 22: 1064–9.

Kawaii, S., and E.P. Lansky. 2004. Differentiation-promoting activity of pomegranate (*Punica granatum*) fruit extracts in HL-60 human promyelocytic leukemia cells. *J Med Food* 7: 13–8.

Khan, M.S.Y., A.A. Siddiqui, and K. Javed. 1990. Chemical investigation of the leaves of *Ficus hispida*. *Indian J Nat Prod*. 6: 14–15.

Kim, J.M., and H.S. Yun-Choi. 2008. Anti-platelet effects of flavonoids and flavonoid-glyco-sides from *Sophora japonica*. *Arch Pharm Res*. 31: 886–90.

Kitajima, J., M. Arai, and Y. Tanaka. 1994. Triterpenoid constituents of *Ficus thunbergii*. *Chem Pharm Bull (Tokyo)* 42: 608–10.

Kuete, V., B. Ngameni, C.C. Simo et al. 2008. Antimicrobial activity of the crude extracts and compounds from *Ficus chlamydocarpa* and *Ficus cordata* (Moraceae). *J Ethnopharmacol*. 120: 17–24.

Kumi-Diaka, J., S. Saddler-Shawnette, A. Aller, and J. Brown. 2004. Potential mechanism of phytochemical-induced apoptosis in human prostate adenocarcinoma cells: Therapeutic synergy in genistein and beta-lapachone combination treatment. *Cancer Cell Int*. 4: 5.

Kuo, Y.H., and H.Y. Lin. 2004. Two novel triterpenes from the leaves of *Ficus microcarpa*. *Helvet Chim Acta* 87: 1071–6.

Lansky, E.P., G. Harrison, P. Froom, and W.G. Jiang. 2005. Pomegranate (*Punica granatum*) pure chemicals show possible synergistic inhibition of human PC-3 prostate cancer cell invasion across Matrigel. *Invest New Drugs* 23: 121–2, Erratum 23: 379.

Lansky, E.P., and R.A. Newman. 2007. *Punica granatum* (pomegranate) and its potential for prevention and treatment of inflammation and cancer. *J Ethnopharmacol*. 109: 177–206.

Lansky, E.P., H.M. Paavilainen, A.D. Pawlus, and R.A. Newman. 2008. *Ficus* spp. (fig): Ethno-botany and potential as anticancer and anti-inflammatory agents. *J Ethnopharmacol*. 119: 195–213.

Li, Y., J. Duan, T. Guo et al. 2009. In vivo pharmacokinetics comparisons of icariin, emodin and psoralen from Gan-kang granules and extracts of *Herba Epimedii*, Nepal dock root, *Ficus hirta* Vahl. *J Ethnopharmacol*. 124: 522–9.

Mandal, S.C., and C.K. Ashok Kumar. 2002. Studies on anti-diarrheal activity of *Ficus his-pida*. Leaf extract in rats. *Fitoterapia* 73: 663–7.

Mandal, S.C., T.K. Maity, J. Das, B.P. Saba, and M. Pal. 2000a. Anti-inflammatory evaluation of *Ficus racemosa* Linn. leaf extract. *J Ethnopharmacol*. 72: 87–92.

Mandal, S.C., B.P. Saha, and M. Pal. 2000b. Studies on antibacterial activity of *Ficus rac-emosa* Linn. leaf extract. *Phytother Res*. 14; 278–80.

Meng, Z., Y. Wang, J. Ji, and W. Zhong. 1996. Studies of chemical constituents of *Ficus carica* L. *Zhongguo Yaoke Daxue Xuebao* 27: 202–4.

Mertens-Talcott, S.U., S.T. Talcott, and S.S. Percival. 2003. Low concentrations of querce-tin and ellagic acid synergistically influence proliferation, cytotoxicity and apoptosis in MOLT-4 human leukemia cells. *J Nutr*. 133: 2669–74.

Michael, J.P. 2007. Indolizidine and quinolizidine alkaloids. *Nat Prod Res*. 24: 191–222.

Michael, J.P. 2008. Indolizidine and quinolizidine alkaloids. *Nat Prod Res*. 25: 139–65.

Nguyen, V.T., V.S. Tran, M.C. Nguyen, B.T. Nguyen, and T.H. Nguyen. 2002. [Study on the chemical constituents of *Ficus semicordata*]. *Tap Chi Hoa Hoc* 40: 69–71.

NIST Chemistry WebBook. 1996. [Online Database]. NIST Standard Reference Database Number 69. The National Institute of Standards and Technology (NIST). URL: http://webbook.nist.gov/chemistry/contact.html [accessed August 14, 2009].

Novak-Hofer, I., S. Cohrs, J. Grünberg et al. 2008. Antibodies directed against L1-CAM synergize with Genistein in inhibiting growth and survival pathways in SKOV3ip human ovarian cancer cells. *Cancer Lett*. 261: 193–204.

Parveen, M., R.M. Ghalib, S.H. Mehdi, S.Z. Rehman, and M. Ali. 2009. A new triterpenoid from the leaves of *Ficus benjamina* (var. *comosa*). *Nat Prod Res*. 23: 729–36.

Peraza-Sánchez, S.R., H.B. Chai, Y.G. Shin et al. 2002. Constituents of the leaves and twigs of *Ficus hispida*. *Planta Med*. 68, 186–8.

Perez, C., J.R. Canal, J.E. Campillo, A. Romero, and M.D. Torres. 1999. Hypotriglyceridaemic activity of *Ficus carica* leaves in experimental hypertriglyceridaemic rats. *Phytother Res*. 13: 188–91.

PubChem. The PubChem Project. 2001. [Online Database]. National Center for Biotechnology Information. U.S. National Library of Medicine. 8600 Rockville Pike, Bethesda, MD 20894. URL: http://pubchem.ncbi.nlm.nih.gov/ [accessed August 16, 2009].

Ragasa, C.Y., E. Juan, and J.A. Rideout. 1999. A triterpene from *Ficus pumila*. *J Asian Nat Prod Res*. 1: 269–75.

Rahuman, A.A., P. Venkatesan, K. Geetha, G. Gopalakrishnan, A. Bagavan, and C. Kamaraj. 2008. Mosquito larvicidal activity of gluanol acetate, a tetracyclic triterpenes derived from *Ficus racemosa* Linn. *Parasitol Res*. 103: 333–9.

Saeed, M.A., and A.W. Sabir. 2002. Irritant potential of triterpenoids from *Ficus carica* leaves. *Fitoterapia* 73: 417–20.

Schmidt, F., C.B. Knobbe, B. Frank, H. Wolburg, and M. Weller. 2008. The topoisomerase II inhibitor, genistein, induces G2/M arrest and apoptosis in human malignant glioma cell lines. *Oncol Rep*. 19: 1061–6.

Sen, A.B., and A.R. Chowdhury. 1971. Chemical investigation of *Ficus glomerata*. *J Indian Chem Soc*. 48: 1165–9.

Serraclara, A., F. Hawkins, C. Perez, E. Dominguez, J.E. Campillo, and M.D. Torres. 1998. Hypoglycemic action of an oral fig-leaf decoction in type-I diabetic patients. *Diabetes Res Clin Pract*. 39: 19–22.

Sowemimo, A.A., F.A. Fakoya, I. Awopetu, O.R. Omobuwajo, and S.A. Adesanya. 2007. Toxicity and mutagenic activity of some selected Nigerian plants. *J Ethnopharmacol*. 113: 427–32.

Sulaiman, M.R., M.K. Hussain, Z.A. Zakaria et al. 2008. Evaluation of the antinociceptive activity of *Ficus deltoidea* aqueous extract. *Fitoterapia* 79: 557–61.

Teixeira, D.M., R.F. Patão, A.V. Coelho, and C.T. da Costa. 2006. Comparison between sample disruption methods and solid–liquid extraction (SLE) to extract phenolic compounds from *Ficus carica* leaves. *J Chromatogr A* 1103: 22–8.

Tsai, I.L., J.H. Chen, C.Y. Duh, and I.S. Chen. 2000. Chemical constituents from the leaves of formosan *Ficus septica*. *Chin Pharm J (Taipei)* 52: 195–201.

Vaya, J., and S. Mahmood. 2006. Flavonoid content in leaf extracts of the fig (*Ficus carica* L.), carob (*Ceratonia siliqua* L.) and pistachio (*Pistacia lentiscus* L.). *Biofactors* 28: 169–75.

Wang, G., H. Wang, Y. Song, C. Jia, Z. Wang, and H. Xu. 2004. [Studies on anti-HSV effect of *Ficus carica* leaves]. *Zhong Yao Cai* [Journal of Chinese Medicinal Materials] 27: 754–6.

Yan, J., D.J. Allendorf, B. Li, R. Yan, R. Hansen, and R. Donev. 2008. The role of membrane complement regulatory proteins in cancer immunotherapy. *Adv Exp Med Biol.* 632: 159–74.

Yang, C.W., W.L. Chen, P.L. Wu, H.Y. Tseng, and S.J. Lee. 2006. Anti-inflammatory mechanisms of phenanthroindolizidine alkaloids. *Mol Pharmacol.* 69: 749–58.

Yang, X.W., B. Xu, F.X. Ran, R.Q. Wang, J. Wu, and J.R. Cui. 2007. [Inhibitory effects of 11 coumarin compounds against growth of human bladder carcinoma cell line E-J in vitro]. *Zhong Xi Yi Jie He Xue Bao* 5: 56–60.

Yarmolinsky, L., M. Zaccai, S. Ben-Shabat, D. Mills, and M. Huleihel. 2009. Antiviral activity of ethanol extracts of *Ficus binjamina* and *Lilium candidum* in vitro. *N Biotechnol.* August 21, 2009. [Epub ahead of print]

Yoon, W.J., H.J. Lee, G.J. Kang, H.K. Kang, and E.S. Yoo. 2007. Inhibitory effects of *Ficus erecta* leaves on osteoporotic factors in vitro. *Arch Pharm Res.* 30: 43–9.

Zand, R.S., D.J. Jenkins, and E.P. Diamandis. 2000. Steroid hormone activity of flavonoids and related compounds. *Breast Cancer Res Treat.* 62: 35–49.

Zhang, L., H.Z. Qu, L. Huang, S.S. Du, D.Y. Hao, and X.P. Wang. 2008. Separation, purification and properties of superoxide dismutase from *Ficus carica* leaves. *Gaodeng Xuexiao Huaxue Xuebao* 29: 1588–91.

Zhang, H.J., P.A. Tamez, Z. Aydogmus et al. 2002. Antimalarial agents from plants. III. Trichothecenes from *Ficus fistulosa* and *Rhaphidophora decursiva*. *Planta Med.* 68: 1088–91.

Zhao, P., H. Zhou, and Y. Wu. 2004. Chemical component analysis of *Ficus carica* L. leaf. *Zhong Cao Yao* 35: 1341–2.

5 Latex

Species of *Ficus*, similar to those of many other higher plant genera, produce a vascular liquid with rubber-like qualities within their vasculatures known as *latex*. Plants again use these complexes for their own purposes, and in the case of latex the purposes undoubtedly are connected with defense, wound healing, and innumerable other circumstances where a flowing inner material could help control and maintain respiration and metabolism.

FIGURE 5.1 *Ficus carica* branch, with the hidden latex flowing within (charcoal by Zipora Lansky).

211

Ficus latex naturally follows our discussion of *Ficus* leaves in Chapter 4, but also of *Ficus* fruits in Chapter 3 since the fruits, when they are immature and green, also contain latex. As the fruit ripens, the latex within it seems to recede or be consumed, but latex lives on in the leaves and stems, and probably also the roots. Latex from immature *Ficus* fruits, leaves, twigs, branches, and stems is known to medicine.

The latex of *Ficus* is notable for the extent of its medical employment through the millennia. This will be summarized in the ethnographic table. However, according to Dioscorides and Pliny, the method of collection is to take the juice from branches of the wild fig tree when they are sapful, before buds appear, by pounding and pressing. Juice is dried in the shade and stored. Traditional uses by indigenous peoples worldwide include external treatments for warts, boils, and dermatitis as well as external applications for rheumatic pain and ophthalmia (inflammation of the eye; Lansky et al. 2008).

Peruvian shamans consider the latex of *Ficus* to be an extremely powerful plant teacher, one that can help teach one medicine and promote the ability to travel "under water," in the spiritual realm. The training is conducted by the aspirant in conjunction with a program of continence and a special diet consisting primarily of plantains and fish, eschewing white sugar, white flour, meat of many different animals, and with numerous other dietary restrictions. The aspirant also practices concurrently with this program a series of chants, in conjunction with the use of *ayahuasca* enhanced by various other "plant teachers." Of the 16 plant teachers used by the shamans, *Ficus* latex, known by them as *renaco*, is considered third in importance, after tobacco and *ayahuasca* (mixture of *Banisteriopsis caapi*, *Psychotria viridis*, and other species) (Luna 1984a,b). Further details of how the shamans intake *renaco*, dosage, etc., are not available at the present time.

HISTORICAL USES OF *FICUS* LATEX

The following tables give an overview of the historical medicinal uses of the latex of different *Ficus* species. For references used in Table 5.2, see Chapter 3, Table 3.6. The scoring system used in Table 5.2 is based on the number of times a reference discusses the use of *Ficus* latex for the purpose indicated. If the number of citations is regular, the reference is given in regular type and scored as "1." If the number of citations is strong, the reference is given in *italic* and scored as "2." If the number of citations is strongest, the reference is given in **bold** and scored as "3." The final score in the last column is the sum of these scores for the individual references. The particular kinds of latex used are starred, but this information does not enter into the calculation of the total score for the function. The indications are shown in the order of importance based on their score values.

CHEMISTRY AND PHARMACOLOGY OF FIG LATEX

Ficus latex is rich in a group of proteolytic enzymes (Gonashvili 1964) known as *ficins* (or *faicins*). This fact, which has been appreciated for over a hundred years, has resulted in some industrial use with promise of many more uses to come (Gaughran

TABLE 5.1
Historical Uses of *Ficus* spp. Latex: Formulas

How to Prepare and Store Latex for Internal and External Applications

Flows from young leaves when broken (Bock 1964); fruit and leaves of cultivated trees yield
  Euphorbia-like white latex if broken before fruits are ripe (Gerard 1633)

Latex storage: Juice taken from branches of the wild fig tree when sapful, before buds appear, by
  pounding and pressing. Juice is dried in the shade and stored (Dioscorides 1902, Pliny 1967–1970)

The latex is taken from the tree at the beginning of spring, before it brings forth fruit, by breaking the
  outside of the bark with a stone (if it is broken deeper, no latex comes out). The drops are gathered in
  a sponge or wool, dried, formed (into pills), and stored in earthenware jars (Dioscorides 1902)

Variations	Notes
Latex	Dropped into wounds caused by the bite of a rabid dog or a scorpion sting; on stings of hornets and wasps; removes warts
	Internally removes kidney stones
	Applied both internally and externally for serpent stings, hard swellings of spleen, and stomachache
	Applied externally, warm, for stomach pain
	Applied in a piece of cotton or wool on the tooth for the relief of toothache
	When rubbed on the spine, stops malarial shivering
Mixed with honey	Removes dimness of eyes; for the beginning stages of catarrh
Mixed with fenugreek flour	As a poultice for gout, leprous sores, itch, and freckles; for stings of venomous animals and dog bites
Mixed with fenugreek flour and vinegar	As a poultice for gout; a depilatory; heals eruptions of eyelids
Mixed with barley flour	Externally for leprosy, sunburns, itch, roughness of skin, skin inflammations, sores, smallpox pustules, measles, freckles, and vitiligo
Mixed with wheat	Externally for scabies, sunburns, and marks on the face
Mixed with grease or wax	Externally, removes warts when spread around them
	As a vaginal suppository, emmenagogue
Mixed with starch	Internally (as a drink) for hardenings of womb, as a laxative and an emmenagogue
Mixed with egg yolk	As a vaginal suppository, emmenagogue
Mixed with crushed bitter almonds	When taken as a drink, opens and relaxes the womb; laxative
Mixed with mustard	Ear drops for earache, tinnitus, and itching of ears; improves loss of hearing

1976). This is undoubtedly behind much of the potent activity of the latex for dissolving neoplastic tumors, especially cutaneous ones such as warts. These enzymes, which have the ability to dissolve proteins, act as a chemical debridement for external tumors, and to benefit digestion insufficiency disorders (Bykov et al. 2000), bearing many similarities to papaya's papain and pineapple's bromelain (Huang et al. 1972, Murachi and Takashi 1970), and may also be utilized as meat tenderizers.

TABLE 5.2

Historical Uses of *Ficus* spp. Latex

Indication	Milk	Milk of Fruit	Milk of Unripe Fruit	Milk of Leaf	Milk of Branches/ Twigs	Milk of Tree	References	Score
Analgesic	*	*			*	*	1, *2*, **6**, 7, 11, 13, 14, **15**, 16, **18**, **20**	19
Against warts	*	*		*	*	*	1, 2, **6**, 11, 13, 14, **15**, 16, 18, 20	17
Laxative	*	*			*	*	*1*, 2, **6**, 7, 11, 13, *14*, **15**, 18, 20	17
Emmenagogue	*	*			*	*	*1*, 2, **6**, 7, 11, *14*, **15**, 16, 17, *18*	15
Against scabies	*	*				*	1, 2, *6*, *11*, *13*, 14, *15*, 16, 20	14
Against tetter	*	*		*		*	*1*, 2, 6, 11, **13**, 14, 15, 18	14
Coagulating milk	*	*		*		*	*1*, 6, 10, 11, 13, 14, *15*, 16, *18*, 20	13
Against poisonous animal bites	*				*	*	1, 2, 6, 11, 14, 16, **18**, **20**	12
Causes ulcers	*	*			*	*	**2**, *6*, 11, 14, *15*, **20**	12
Against scorpion sting	*	*				*	*1*, 2, 6, 7, 10, 11, *14*, *15*	11
Against toothache	*	*			*	*	1, 2, 6, 13, 14, *15*, 16, 18, 20	10
Against wounds	*					*	2, 6, 7, 11, 15, 17, *18*, 20	10
Opens the mouths of the veins (causing bleeding)	*	*		*	*	*	*6*, 11, 13, 14, **15**, 18, 20	10
Against rabid dog bite	*	*			*	*	14, *15*, 16, **18**, 20	9
Closing [wounds, etc.]	*					*	2, 6, 7, 11, 15, 17, *20*	8
Against freckles	*	*		*			1, 2, 6, 11, **18**	7
Against gout	*					*	*1*, 2, 6, 11, *14*	7
Against marks on skin	*					*	15, 16, **18**	7
Against scars	*	*				*	7, 13, *15*, *18*	7
Against ulcers in the head		*			*	*	*13*, **15**, 16, 20	7

TABLE 5.2 (continued)
Historical Uses of *Ficus* spp. Latex

Indication	Milk	Milk of Fruit	Milk of Unripe Fruit	Milk of Leaf	Milk of Branches/ Twigs	Milk of Tree	References	Score
Dissolving coagulated milk	*	*				*	10, 11, 13, 15, 16, 20	7
Against insect stings (wasp, hornet)	*					*	1, 14	6
Against leprosy	*	*		*	*		1, 14, 18, 20	6
Against putrid ulcers		*			*	*	13, 15, 20	6
Against swelling, tumors (general)	*					*	6, 7, 11, 15, 17	6
Against dandruff	*	*			*	*	2, 13, 15, 20	5
Against malarial shivering	*					*	2, 6, 7, 11, 15	5
Against snakebite	*					*	2, 6, 7, 11, 15	5
Against vitiligo	*				*		2, 6, 11, 20	5
Softening (in general)	*					*	2, 6, 7, 11, 15	5
Against all swellings/ tumors of spleen	*					*	6, 7, 11, 15	4
Against colic	*					*	2, 6, 11, 15	4
Against hard swelling, thickening of spleen	*					*	6, 7, 11, 15	4
Against hardness, closure of womb	*						2, 6, 11, 14	4
Against ordinary dog bite	*						1, 2, 6, 11	4
Against shrew mouse bite	*				*	*a	1, 7, 20	4
Opens the womb	*	*				*	1, 15, 16	4
Spicy, sharp	*		*	*	*		11, 20	4
Against allergic skin reactions		*		*	*		18, 20	3
Against skin problems		*			*	*	15, 20	3
Burning, stinging	*			*	*		6, 20	3
Causes intestinal ulcers				*	*	*	13, 20	3
Cleaning, absterging	*			*	*		6, 20	3
Cleans womb	*						6, 7, 11	3
Prevents eye cataract	*						6, 7, 17	3
Warming				*	*	*	6, 20	3
Against all swellings/ tumors of womb	*						6, 11	2
Against coagulated milk in stomach (poisoning)				*		*	18	2

(continued on next page)

TABLE 5.2 (continued)
Historical Uses of *Ficus* spp. Latex

Indication	Milk	Milk of Fruit	Milk of Unripe Fruit	Milk of Leaf	Milk of Branches/ Twigs	Milk of Tree	References	Score
Against hardness, hardenings of the womb	*						6, 11	2
Against marks in eyes	*						6	2
Against measles			*		*		18	2
Against scabies/eruptions in eyelids	*						1, 14	2
Against scurvy			*		*		18	2
Against smallpox			*		*		18	2
Against spreading ulcers			*		*		18	2
Bitter					*		20	2
Coagulating blood	*						7, 10	2
Corrects discoloring	*				*		14, 20	2
Corroding			*			*	15	2
Depilatory	*					*	1, 14	2
Dissolving	*					*	7, 11	2
Dissolving coagulated blood	*						7, 10	2
Hemostatic	*						7, 10	2
Kills parasitic worms in whole body	*						13, 15	2
Opening hemorrhoids	*						6, 18	2
Softens skin			*		*		18	2
Against "covering of the eye"	*						7	1
Against all hard swellings/tumors, hardenings						*	20	1
Against all swellings/ tumors of liver	*						2	1
Against chilblains	*						7	1
Against consumption of gypsum					*		20	1
Against deafness					*		20	1
Against ear itch					*		20	1
Against earache					*		20	1
Against headache						*	20	1
Against itch	*						2	1

TABLE 5.2 (continued)

Historical Uses of *Ficus* spp. Latex

Indication	Milk	Milk of Fruit	Milk of Unripe Fruit	Milk of Leaf	Milk of Branches/ Twigs	Milk of Tree	References	Score
Against kidney stones		*					7	1
Against morphea					*		20	1
Against plague						*	20	1
Against ringworm						*	16	1
Against skin inflammations	*						17	1
Against spleen disease						*	20	1
Against stomach pain	*						6	1
Against sunburns						*	16	1
Against swellings/tumors in muscles	*						7	1
Against thickness of lens	*						7	1
Against tinnitus					*		20	1
Against torsion of eye	*						6	1
Against tuberculosis, scrofula	*						7	1
Against urinary problems	*						7	1
Against weakness of sight	*						6	1
Against worry, obsession						*	20	1
Bad for stomach	*						13	1
Benefits body						*	20	1
Caustic		*					11	1
Cleans kidney	*						7	1
Cleans kidneys of materia [excluding stones]	*						7	1
Cleans ulcers						*	16	1
Diuretic	*						7	1
Good taste	*						1	1
Opening; against obstruction	*						2	1
Rarefying						*	6	1
Tenderizes meat (in cooking)	*						20	1

[a] From the breaking point of the leaf.

Enzymological investigations have focused on ficin's physical properties, isolation, and characterization, progress generally following advancement of the state of the art (El-Fekih and Kertesz 1968; Fadýloğlu 2001; Husain and Lowe 1970; Kramer and Whitaker 1964, 1969a; Li et al. 1981; Liener and Friedenson 1970; Porcelli 1967). Also, the role of ficin as an enzymatic catalyst has been explored relative to hydrolysis of α-*N*-benzoyl-l-arginine ethyl ester and α-*N*-benzoyl-l-argininamide (Kramer and Whitaker 1969b), as well as the internal plant metabolism from one ficin subtype to another (Kramer and Whitaker 1969c).

The latex also contains other pharmacologically important proteins, including one recently discovered in the "Inzhir tree" (*Ficus carica*) and found to have a molecular weight of 6481 and antifungal activity (Mavlonov et al. 2008). This activity is linked to the presence of at least three different chitinases in the ficin, enzymes that bind to the chitin in the fungal cell walls so that the enzyme can act to dissolve it (Taira et al. 2005). Refinement of these proteases and combining with cysteine and saline is claimed to restimulate the growth process in aged hens and protect birds from influenza (Park 2008). Ficin proteases also possess potent anthelmintic properties, the mechanism of action being related to an attack on the worm's cuticle (Stepek et al. 2005). Cysteine proteases from the latex of *F. virgata* were very toxic to lepidopteran larvae and led to the idea of their evolution being partly a defense to inactivate herbivorous insects (Konno et al. 2004).

Another component of the ficin recently purified from the latex of the Shandong fig trees in China is inhibited by heavy metal ions and promoted by l-cysteine, with a half-life at 65°C of over an hour. This purified ficin was shown to be a cysteine proteinase (Huang et al. 2008), such as papain and bromelain, occurring in papaya, pineapple, and kiwi. Also used in meat tenderizers (Bada and Kato 1987), cysteine proteases, including ficin, can induce hemostasis via Human Factor X (Richter et al. 2002) and may also prove extremely important in regulating malignancy, since protease-activated receptors are linked to tumor progression (Borensztajn and Spek 2008) and they figure prominently in hypercoagulation states as well as cancer (Ten Cate and Falanga 2008). One cysteine proteinase purified ficin, considered here as a single linear peptide chain (a purified single protein), isolated from the latex of *F. carica* (Devaraj et al. 2008a), and a related aspartic protease (that targets aspartate linkages rather than cysteine linkages in peptides) from the latex of *F. racemosa*, showed maximum activity at 60°C, with a midpoint of thermal inactivation of 70°C that demonstrated exceptional thermostability, relative to the ficin from *F. carica* (Devaraj et al. 2008b). Other earlier purification efforts of ficin yielded in one case from *F. carica* a three-protein complex that was optimally active at 50°C and pH 7 (Kim et al. 1986). In another case, a serine-centered (as opposed to cysteine-centered) protease was isolated from *F. elastica*. The enzyme was optimally active at pH 6 (Lynn and Clevette-Radford, 1986).

Proteolytic enzymes have also been isolated from the latex of *Ficus pumila* L. (Perello et al. 2000), *F. hispida* (Chetia et al. 1999), and *F. glabrata* (Jones and Glazer 1970, Malthouse and Brocklehurst 1976). Six discrete ficins were isolated from *F. glabrata*, and it was concluded that autodigestion of the ficins during collection and storage did not occur (Kortt et al. 1974). One ficin isolated from *F. carica* was most active at pH 8 and 60°C and contained a sugar moiety bound to the protein (Sugiura and Sasaki 1974), but was otherwise similar to the ficins earlier purified from the same species by the same team (Sugiura and Sasaki 1973). In a screening of the proteolytic activity of the latex from 24 latex-bearing plants, the highest activity was for *F. glomerulata* (990 units), with the second highest being for *Euphorbia tirucelli* (700 units) (Chary and Reddy 1983); and from a group consisting of *F. carica, Carica papaya, Cryptostegia grandiflora, Calotropis procera,* and *Calotropis gigantea, F. carica* had the highest proteolytic activity (measured as tyrosine release from casein) and an associated time of 45 min, while *Carica papaya* had one of 75 min, *Cryptostegia grandiflora* 120 min, *Calotropis procera* 150 min, and *Calotropis gigantea* 210 min (Pant and Srivastava 1966). Nevertheless, there is homology in amino acid sequence between the enzymes in *Ficus* latex and those of other latex-producing plant species such as *Hevea brasiliensis* (Tata et al. 1976). In an earlier comparative study of the proteolytic activity of the latex from 46 different *Ficus* species, only 13 species showed appreciable activity, with the highest from the latex of *F. stenocarpa* followed closely by the latexes of *F. carica* and *F. glabrata* (Williams et al. 1968). Other work looked at two different proteins within (presumably) *F. carica* ficin and found only one that effects protein hydrolysis (Brovetto and Porcelli 1967, Porcelli 1967). Diversity in the number of proteolytic components among the latexes of 10 varieties of *F. carica* was observed. Ten active components were identified, all of which occurred in variety Kadota. Brown Turkey, Stanford, and Adriatic had 8 components each; King, 7; Beall, Black Mission, Conadria, 6; and Calimyrna, Blanquette, 4. The influence of genetic and ecological factors was considered in the evolution of the diversity (Sgarbieri 1965).

Preparation of latex for pharmaceutical purposes, at least isolation of the proteolytic enzyme complex, may be facilitated by lowering viscosity to 1/6 of the original, allowing easier subsequent filtration and centrifugation, by raising the pH of the natural *Ficus* latex from 4 to 6 or 7 (Walti 1954). Diversity of the latex among 92 species of *Ficus* in terms of color, rubber content, and toxicity has been charted (de Wildeman 1949). The protease action of ficin, as well as of papain, may be activated with sodium thiosulfate, hydrogen cyanide, or cysteine (Jaffe 1945). The latex of *F. elastica* is stable over time and does not putrefy, but eventually creams and finally develops an irreversible clot (Daniel et al. 1938, Stevens 1939), and this clotting is more likely as temperature increases (Cornish and Brichta 2002). Stabilization of the ficin crystal from *Ficus* latex can be accomplished by filtration in an acidulated aqueous solution of 0.001 N HCl, freezing the wet crystals thus obtained, and removing the frozen solvent under high vacuum (Major and Walti 1939). *Ficus* latex has long been known to break down proteins under both acid and alkaline conditions, though the process is much more efficient under mild acids in some situations and, more recently, it has been recognized also in neutral conditions (Devaraj et al. 2008b). *Ficus* latex has a similar digestive effect on starches, and also coagulates milk. Treatment under too high amounts of HCl will inactivate the proteolysis, though sodium carbonate, which inactivates pepsin, does not (Hansen 1886).

The modern use of employing *Ficus* latex in a transdermal patch, ointment, or cream gel for broad anti-inflammatory action has been described (Santana 2008). Compounds other than proteolytic enzymes that probably occur in *Ficus* latex (as well as in the latex of many other plants) include amino acids, vitamins, carbohydrates, lipids, cardiac glycosides, and possibly also alkaloids (Shukla and Krishna Murti 1971). Other enzymes in the latex include diastase, esterase, and lipase (Sapozhnikova 1940). For collecting ficin from *F. carica* in Khorasan province of Iran, the ideal time for collection was found to be the fall, the ideal organ of the tree the branch ends, calcium ion solutions were found to help the ficin to develop in situ, and the ficin was found to be 50–100 times more active as a protease than papain (Rajabi et al. 2006). Purified, enzymatically active rubber particles from *F. elastica* are able to produce rubber (a component of the latex on the order of 25% of the total latex from this species) in vitro in the presence of a chemical initiator and a divalent magnesium cofactor (Espy et al. 2006). A crude ficin was also produced from cultures of *F. carica* cells (calluses), though the protease differed from that of the latex from figs grown on trees (Nassar and Newbury 1987). Near-infrared (NIR) spectroscopy has been successfully applied to the quantification of latex in samples of *F. elastica* (Cornish et al. 2004).

Production of latex by *F. elastica* has been proved to be affected by both plant hormones and abiotic stress such as wounding, in that latex-producing genes are induced following injury (Kim et al. 2003). The divalent metal cofactor contained within the latex is an important element in this regulation (Kang et al. 2000).

Other compounds have been isolated from fig latex and found to inhibit proliferation of cancer cells. Among these are a class of 6-O-acyl-β-d-glucosyl-β-sitosterols discovered by Mechoulam and colleagues (Mechoulam et al. 2002, Rubnov et al. 2001). In this series, the acyl group (R) may consist of palmitoyl, linoleyl stearyl or oleyl moieties as shown in Figure 5.3, with the palmitoyl and linoleyl groups the most prevalent. A more recent study (Wang et al. 2008) showed the complex fresh

FIGURE 5.2   *Ficus elastica* (27.7.2009, Binyamina, Israel, by Helena Paavilainen).

FICIN

Ficin, sometimes also written Ficain, is the name given for a hundred years to the more-or-less purified protein segment of the *Ficus* latex. This ficin is a potent cysteine proteinase. It catalyzes the catalysis of proteins and peptides, and its active site on the amino acid chain is occupied by a cysteine molecule. Cysteine proteinases also include bromelain and papain, and occur in papayas, pineapple, and kiwi. Ficin is much more potent as a cysteine proteinase, but much less studied. Still, it is an article of commerce as a food additive for the food-manufacturing industry.

Modern work (Devaraj et al. 2008 a,b) has purified a single peptide within the mix of both "active" and "inactive" proteins and peptides that make up the classical ficin, and the authors have called this single peptide, which apparently has exceptional activity relative to other compounds in the mix, also "Ficin." So, depending on your temporal frame of reference, *Ficin* may refer either to a mix of peptides including some cysteine proteinases, likely other proteinases with other amino acids at their target sites, and some "inactive" compounds that may nevertheless contribute in some ways, possibly still mysterious, to the activity, or it may refer to a highly purified single enzyme, Ficin. Single-molecule Ficin is not necessarily, or rather, it almost certainly is not, more potent than a natural cysteine protease complex to which the pure compound belongs.

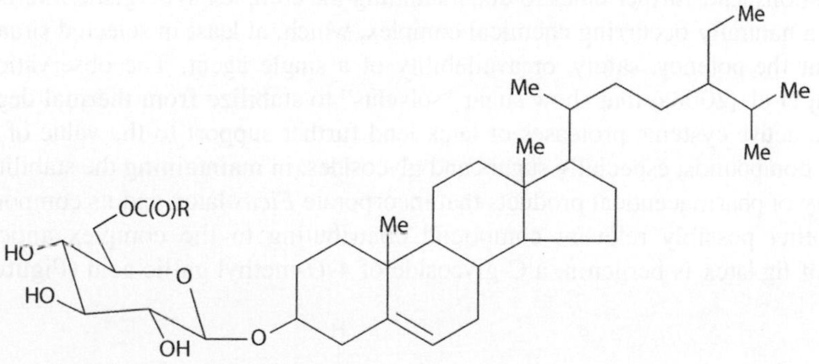

FIGURE 5.3 6-*O*-acyl-β-d-glucosyl-β-sitosterol isolated from latex of *Ficus carica*. (Reprinted with permission from Mechoulam et al. 2002. 6-*O*-acyl-β-d-glucosyl-β-sitosterols from fig (*Ficus carica*) resin: Isolation, structure elucidation, synthesis and their use as antiproliferative agents. PCT Int. Patent Application WO 2002-IB918 20020326, 31 pp, 2002.) [I; R = $(CH_2)_{14}Me$, $(CH_2)_{16}Me$, $(CH_2)_7CH:CH(CH_2)_7Me$, $(CH_2)_7CH:CHCH_2CH:CH(CH_2)_4Me$].

fig latex from *F. carica* to have significant effects in inducing apoptosis as measured by staining with a mixture of 100 µg/mL acridine orange and 100 µg/mL ethidium bromide in human *glioma multiforme* and hepatocellular carcinoma cells, while normal human fetal liver cells were unaffected. The effect, based on the values obtained by Rubnov et al. (2001), was stronger than that obtained from the pure compounds, suggesting that these sitosterol glycosides contribute to, but are not solely responsible for, the anticancer effect of the latex. Wang et al. (2008) also succeeded in linking the proapoptotic and antiproliferative effect of the latex in the cancer cells to disruption of cell cycle and colony formation, again pointing to a multifactorial mechanism of cancer suppression, which, nonetheless, is highly specific to malignant species, sparing normal cells.

Two triterpenoid compounds from the latex of *Ficus sur* L. were identified as pentacyclic structures of the oleanane and ursene types (Feleke and Brehane 2005). Oleanane terpenoids (Deeb et al. 2009) and their saponins (Gauthier et al. 2009) are cytotoxic to cancer cells through targeting of prosurvival Akt and mTOR pathways. Significant cytotoxicity by an ursene triterpenoid saponin has also recently been described in MDA-MB-231 estrogen receptor negative human breast cancer cells in vitro (Calabria et al. 2008).

Coumarins, α-amyrin and bergapten, were found in the latex of *F. sycomorus*, *F. benjamina*, and *F. elastica*, with imperatorin found only in *F. sycomorus* and *F. benjamina*, and xanthotoxin only in *F. sycomorus* (Abdel-Wahab et al. 1989). Bergapten is a powerful inducer of differentiation and γ-globin gene expression in erythroid cells (Guerrini et al. 2009), differentiation being an important mechanism for recovery from cancer (Kawaii and Lansky 2004). Amyrins occur in other settings with triterpenoids, including a lipophilic extract from *Viscum album* (Urech et al. 2005), a methanolic extract of *Celastrus rosthornianus* (Wang 2007), and the Korean folk medicine *Dendropanax morbifera* Leveille (Araliaceae) (Chung et al. 2009), and in all cases suggestively augment the effect of more active triterpenoids. All of these observations lend further clues to understanding the complex synergistic interactions within a naturally occurring chemical complex, which, at least in selected situations, augment the potency, safety, or availability of a single agent. The observations of Devaraj et al. (2008a) that show sugar "solvents" to stabilize from thermal degradation the active cysteine proteases of latex lend further support to the value of many "inert" compounds, especially sugars and glycosides, in maintaining the stability and integrity of pharmaceutical products that incorporate *Ficus* latex and its components.

Another possibly relevant compound contributing to the complex anticancer effect of fig latex is bergenin, a C-glycoside of 4-*O*-methyl gallic acid (Figure 5.4),

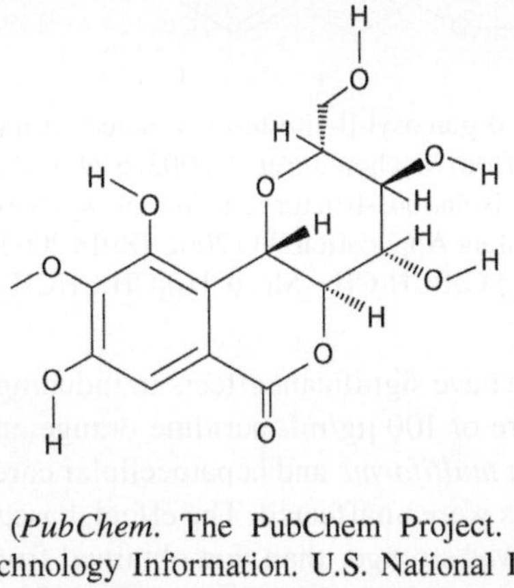

FIGURE 5.4 Bergenin. (*PubChem*. The PubChem Project. 2001. [Online Database]. National Center for Biotechnology Information. U.S. National Library of Medicine. 8600 Rockville Pike, Bethesda, MD 20894. URL: http://pubchem.ncbi.nlm.nih.gov/ [accessed August 16, 2009]).

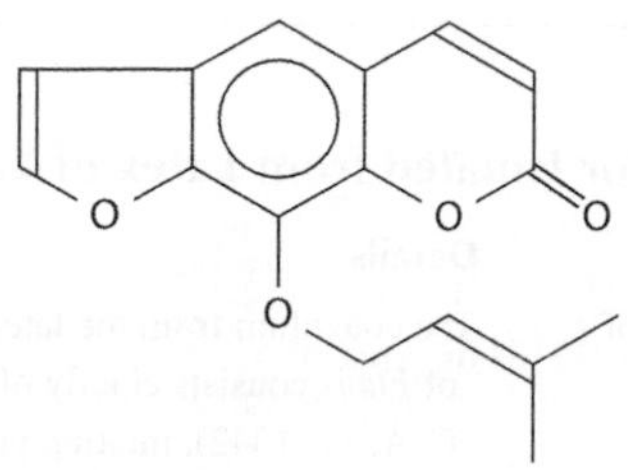

FIGURE 5.5 Imperatorin. (*NIST Chemistry WebBook*. 1996. [Online Database]. NIST Standard Reference Database Number 69. The National Institute of Standards and Technology (NIST). URL: http://webbook.nist.gov/chemistry/contact.html [accessed August 14, 2009].)

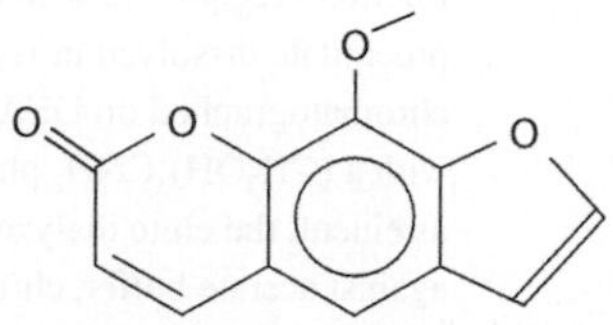

FIGURE 5.6 Xanthotoxin. (*NIST Chemistry WebBook*. 1996. [Online Database]. NIST Standard Reference Database Number 69. The National Institute of Standards and Technology (NIST). URL: http://webbook.nist.gov/chemistry/contact.html [accessed August 14, 2009].)

isolated, along with lupeol acetate and β-sitosterol, from the latex of *Ficus glomerata* Roxb. (Hai et al. 1991). Bergenin at oral doses of 5, 10, 20, 40, and 80 mg/kg per dose dependently inhibited production of pro-inflammatory Th1 cytokines (IL-2, IFN-γ, and TNF-α) while potentiating anti-inflammatory Th2 cytokines (IL-4 and IL-5) in the peripheral blood of adjuvant-induced arthritic balb/c mice, suggesting an overall anti-inflammatory effect based on rebalancing of pro-inflammatory and anti-inflammatory mediators (Nazir et al. 2007), given the pivotal role inflammation plays in cancer progression (Lansky and Newman 2007, Lansky et al. 2008). Bergenin, which exhibits marked antioxidant activity as well (De Abreu et al. 2008), is presently being touted as a weight loss solution and is the subject of intensive ongoing investigation regarding its intestinal absorption and pharmacokinetics (Wang et al. 2009, Yu et al. 2009).

HYPERTENSION

Other possible drugs from *Ficus* latex are peptides/enzymes and small secondary metabolites, such as coumarins, flavonoids, sterols, and terpenoids. Examples from the first group from the latex of *F. carica* are peptides that are inhibitors of angiotensin I-converting enzyme (ACE; Maruyama et al. 1989), an important target in hypertension. Petrov et al. (2000) showed one type of *protease inhibitor*, a leucine aminopeptidase inhibitor, bestatin, to stimulate activity of ACE in T-lymphocytes, while the serine protease inhibitor leupeptin inhibited its activity in a dose-dependent manner. The chymase inhibitor, chymostatin, was without effect.

TABLE 5.3
Chemicals Identified and/or Isolated from Latex of Various *Ficus* spp.

Ficus Species	Compounds	Details	Reference
alba	Wax consisting of stearic acid, β-amyrin, and lupeol (free and as the acetate, lupeol also as benzoate)	The coagulum from the latex of varieties of *Ficus* consists chiefly of a wax (cf. C. A. 17, 1342), melting point approximately 60°C. By hydrolysis, this yields stearic acid, β-amyrin, and lupeol.	Ultee 1922b
anthelminthica	Ficin; tryptophan	*F. anthelminthica* latex was extracted with 0.01 N HCl, dialyzed, proteolytic fraction precipitated with $(NH_4)_2SO_4$, precipitate dissolved in H_2O, the solution chromatographed on DEAE-cellulose with a $(CH_2OH)_3CNH_2$ phosphate buffer as eluent, the elute dialyzed in vacuo against acetate buffer, chromatographed on CM-cellulose with an acetate buffer as eluent, the eluate concentrated by dialysis or lyophilization, and the concentrated solution allowed to stand at pH 5 to give crystallized ficin.	Bettolo et al. 1963
benghalensis	Rubber (17%)	13C NMR analysis of samples from latex of *F. benghalensis* prepared by successive extractions with acetone and benzene confirmed that the benzene-soluble residues were natural rubber, cis-1,4-polyisoprene. The rubber content in the latex was approximately 17%. Gel permeation chromatography revealed that the molecular mass of the natural rubber from *F. benghalensis* was approximately 1500 kDa.	Kang et al. 2000
benghalensis	12% Rubber; uncrystallizable steroids; α-amyrin acetate; traces of wax containing stearic acid	Fresh latex containing 12% rubber was coagulated with EtOH. Uncrystallizable steroids, α-amyrin acetate (characterized), and traces of wax containing stearic acid were found in the coagulum.	Santhakumari and Pillay 1959
benjamina	Coumarins: α-amyrin, bergapten, imperatorin	Coumarins were isolated from the latex of *F. benjamina* by silica gel chromatography.	Abdel-Wahab et al. 1989

TABLE 5.3 (continued)
Chemicals Identified and/or Isolated from Latex of Various *Ficus* spp.

Ficus Species	Compounds	Details	Reference
carica	Mixture of 6-*O*-acyl-β-d-glucosyl-β-sitosterols, the acyl moeity being primarily palmitoyl and linoleyl with minor amounts of stearyl and oleyl	A mixture of 6-*O*-acyl-β-d-glucosyl-β-sitosterols, the acyl moeity being primarily palmitoyl and linoleyl with minor amounts of stearyl and oleyl, was isolated from *F. carica* latex, and its identity was established by spectroscopic methods (NMR, MS) and confirmed by chemical synthesis.	Rubnov et al. 2001
carica	Ficin		Nassar and Newbury 1987
carica	Ficin	Chromatographic separation of ficin derived from *Ficus carica* yielded six proteolytic fractions with a different specificity toward FX. Ficin from *F. carica* latex was found to be heterogeneous by electrophoresis, $(NH_4)_2SO_4$ and EtOH precipitation, and chromatography on carboxymethyl cellulose at pH 7.0. The number and relative amounts of the active components of *F. carica* varied among the varieties studied. *F. carica* var. *Kadota* and var. *Calimyrna* had 10 and 4 proteolytically active components, respectively.	Richter et al. 2002, Sgarbieri et al. 1964
carica	Ficins C and D	Ten proteolytic enzyme components were separated from *F. carica* var. *Kadota* latex by chromatography on carboxymethyl cellulose. Ficins C and D were crystallized. All 10 components were able to act on casein, α-benzoyl-l-argininamide, and milk, but there were large relative differences among the components.	Kramer and Whitaker 1964

(continued on next page)

TABLE 5.3 (continued)

Chemicals Identified and/or Isolated from Latex of Various *Ficus* spp.

Ficus Species	Compounds	Details	Reference
carica var. *Horaishi*	Ficins A, B, C, and D; ficin S (sugar-containing proteinase)	A sugar-containing proteinase, Ficin S, from *F. carica* var. *Horaishi* was purified by CM-cellulose and CM-Sephadex C-50 and crystallized. The purified Ficin S was electrophoretically homogeneous. The sugar content of Ficin S was determined to be 4.8% by the phenol-H_2SO_4 method. The enzyme was most active at pH 8.0 and 60°C, and stable over a pH range 2.0–8.0 at 4° for 20 h and below 60°C for 30 min. The enzyme was activated by cysteine and mercapto-ethanol, but inhibited especially by $HgCl_2$ and *p*-chlomercuribenzoate. Ficin S differs only in isoelectric point and sugar content from Ficin A, B, C, and D. Ficins A, B, C, and D from the latex of *F. carica* var. *Horaishi* were similar in sedimentation constant, partial specific volume, specific rotation, extinction coefficient, and intrinsic viscosity, but differed in isoelectric point. Each of these enzymes apparently consisted of a single polypeptide chain with an N-terminal leucine. They differed in amino acid composition, particularly in the content of lysine, threonine, valine, and leucine, and Ficin B and C contained one buried SH group.	Sugiura and Sasaki 1973, 1974
carica	Enzymes: diastase, esterase, lipase, and protease		Sapozhnikova 1940
carica	Proteolytic components	Ficin from *F. carica* latex was found to be heterogeneous by electrophoresis, $(NH_4)_2SO_4$ and EtOH precipitation, and chromatography on carboxymethyl cellulose at pH 7.0. The number and relative amounts of the active components of *F. carica* varied among the varieties studied. *F. carica* var. *Kadota* and var. *Calimyrna* had 10 and 4 proteolytically active components, respectively.	Sgarbieri et al. 1964, Kramer and Whitaker 1969b

TABLE 5.3 (continued)
Chemicals Identified and/or Isolated from Latex of Various *Ficus* spp.

Ficus Species	Compounds	Details	Reference
carica (continued)		The effect of pH on the hydrolysis of $N\alpha$-benzoyl-l-arginine Et ester (BAEE) and $N\alpha$-benzoyl-l-argininamide (BAA) by a proteolytic enzyme component purified from *F. carica* var. *Kadota* latex was studied over pH 3–9.5. The pH optimum is 6.5 for both substrates. Km (app) values for BAEE and BAA were 3.32×10^{-2}M and 6.03×10^{-2}M, respectively, over the pH range 3.9–8.0.	
carica	Peroxidase (POX); trypsin inhibitor (TRI); basic class I chitinase (CHI)	Wounding treatment strongly induced the expression of the three stress-related genes in *F. carica*, peroxidase (POX), trypsin inhibitor (TRI), and a basic class I chitinase (CHI). Different expression of the stress-related genes following various abiotic stress or plant hormone treatments suggests that a cross talk exists between the signal transduction pathways elicited by abiotic stresses and hormones in plants.	Kim et al. 2003
coronata	Lipase; proteolytic diastase		Gerber 1913
domestica	Anticoagulant substance	The latex prevented blood clotting at a dilution of 0.1% in vitro.	Echave 1954
elastica	α-Amyrine; small amounts of β-amyrine; lupeol	*F. elastica* latex, cleaned by brief centrifuging, had d27 0.950 and total solids 36.11%. Crepe prepared from this latex showed 5.97% acetone extract (4.85 unsaponifiable), 0.18 alcohol KOH extract, 0.31 aqueous extract, 0.30 ash, 0.11 N, and 0.0008 Cu. The unsaponifiable part of the acetone extract contained chiefly α-amyrine with small amounts of β-amyrine and lupeol.	van der Bie 1946
elastica	Coumarins: α-amyrin; bergapten	Coumarins were isolated from the latex of *F. elastica* by silica gel chromatography.	Abdel-Wahab et al. 1989

(continued on next page)

TABLE 5.3 (continued)
Chemicals Identified and/or Isolated from Latex of Various *Ficus* spp.

Ficus Species	Compounds	Details	Reference
elastica	Ficin E (a serine-centered protease)	A protease of molecular weight 50,000 was purified from the latex of *F. elastica* by BioGel P chromatography and HPLC. The enzyme had a pI (isoelectric point) of 3.7, and optimum pH was 6.0. Its activity was dependent on serine and histidine residues.	Lynn and Clevette-Radford 1986
fulva	Wax containing stearic acid; a little rubber	The latex contains large quantities of a wax, which, upon hydrolysis, yields stearic acid.	Ultee 1922a
glabella	Ester of cinnamic acid		Ultee 1949
glabrata	Ficin; nine proteolytic components	Ficin from *F. glabrata* latex was found to be heterogeneous by electrophoresis, $(NH_4)_2SO_4$ and EtOH precipitation, and chromatography on carboxymethyl cellulose at pH 7.0. By chromatography on carboxymethyl cellulose, *F. glabrata* latex was found to contain nine components with proteolytic activity. Crystallized ficin was found to be the largely active Component G of the latex.	Sgarbieri et al. 1964
glabrata	Ficin: a mixture of ficins I–IV and ficin G, in which ficins II and III predominated	Fully active ficin (EC 3.4.22.3) containing 1 mol of thiol with reactivity toward 2,2′-dipyridyl disulfide at pH 4.5 per mol of protein, was prepared from the dried latex of *F. glabrata* by covalent chromatography on a Sepharose-glutathione-2-pyridyl disulfide gel. The enzyme preparation was a mixture of ficins I–IV and ficin G, in which ficins II and III predominated.	Malthouse and Brocklehurst 1976
glabrata	Five sulfhydryl endopeptidases, one of them corresponding to ficin	Five sulfhydryl endopeptidases were purified to homogeneity from commercial preparations of *F. glabrata* latex. Each of these enzymes appeared to consist of a single polypeptide chain, of 25,000 to 26,000 molecular weight, with amino-terminal leucine. They displayed very similar specificity and kinetic properties toward the B chain of oxidized insulin, casein, and synthetic ester substrates.	Jones and Glazer 1970

TABLE 5.3 (continued)
Chemicals Identified and/or Isolated from Latex of Various *Ficus* spp.

Ficus Species	Compounds	Details	Reference
glabrata	N-acetyl-β-d-hexosaminidase	N-acetyl-β-d-hexosaminidase (EC 3.2.1.52), active against both p-nitrophenyl-N-acetyl-β-d-glucosaminide and p-nitrophenyl-N-acetyl-β-d-galactosaminide, is present in latex of *F. glabrata*. Substrate competition, competitive inhibition studies, and Arrhenius plots confirm that, in the hexosaminidase, only one kind of active site is responsible for both activities. This active site may bind the N-Ac moiety of the substrate, with a hydrophobic interaction of the Me group involved.	Orlacchio et al. 1985
glomerata	Bergenin I; lupeol acetate; β-sitosterol		Hai et al. 1991
platyphylla	The prepared latex of the fruit contained aspartate, hydroxyproline, tryptophan, tyrosine, arginine, lysine, alanine, histidine, and two unidentified acids.		El-Sayyad et al. 1986
racemosa	An aspartic protease (ficin)	A protease from the latex of *F. racemosa* was purified to homogeneity and was a single polypeptide chain of approximately 44.5 kDa molecular weight as determined by MALDI-TOF. The enzyme exhibited a broad spectrum of pH optima at pH 4.5–6.5 and showed maximum activity at 60°C. The enzyme activity was completely inhibited by pepstatin-A, indicating that the purified enzyme was an aspartic protease. Its enzymic specificity studied using the oxidized B chain of insulin indicated that the protease preferably hydrolyzed peptide bonds C-terminal to Glu, Leu, and Phe residues (at the P1 position). The broad specificity, pH optima, and elevated thermostability indicated that the protease was distinct from other known ficins.	Devaraj et al. 2008b

(continued on next page)

TABLE 5.3 (continued)
Chemicals Identified and/or Isolated from Latex of Various *Ficus* spp.

Ficus Species	Compounds	Details	Reference
sur	Two pentacyclic triterpenoids of oleanane and ursene structures	The compounds isolated from the latex are naturally acetylated in the 3-position, and their structures have been elucidated on the basis of spectroscopic studies.	Feleke and Brehane 2005
sycomorus	Coumarins: α-amyrin; bergapten; imperatorin; xanthotoxin	Coumarins were isolated from the latex of *F. sycomorus* by silica gel chromatography.	Abdel-Wahab et al. 1989
Ficus spp.	Diastase, esterase, lipase, protease		Sapozhnikova 1940

As in the earlier scenario regarding blood coagulation, proteases, which may also act as competitive binding inhibitors of other proteases, can affect the activity of ACE in paradoxical ways, indicating that within the complex milieu of chemicals in latex are proteases that can both stimulate and inhibit the activity of ACE. Careful bench research should help in elucidating the distinctions and planning therapeutic interventions for either purpose: to lower or to increase blood pressure through modulation of the enzyme ACE, using, as raw material, enzymes or different subfractions of mixed enzymes, derived from the protease pool in *Ficus* latex and in ficin.

SIGNAL TRANSDUCTION MODULATION

The complexity of proteases within the natural *Ficus* latex undoubtedly gives rise to a very wide spectrum of biologically active compounds. Proteases, that is, enzymes specialized to digest bonds between amino acids on protein strands, exert a great multiplicity of physiological effects, not only from direct digestion by proteases but much more so by the actions of proteases in signaling cascades. Thus, the specific protein digestion acts as a signal to initiate further "downstream" biochemical actions. Examples of cascades where proteases are known to play an important role include the complement system (discussed in Chapter 3 relative to lectins), blood coagulation and tumor regulation.

Borensztajn and Spek (2008) note:

Protease-activated receptors (PARs) are G-protein-coupled receptors (GPCRs) that are activated by a unique proteolytic mechanism. Besides the important role of blood coagulation factors in preventing bleeding after vascular injury, these serine proteinases actively engage target cells, thereby fulfilling critical functions in cell biology. Cellular responses triggered by coagulation factor-induced PAR activation suggest that PARs play an important role in proliferation, survival, and/or malignant transformation of tumor cells. Indeed, PAR expression correlates with cancer malignancy, and clinical studies show that anticoagulant treatment is beneficial in cancer patients.

Note that the proteases referred to earlier are described as serine proteases, while the main ones found in *Ficus* latex are cysteine proteases. Still, the proteases are of the same general class and may competitively interfere with the others' binding. Recently, even an aspartic protease has been isolated from the latex of *F. glomerata* (Devaraj et al. 2008b), further highlighting the diversity of proteases within *Ficus* latex, particularly when looking across the entire genus. Indeed, the possibility of a veritable treasure chest of therapeutic proteases found within the rich protease milieu that is traditional Ficin, acting through signaling effects on biochemical cascades, should not be overlooked.

BLOOD COAGULATION

Studies of Ficin in regard to blood coagulation have yielded paradoxical, or perhaps *biphasic* or even amphoteric results. Early work introduced latex of *Ficus glabrata* H.B.K. (Azevedo 1949) and a later one that of *F. domestica* L. (Echave 1954) into human blood in vitro and observed suppression of clotting. Then, in 2002, Richter et al. investigated a mixed-protease Ficin derived from the latex of *F. carica*, and studied as well the effect of certain mixed proteinase fractions of the whole on factor X (FX). Six proteolytic fractions separated by chromatography were checked, and two of those, with molecular masses of 23.2 and 23.5 kD, had the highest specificity toward FX. Factor X was converted to activated FXβ by consecutive proteolytic cleavage in the heavy chain between Leu178 and Asp179, Arg187 and Gly188, and Arg194 and Ile195 (FX numbering system) with concomitant release of a carboxy-terminal peptide. "These data suggest the hemostatic potency of Ficus proteases is based on activation of human coagulation factor X." (Richter et al. 2002)

Looking back on these combined findings over the past half a century, it appears that the proteases in fig latex *are* active in modulating coagulation. However, depending upon dosage, dosage form, and undoubtedly other factors that we do not yet understand, the latex may either activate or downregulate coagulation. More work in this basic area of research is required before treatment can begin, but there is great potential within *Ficus* latex for novel complex products to either facilitate or attenuate clotting activity via coagulation factor X and/or other coagulation factors.

CHITINASE-RELATED ACTIVITIES

Chitin, a repeating polymer comprising repeating units of *N*-acetylglucosamine, first appeared in the exoskeletons of *Cambrian* arthropods (e.g., trilobites). The oldest preserved chitin dates to the *Oligocene*, about 25 million years ago (Briggs 1999). Chitin is roughly synonymous with *cuticle* and is the basic protection in fungal walls as well as in exoskeletons and nematodes. Thus, enzymes that can break these linkages and disrupt specifically the walls of pathogenic fungi or parasitic worms would be highly advantageous, especially if these fungi and parasites were keen to infiltrate your (the *Ficus'*) defenses.

These enzymes, chitinases, also occur in the milieu of the *Ficus* latex, if not the Ficin as well, and serve as a defense first against fungi. The potencies of these chitinases, isolated from the latex of *F. microcarpa*, were improved in vitro when the ionic strength of the media was increased (Taira et al. 2005).

Chitinases also, as mentioned, initiate a defense against nematodes, and the action of *Ficus* spp. latex fractions in this context has been extensively studied and will be briefly summarized here. Cysteine proteinases, as noted earlier, are the main types of proteases (protease and proteinase are roughly synonymous) found in Ficin and *Ficus* latex, as well as in papaya, pineapple, and kiwi. These compounds, in vitro, effected digestion of the cuticle against two sedentary plant parasitic nematodes of the genera *Meloidogyne* and *Globodera*, and further, the action was blocked by the specific cysteine protease inhibitor, E-64, reinforcing the case for a primary role of cysteine proteinases in the nematicidal action of *Ficus* latex (Stepek et al. 2007b). These authors had earlier also elaborated a similar action for these compounds against the gastrointestinal nematode, *Heligmosomoides polygyrus* (Stepek et al. 2005) and the rodent stomach nematodes *Protospirura muricola* (Stepek et al. 2007a) and *Trichuris muris* (Stepek et al. 2006) both in vitro and in vivo.

A clinical trial utilizing doses of the latex from *Ficus glabrata* H.B.K. down to 0.05% in physiological saline was used to control intestinal helminthiasis in Indian and non-Indian rural populations in three Amazonian villages. The study on 181 persons has resulted in a recommended dosage of 1.0 cm^3 of prepared latex/kg per day for 3 days to be repeated every 3 months as a preventive or maintenance treatment. The treatment was effective against these organisms, ranked according to their occurrence clinically in these settings, *Ascaris* 68%, *Strongyloides* 42%, *Trichuris* 41%, *Ancylostoma/Necator* 26%, and *Taenia* 1%, with an overall occurrence of 92% (Hansson et al. 1986).

Nearly 20 years later, the same group issued a caution related to the medicinal use of the latex from *F. insipida*. It seemed that, in one case, an acute toxic reaction was noted in a 10-year old girl who had been treated with a dose of *oje*, the *F. insipida* latex, not much higher than the recommended one. This led to an investigation of 39 cases from the area over a 12-year period that required hospitalization for toxic reactions to treatment with *oje*, most probably due to overdose. The mortality rate was estimated to have been 0.01%–0.015% among patients supposedly treated with *oje* in the area. Severe intoxication led to symptoms of cerebral edema. The main treatment was osmotic diuresis with mannitol, which had been instituted in 1996. The recommended dose was defined as 1 cc per kg, and the toxic dose as 1.5 cm/kg, indicating obviously an extremely narrow and worrisome therapeutic index for this treatment. From the 39 cases, one fatality occurred in hospital and two outside of the hospital (Hansson et al. 2005). In addition to this human work, one dog study of favorable effects of *Ficus* latex against the nematode *Ascaris* sp. in dog is on record (Nagaty et al. 1959).

WARTS

Warts are small rough, benign tumors caused by a virus, specifically, the human papilloma virus, or HPV. *Ficus* latex has long been used as a local appliqué as a means of speeding up their resolution. Although the mechanism of action is incompletely understood, two Iranian studies, one in humans and one in cattle, have highlighted the efficacy of this specific therapy.

In the bovine study, 12 cows suffering from teat papillomatosis were divided into three groups. One group was treated once every 5 days with local application of latex from *F. carica*, one every 5 days with 10% salicylic acid, a common conventional treatment, and one group kept as a control. In the two treatment groups, de-epithelization and shrinking of the warts began after 5 days. In the control group, warts in some of the animals began to disappear spontaneously only after 15 days. The conclusion reached from the study was that the *Ficus* latex had a therapeutic effect roughly on par with the salicylic acid (Hemmatzadeh et al. 2003).

In the human study, 25 patients with common warts were recruited from a hospital outpatient clinic. Half the patients were instructed to apply *F. carica* latex to the warts on one side of the body. Warts on the contralateral side were treated with standard cryotherapy. After 6 months, 11 patients (44%), enjoyed complete resolution of the *Ficus* latex-treated warts. The remaining 14 patients (56%) had a complete cure following cryotherapy. Two patients had complete remission on both sides of the body, and. two failed to respond to either cryotherapy or fig tree latex. Although fig tree latex therapy was marginally less effective than cryotherapy, adverse effects were observed only in the cryo-treated warts. Overall, fig tree latex therapy of warts was judged to have offered several benefits, namely, short-duration therapy, no reports of any side-effects, ease-of-use, patient compliance, and a low recurrence rate (Bohlooli et al. 2007).

Although the mechanism for the action of the latex in these cases was not known, it is easy to assume that the efficacy is related to the proteolytic effects of the enzymes. Nevertheless, the extent to which the proteases act directly by "dissolving the wart" and the extent that they act as secondary signalers of more complex cascades will require further careful studies to enable us to delineate the mechanism in a definitive manner.

CANCER

The detailed direct use of the latex in rodent models pioneered by Ullman and coworkers in the 1940s and 1950s at the Hebrew University in Jerusalem set many precedents in both toxicity and anticancer efficacy. Some modern work has gone forward both with in vitro confirmation, identification of some novel compounds that may be contributory to the effect, purification of nontoxic fractions, and verification of their specific anticancer activity, for example, against angiogenesis.

Ullman et al. (1945) initially tested the *F. carica* latex, dissolved in water and subsequently extended into Ringers lactate solution, by both intravenous and then subcutaneous injections into rabbits. Intravenous injection of 0.2 cc of the original juice, so dissolved, caused death in 10–20 s, with symptoms of "violent intoxication" consisting of bleeding from the mouth and nose, tonic/clonic convulsions, crying, and other expressions of violent pain. Intravenous injection of 0.05 cc of the original juice diluted in the same manner caused death in 1–2 min with symptoms of paresis of the legs, especially of the hind legs, and dyspnea. Postmortem examination in all cases revealed anemia of the skin and eyes, hemorrhages in the muscles, hyperemia in the intestines, and congestion of the right atrium. Death typically occurred in diastole, preceded by a bradycardic period. Doses less than 0.02 cc, however, caused no visible toxic reaction. In albino rats, doses greater than 0.02 cc caused a similar picture of death within 2 min. Injection of daily subactive doses for several days followed by a single injection a week later resulted in severe shock. It was also noted that significant and profound variation in this toxicity was exhibited by latex from different trees, presumably all different examples of *F. carica* from varying growing conditions, cultivars, or individuals. The nature of the relationships among the different trees, or of the consistency of different doses, etc., from single or different trees as used in this study, was not further specified.

Subcutaneous injections of the latex, in doses greater than 0.1–0.12 cc, into albino rats of 70–80 g caused death in 3–5 min. Other than "anemia of the skin and hyperemia of the intestines" there were no other striking postmortem findings; however, by beginning with 0.04 cc, a nonactive dose, and gradually doubling this on a daily basis, 0.4 cc could be reached prior to a lethal effect. Such subcutaneous conditioning, however, did not desensitize the animals to a subsequent intravenous injection. Intramuscular injection of up to 0.25 cc caused local necrosis, but no systemic signs of toxicity. Further, all injections, both subcutaneous and intramuscular, were accompanied by local necrosis, though local inunction of the latex to the bare skin did not. This problem became a severe limiting factor in the amount of injections a single animal could tolerate. Sublethal injections of the latex on successive days also resulted in profound anemia; in one case, the erythrocyte count went from about 6K to 3K after a week of treatments.

An anticancer exploration was then launched by subcutaneously transplanting the albino rats with benzpyrene sarcoma 616, following which 70%–80% of the animals developed palpable tumors 0.5–1 cm in diameter. The rats received daily injections of 1/25 solutions of the latex in water, starting at 1.0 cc and advancing toward 2.0 cc; the injections were discontinued after 4–9 days due to local necrosis at the injections sites as described earlier. It was further found that inclusion of 2% sodium carbonate or sodium bicarbonate in the injection medium attenuated the local necrotic effect but also weakened the anticancer effect, which is described in the following text. Heating the solution, on the other hand, to 60°–70° (presumably centigrade), diminished the local necrotic effect but not the anticancer power. Heating to over 78° resulted in a heavy precipitate and the simultaneous loss of efficacy.

Daily doses of 0.02 cc of the original juice were without effect, but 0.04–0.08 cc/day subcutaneously resulted in complete regressions in many cases of small tumors under about 1.0 cm; larger tumors from 1–2 cm remained stationary, and then regressed or opened after 10–14 days. Attempts to augment the effect through combinations with other agents of known growth inhibiting properties including methylene blue, heptaldehyde, and cholinbromide were unsuccessful. Rats not receiving the injections showed neither regression nor attenuation of their tumors.

Next, initial trials were undertaken to separate major phases of the fresh latex. The latex, when filtered, yielded a "brownish-yellow liquid with a sweet odor." This liquid was then subject to dialysis against distilled water. Dialysis involves the movement of one liquid, the dialyzed (here, the filtered *F. carica* latex), across a semipermeable membrane (here, an unspecified type of (presumably laboratory-grade) cellophane), which has contact with, here, ten times its volume of distilled water. The smaller compounds go across, and the larger ones get left behind.

A clear yellow dialyzate went across the membrane, and tested positive by the Boucharda–Meyer–Dragendorff–Sonnenschein–Marme method for alkaloids, although neither alkaloids nor alkaloid-rich fractions were separated or identified further. A residual liquid, during which sediment formed during dialysis, was filtered off and also tested positive for alkaloids. The precipitate, which could be augmented by exposure of the fresh latex to $(NH_4)_2SO_4$, dissolved in Ringer's but not in water, and was judged by several tests to be a globulin (i.e., proteins). The juice also yielded "negligible" amounts of extract when in contact with ether, $CHCl_3$, C_6H_6, petrol ether, carbon tetrachloride, or CS_2, either as such, or after acidification or alkalization. The method of Stass-Otto for extracting alkaloids was applied, but proved nonproductive.

The dialysate did not possess the acutely toxic and necrosis-inducing properties of the whole material, though it did possess the tumor-regression-inducing property, as well as, unfortunately, also the anemia-inducing property. Up to ten times the volume of dialysate as a lethal dose in volume of the full latex could be injected without inducing acute symptoms. And, in "many cases," 100% remission of tumors occurred following 1 to 2 weeks of daily subcutaneous injections of dialysate, up to 2 cc per dose. In all cases, injections of either the dialysate or fresh juice were directed to areas as remote as possible from the palpable or visible tumors.

The undialyzable liquid exhibited the toxic and necrotizing effects of the full latex, but did not have the effect of promoting tumor regression. It also exhibited the proteolytic effects of the full latex on gelatin, and tested positive with the Biuret reaction to detect peptide bonds.

The sediment, after washing with water, was insoluble in Ringer's solution or weak base, but was soluble in 0.05 N HCl or the dialysate. Injection of the sediment, suspended in Ringers, in no cases effected tumor regression but in many cases caused tumors to open, with growth continuing only at their peripheries. When redissolved in the dialysate, the combination was without toxicity or necrotizing influence, and also effected a nice rate of tumor regression. The combination also facilitated an additive effect on tumor regression when subcutaneous injections of it were used in conjunction with radiotherapy. All told, in a series of 134 rats with tumors treated with the combination or the dialysate alone, 58 showed tumor regression, while only 2 rats in a control group of 142 exhibited tumor regression. Percutaneous rubbing of the skin with any of these fractions failed to elicit regression of underlying tumors. No toxic effects were observed from per os administration.

In further work (Ullman 1952), the process of preparing the experimental medicine was refined. An initial extraction of the fresh latex using 96% ethanol could be instituted to remove the putative anemia-producing fraction. To prepare the solution for experimental subcutaneous injections, fresh, filtered latex was dialyzed against a tenfold quantity of distilled water. This dialysate was then filtered, to it was added 0.5% phenol, and then the solution was heated on three alternate days for 1 hour to 68°C for the purpose of achieving sterilization. Adult mice, 30 g, could be sensitized to increasing amounts of the solution to tolerate up to 0.6–0.7 cc. Latex prepared from trees in Jerusalem caused less mortality than latex from California, though no reason was found for this. There was some problem also with the phenol additive (used as a preservative) causing seizures in the mice, though not in the rats. For mammary cancer, treatments were given either prior to development of tumors, at the time of first noticing the tumor, or at the time the tumor was already at least 0.5 cm in diameter. Other tumors born in mice against which the product was tested included myogenic leukemia, malignant melanoma, sarcoma, and lymphosarcoma. For most of these, the latex product resulted in a delaying of tumor onset if used at an early stage. Tumor regression, mainly of tumors less than 0.5 cm, was observed.

In a subsequent series of studies (Ullman et al. 1952), two different treatment protocols were used in treatment of spontaneous mammary tumors and CFW and db noninbred and inbred mice, from Rockland Farms and Jackson Labs, respectively. The two methods were (1) five injections/week, and (2) one or two injections a day, 7 days a week, for a month. Favorable changes documented after months of treatment included fewer metastases, less penetration through skin, less lymph node invasion, favorable changes in vaginal epithelium via vaginal and cervical smears, increase of polymorphonuclear leukocytes, and also lymphocytes and basophiles. Tumors of treated animals were impossible to transplant into other animals of the same strain. Vaginal epithelium cornification, indicating estrogenic potentiation, was attenuated in both tumor-free and mammary tumor-bearing mice, indicating an overall antiestrogenic effect of a purified fraction R3, determined to be free of necrotizing and toxic properties. Tumor-bearing mice treated with this latex fraction had relatively higher counts of lymphocytes relative to leukocytes than the untreated or saline-treated tumor-bearing mice. These changes were hypothesized as being related to immunological forces being mobilized for the overcoming of the tumor. Other changes reported from treatment were spontaneous necrosis in tumor tissue specifically, and increased size of adrenals, liver, kidneys, thymus, and spleen, again hypothesized to be linked to immune system (or reticuloendothelial system) mobilization. Treated animals that had spontaneous mammary tumors removed by operation had fewer recurrences than untreated animals, suggesting a chemoprotective effect.

Taking Ullman's work as a starting point, Rubnov et al. (2001), as mentioned earlier, isolated a series of 6-*O*-acyl, β-d-glucosyl-β-sitosterols (AGS). These compounds were tested for their effects in limiting proliferation in several different cancer cell lines, namely, estrogen receptor positive MCF-7 human breast cancer, DU 145 androgen receptor negative human prostate cancer, Raji and DG-75 Burkitt B cell lymphomas, and Jurkat and HD-MAR T-cell leukemias. Dose-dependent inhibitory activity by AGS was seen in all the cell lines tested at 25 µg/mL and 50 µg/mL. Synthetically prepared versions of these compounds had the same activity as the naturally extracted ones. There were considerable problems in getting the compounds into water or DMSO or many other common solvents because of the compounds' inherent insolubility in multiple media.

In 2008, Wang et al. published their study on the cytotoxicity of *Ficus* latex in a number of additional cell lines, looking more deeply at the mechanism of suppression, including methods to assess both cell cycle and apoptosis, as well as comparative studies in both normal and cancer cells. Human glioblastoma, hepatocellular carcinoma, and normal liver cells were used for testing the *Ficus* latex. The tests included those for cytotoxicity, colony formation inhibition, Brdu incorporation, acridine orange/ethidium bromide (AO/EB) staining for apoptotic cells, cell cycle distribution through flow cytometry (FCM), and ADP-ribosyltransferase (NAD+; poly(ADP-ribose) polymerase)-like 1 (ADPERL1) mRNA expression through RT-PCR in response to treatment. Following the treatment, proliferation, colony formation, and Brdu labeling indices of cancer cells decreased ($P < 0.05$), while the AO/EB-stained apoptotic cells increased ($P < 0.05$). By FCM analysis, an increase of G(0)/G(1) phase cell population and decrease of S and G(2)/M phase cells were observed ($P < 0.01$), while both ADPRTL1 mRNA expression and apoptotic indices increased ($P < 0.01$). All told, the results provide reinforcement regarding the ability of the latex to inhibit cancer cell growth and progression and highlight the roles of cell cycle disruption and apoptosis in mediating these effects.

In another work that directly builds on Ullman's pioneering efforts, Pawlus et al. (2008) studied latex pressed from immature fruits of several different varieties of *F. carica* in its ability to interfere with angiogenesis. Several novel ex vivo human models were used to achieve direct visualization of blood vessels, specifically derived from placental veins and human tumor biopsies. Antiangiogenic actions in these assays by fractions prepared from the latexes of three out of six of the *F. carica* varieties studied were confirmed using a human umbilical vein (HUVEC) tubule formation assay, though the fractions were nontoxic as measured by an MTT cell proliferation assay. These latex fractions also downregulated vascular endothelial growth factor (VEGF) expression in the HUVEC cells, and also downregulated expression of VEGF receptors. These changes are consistent with suppression of angiogenesis. The fraction was also studied with LC/MS, and rutin was found to be an important constituent.

F. carica latex is, in short, a natural product that is expected to be increasingly studied for its anticancer activity. In spite of its potential toxicity, its association with a safe and edible fruit lowers the barriers for its acceptance as an important matter for further investigation (Amara et al. 2008).

PSORIASIS

Mavromoustakos et al. (2008), presented at the same Greek venue as the preceding study, focused on the latex extracted from branches of *F. sycomorus* utilizing a variety of solvents. A fraction of midpolarity was found to be rich in furocoumarins, including psoralen. The identity of psoralen was subsequently confirmed with 1H NMR. The crude latex extract was also tested by direct topical application to psoriatic skin lesions in human volunteers. Photos taken before and after the treatments confirmed that the treated areas decreased in size.

HEPATOTOXICITY

The oral and intraperitoneal effects of *F. carica* latex on lipid peroxidation and carbon tetrachloride-induced lipid peroxidation in liver homogenates of female rats were investigated. The oral treatment had no effect, but intraperitoneal injection resulted in hepatic lipid peroxidation. Given prior to the carbon tetrachloride, the latex produced no hepatoprotective effect. On the contrary, addition of the whole latex to the incubation mixture produced a dose-dependent increase in lipid autoxidation, while the chloroform and ether extracts of the latex, as well as heated latex, had no effect on hepatic lipid autoxidation (Al-Bayati and Alwan 1990). This work indicates the presence of hepatotoxic and pro-oxidative elements within the latex that may be eliminated upon simple processing such as heating, or by extraction with solvents.

ALLERGY

Latex from *Ficus* spp. is a classic allergen, and reactions to its exposure have been extensively investigated or at least clinically observed, commented upon, and treated. In addition to direct contact of skin with the plants resulting in contact urticarias (Guillet and Guillet 2004), the *Ficus*, especially houseplants of *F. benjamina* (the weeping fig), act as an airborne IgE allergen and can produce symptoms ranging from rhinoconjunctivitis, asthma, angioedema, and pruritus to anaphylactic shock (Bircher et al. 1993; Brehler and Theissen 1996, Charpin and Vervloet 1997, Chełmińska, M. 2004, Díez-Gómez et al. 1998, Pradalier et al. 2004, Rudack et al. 2003, Schenkelberger et al. 1998, Werfel et al. 2001). Gardeners and other caretakers of plants are susceptible, though most of the patients are atopic individuals, but cases have been described of nonoccupational, indoor-related rhinoconjunctivitis in a nonatopic patient. Prick-test and RAST to FB latex were positive and removal of the ficus plant resolved their symptoms, confirming the etiologic role of the plant (Pradalier et al. 2004).

In 1990, Axelsson et al. from the Department of Clinical Immunology, Karolinska Hospital in Stockholm, Sweden, examined the allergen composition of *F. benjamina* latex with sodium dodecyl sulfate–polyacrylamide gel electrophoresis and immunoblotting in 11 occupationally exposed plant keepers, of whom 7 were nonatopic, and in 9 nonoccupationally exposed atopic patients with a positive radio-allergosorbent test to weeping fig. A total of 11 allergenic components were identified. Three of them were found to be major allergenic components with molecular weights of approximately 29,000, 28,000, and 25,000 Da, respectively. The major allergenic components were denatured by heat in the temperature range of 60°C–90°C.

Allergy to *F. benjamina* latex, which is assumed to be the major source of the allergens, even those airborne, is cross-reactive with natural rubber allergens, both fresh and dried common figs, and also fruits or foods imbued with similar latex, including avocado, kiwi, pineapple, papaya, breadfruit, and banana (Antico et al. 2003, Brehler et al. 1997, De Greef et al. 2001, Erdmann et al. 2004, Focke et al. 2003, Hemmer et al. 2004). The potential life-threatening nature of allergies to *Ficus* and its cross-reactivity is an area of growing concern, such that increasing surveillance to this potential problem utilizing IgE skin tests or in vitro methods such as CAP or ELISA may be indicated (Ebo et al. 2003, Stevens 2000). Sensitization to *F. benjamina* latex is found in 2.5% of atopic individuals and mostly occurs independently of *Hevea* latex allergy. Cross-reactivity to other fruits and rubber latex is mediated at least in part by thiolproteases (Hemmer et al. 2004).

TABLE 5.4
Pharmacological Studies of *Ficus* spp. Latex

Ficus Species	Pharmacological Action	Details	Reference
anthelminthica	Proteolytic	Latex was extracted with 0.01 N HCl, dialyzed, proteolytic fraction precipitated with $(NH_4)_2SO_4$, precipitate dissolved in H_2O, the solution chromatographed on DEAE-cellulose with a $(CH_2OH)_3CNH_2$ phosphate buffer as eluent, the elute dialyzed in vacuo against acetate buffer, chromatographed on CM-cellulose with an acetate buffer as eluent, the eluate concentrated by dialysis or lyophilization, and the concentrated solution allowed to stand at pH 5 to give crystallized ficin.	Bettolo et al. 1963
benjamina	Causes respiratory allergy	New type of allergy caused by airborne allergens from nonpollinating plants; cross-reactions to latex.	Rudack et al. 2003
benjamina	Allergen	Case of perennial rhinoconjunctivitis caused by allergic reaction to *F. benjamina*.	Werfel et al. 2001
benjamina	Allergen	Case of nonoccupational, indoor-related rhinoconjunctivitis caused by *F. benjamina* in a nonatopic patient: prick test and RAST to FB latex were positive, and removal of the ficus plant resolved their symptoms, confirming the etiologic role of the plant. The patient did not demonstrate sensitization to other common allergens (except weeds) involved in respiratory and food allergies.	Pradalier et al. 2004
benjamina	Allergen	Latex-allergic patients are at higher risk of becoming sensitized to *Ficus*. Hev b 6.02 in natural rubber latex is a major cross-reactive allergen, and its counterpart in *F. benjamina* is an acidic protein with a molecular weight of about 45 kDa and a hevein-like N-terminal domain.	Chen et al. 2000

(continued on next page)

TABLE 5.4 (continued)
Pharmacological Studies of *Ficus* spp. Latex

Ficus Species	Pharmacological Action	Details	Reference
benjamina	Allergen	The number of *F. benjamina* sensitizations is increasing in Germany, which can partly be explained by the cross-reactivity between *Hevea brasiliensis* latex and *F. benjamina* and the rapidly increasing number of mostly occupational latex allergies. The plant itself is also a potential sensitizer as an ornamental plant. For diagnosis, prick tests with *F. benjamina* latex seem to be more sensitive than in vitro methods (RAST, CAPRAST).	Schenkelberger et al. 1998, Axelsson et al. 1990
		The allergen composition of crude extract from latex of *F. benjamina* was investigated by SDS-PAGE and immunoblotting. The allergen–antibody complexes were visualized by rabbit anti-IgE and β-galactosidase-labeled sheep anti-rabbit IgG, using a chromogenic insoluble substrate. A total of 11 allergenic components were identified. Three of them were major allergenic components, identified by >50% of the investigated sera. These three IgE-binding components had molecular weights of approximately 29,000, 28,000, and 25,000 Da, respectively.	
carica	Milk coagulating	The milk-clotting enzyme from latex of *F. carica* can be substituted for animal rennet to yield Cheddar and processed cheese.	Krishnaswamy et al. 1961

TABLE 5.4 (continued)
Pharmacological Studies of *Ficus* spp. Latex

Ficus Species	Pharmacological Action	Details	Reference
carica	Hemostatic; activates human coagulation factor X proteases; shortens the activated partial thromboplastin time and the prothrombin time of normal plasmas and plasmas deficient in coagulation factors, except plasma deficient in factor X (FX) and generated activated FX (FXa) in defibrinated plasma	Ficin derived from *F. carica* shortened the activated partial thromboplastin time and the prothrombin time of normal plasmas and plasmas deficient in coagulation factors, except plasma deficient in factor X (FX) and generated activated FX (FXa) in defibrinated plasma. Chromatography separation of ficin from *F. carica* yielded six proteolytic fractions with a different specificity toward FX. Two factor X activators, with molecular masses of 23.2 and 23.5 kDa, caused purified human FX to be converted to activated FXβ by consecutive proteolytic cleavage in the heavy chain between Leu178 and Asp179, Arg187 and Gly188, and Arg194 and Ile195 (FX numbering system) with concomitant release of a carboxy-terminal peptide. The cleavage pattern of FXa degradation products in the light chain was influenced by Ca^{2+} and Mn^{2+}.	Richter et al. 2002
carica	Anticancer	A mixture of 6-O-acyl-β-d-glucosyl-β-sitosterols, the acyl moeity being primarily palmitoyl and linoleyl with minor amounts of stearyl and oleyl, was isolated as a potent cytotoxic agent from *F. carica* latex. Both the natural and the synthetic compounds showed in vitro inhibitory effects on proliferation of various cancer cell lines.	Rubnov et al. 2001

(continued on next page)

TABLE 5.4 (continued)
Pharmacological Studies of *Ficus* spp. Latex

Ficus Species	Pharmacological Action	Details	Reference
carica	Antiangiogenic	Latex from immature fruits of several different *F. carica* varieties were tested for antiangiogenic activity in a unique system involving the use of segments of placental veins and human tumor biopsies. The nontoxic, antiangiogenic activity was confirmed using a human umbilical vein endothelial cell (HUVEC) tubule formation assay and a MTT antiproliferation assay. In HUVEC cells, there was a marked concentration-dependent decrease in both the content of vascular endothelial growth factor (VEGF) as well as downregulation of VEGF receptor.	Pawlus et al. 2008
carica	*Antimetastatic; antitumor (mammary tumors) *Anticarcinoma; antitumor; antiadenocarcinoma; antileukemia; antilymphosarcoma; antisarcoma	In mice bearing spontaneous mammary tumors and treated with fraction R3 of the latex of *F. carica*, there were fewer metastases before or after surgery, less invasion of lymph glands, and less penetration of the skin than in control animals not receiving the fraction. The tumors in the treated animals were movable and not attached to the skin or surrounding tissues; when transplanted to animals of the same strain, these tumors did not take. The treated animals showed a change in vaginal smears, an increase in polymorphonuclear leucocytes and lymphocytes, and an activation of the adrenal cortex with resulting activation in liver, spleen, and bone marrow.	Ullman 1952, Ullman et al. 1952

TABLE 5.4 (continued)

Pharmacological Studies of *Ficus* spp. Latex

Ficus Species	Pharmacological Action	Details	Reference
		Some fractions of the latex of unripe fruits of *F. carica* markedly inhibited the growth of spontaneous mammary carcinoma in mice, induced necrosis in spontaneous mammary tumors, delayed the takes in transplantable adenocarcinoma, myeloid leukemia, lymphosarcoma, and sarcoma, and caused regressions of sarcoma.	
carica	Inhibits the growth of subcutaneously transplanted benzopyrene sarcomas B, 616, and 2192; produces regression of some intraperitoneal and subcutaneous sarcoma transplants; necrotic action on the skin; intravenously toxic; produces anemia; increases susceptibility of benzopyrene sarcoma 616 to the therapeutic action of x-rays; tumor inhibiting; proteolytic; produces capillary lesions on intravenous administration	Latex from *F. carica* L., given subcutaneously, inhibited the growth of subcutaneously transplanted benzopyrene sarcomas B, 616, and 2192 in rats. An alkaloid fraction (A) of the latex, and a nondialyzable, nonprotein, nonalkaloid fraction (C), injected subcutaneously into rats, rendered them practically nonsusceptible to subsequent subcutaneous transplantation of benzopyrene sarcoma 616. The same fractions, given subcutaneously and intravenously, produced regression of some intraperitoneal and subcutaneous sarcoma transplants. The unfractionated latex, and a globulin fraction, had a strongly necrotic action on the skin, and when injected intravenously were very toxic. An alc.-sol. fraction produced anemia. The susceptibility of benzopyrene sarcoma 616 to the therapeutic action of x-rays, which normally is very low, was increased considerably by subcutaneous injection of fractions A and C.	Ullmann et al. 1945
carica	Proteinase		Sugiura and Sasaki 1973, 1974

(continued on next page)

TABLE 5.4 (continued)
Pharmacological Studies of *Ficus* spp. Latex

Ficus Species	Pharmacological Action	Details	Reference
carica	Proteolytic	Latex from leaves and tender shoots of *F. carica* was iced and tested for proteolytic activity by measuring the amount of tyrosine released from a casein substrate. *Ficus* latex showed maximum activity with an optimum time of 45 min. Latex (0.5 mL) was diluted to 2 mL With 0.2 N Na_2CO_3, mixed with 2 mL of 2% casein solution and 2 mL phosphate buffer, pH 7, and the mixture was stirred and held at 37°–38° (presumably) C. The reaction was stopped by adding 4 mL Cl_3CCO_2H 0.25 M. After 10 min, contents were centrifuged and the clear supernatant was used for the assay: 0.1–1.0 mL of supernatant was diluted to 2 mL with H_2O, α-nitroso-β-naphthol in EtOH (0.05 mL, 1 mg. wt./vol.) was added followed by 2 mL $FeNH_4(SO_4)_2$. The mixture was carefully boiled, allowed to stand 1 h at room temperature, and the color read at 400 μm. Activity was expressed in mg. tyrosine liberated from 100 mg latex at pH 7. The proteolytic action of *F. carica* latex was studied in vitro by the hydrolysis of proteins (casein, fibrin, and gelatin) and in vivo by the anatomical and physiological effect of injections in rats and guinea pigs. Its fibrinolytic action was very strong. Paper electrophoresis showed the presence of three different enzymes. Ten proteolytic enzyme components were separated from *F. carica* var. *Kadota* latex by chromatography on carboxymethyl cellulose. In rechromatography, ficins C and D were crystallized. All ten components were able to act on casein, α-benzoyl-l-argininamide, and milk, but there were large relative differences among the components.	Pant and Srivastava 1966, Eristavi et al. 1963, de Cutinella et al. 1964, Kramer and Whitaker 1964, Williams et al. 1968

TABLE 5.4 (continued)
Pharmacological Studies of *Ficus* spp. Latex

Ficus Species	Pharmacological Action	Details	Reference
		Latex of *F. carica* has a high proteolytic activity, with the different varieties containing multiple proteolytic enzymes differing as detected by free boundary electrophoresis although the proteolytic activity of the latexes was reasonably constant.	
carica	Antitumor	Latex of *F. carica* was tested in vivo using three subsequent bioassays: the BST (Brine Shrimp Toxicity bioassay), AWD (Agar well diffusion antimicrobial bioassay), and AtPDT (Agrobacterium tumefaciens Potato Disc Tumor bioassay). It showed high antitumor activity.	Amara et al. 2008
carica	Fruit latex: proteolytic enzyme	Proteolytic enzymes were obtained from the latex of fruits of *F. carica*. Enzymes were extracted from 1 g ground dried material with 40 mL H_2O or 20 mL 5% NaCl at room temperature for 14–16 h. Proteolytic activities were detected on casein (I) and gelatin (II) solutions.	Gonashvili and Gonashvili 1968
carica	Fibrinolytic	The latex of *F. carica* caused a complete hydrolysis of 1 mg of fibrin with 0.6 mL of a 1:1000 freshly prepared aqueous dilution of latex. Solutions prepared the previous day were fully effective, even at 0.2 mL. However, on standing for 2 weeks, virtually all the fibrinolytic activity was lost. Storage for 24 h at 4°C had no observable effect. Boiling for 10 min destroyed the enzyme. The addition of 0.5 mL of 1:1000 dilution of latex to 0.5 mL of normal dog plasma hydrolyzed the fibrinogen to the point that no fibrin could be formed. (Eristavi et al. 1963)	Eristavi et al. 1963, de Cutinella et al. 1964

(continued on next page)

TABLE 5.4 (continued)

Pharmacological Studies of *Ficus* spp. Latex

Ficus Species	Pharmacological Action	Details	Reference
carica (continued)		The proteolytic action of *F. carica* latex was studied in vitro by the hydrolysis of proteins (casein, fibrin, and gelatin) and in vivo by the anatomical and physiological effect of injections in rats and guinea pigs. Its fibrinolytic action was very strong. Paper electrophoresis showed the presence of three different enzymes (de Cutinella et al. 1964).	
carica	Hemostatic proteases	Ficin derived from *F. carica* shortened the activated partial thromboplastin time and the prothrombin time of normal plasmas and plasmas deficient in coagulation factors, except plasma deficient in factor X (FX), and generated activated FX (FXa) in defibrinated plasma. Chromatographic separation of ficin from *F. carica* yielded six proteolytic fractions with a different specificity toward FX. Data suggest that the hemostatic potency of *Ficus* proteases is based on activation of human coagulation factor X.	Richter et al. 2002
carica	Anticancer	Fig fruit latex (FFL) contains significant amounts of polyphenolic compounds and can serve as a source of antioxidants. It caused a decrease in proliferation, colony formation, and Brdu labeling indices of human glioblastoma and hepatocellular carcinoma cells ($P < 0.05$), while the AO/EB stained apoptotic cells increased ($P < 0.05$). By FCM analysis, an increase of G(0)/G(1) phase cell population and decrease of S and G(2)/M phase cells were observed ($P < 0.01$), while both ADPRTL1 mRNA expression and apoptotic indices increased ($P < 0.01$). Thus, FFL exhibited potent cytotoxicity in some human cancer cells with little effect in normal cells at certain concentration. The mechanism for such effects might be associated with the inhibition of DNA synthesis, induction of apoptosis, and cell cycle arrest of cancer cells.	Wang et al. 2008

TABLE 5.4 (continued)
Pharmacological Studies of *Ficus* spp. Latex

Ficus Species	Pharmacological Action	Details	Reference
carica	Anthelmintic	0.1 per cent suspension can be quite effective as an anthelmintic	Nagaty et al. 1959
carica	Lipase, diastase, amylase	The latex of fig contains a lipase that is 1/12 as active in a neutral medium as that of the paper mulberry and more active in acid than in neutral media. It has starch-splitting properties, and is 1/3 as strong as the latex of paper mulberry. The diastase is more resistant to heat than the accompanying lipase. The activity of the fig amylase is only slightly diminished by dialysis in distilled water, is not increased by NaCl, and is greatest in a slightly acid medium. It is activated by CaCl$_2$. Fig latex has very pronounced proteolytic properties, its milk-coagulating power being 100 times greater than that of the paper mulberry. This coagulating power is lost in large measure on dialysis in distilled water.	Gerber 1912
carica	Causes hydrolysis of tributyrin and monobutyrin	The latex of *F. carica* has no action on triolein but causes hydrolysis of tributyrin and monobutyrin.	Visco 1924
carica	Casein digesting, milk clotting	The casein-digesting and milk-clotting activity of *F. carica* latex was determined. At −25°C very little enzyme activity was lost after storage for 60 days. Freezing and thawing did not affect activity. Ficin became inactivated at room temperature because of the oxidation of the essential sulfhydryl group and also because there was an irreversible loss in activity of the enzyme.	Whitaker 1958
carica	Bacteriolytic/lysozymelike activity against *Micrococcus lysodeikticus*	The bacteriolytic/lysozymelike activity of the latex of *F. carica* against *Micrococcus lysodeikticus* (in units/mL) was 213	Shukla and Krishna Murti 1961

(continued on next page)

TABLE 5.4 (continued)
Pharmacological Studies of *Ficus* spp. Latex

Ficus Species	Pharmacological Action	Details	Reference
carica	Antifungal	A low-molecular-weight protein with antifungal activity was isolated from freshly collected latex of *F. carica* L. by successive affinity chromatography over chitin, cation-exchange chromatography over SP-Sephadex C-50, and reverse-phase HPLC.	Mavlonov et al. 2008
carica	Inhibits [3H]benzo[*a*]pyrene[3H-B(a)P] binding to rat liver microsomal protein	Milk latex from *F. carica* produced dose-dependent inhibition of [3H] benzo[*a*]pyrene[3H-B(a)P] binding to rat liver microsomal protein. The activity decreased following centrifugation to two layers (upper and lower layers) and also after extraction with chloroform or ether and when heated at 100°C.	Alwan and Al-Bayati 1988
carica	Inhibitory peptides of angiotensin I-converting enzyme (ACE)	Three inhibitory peptides of angiotensin I-converting enzyme (ACE) were isolated from fresh latex of *F. carica*. The amino acid sequences of these peptides, identified by the Edman procedure and carboxypeptidase digestion, were Ala-Val-Asn-Pro-Ile-Arg, Leu-Tyr-Pro-Val-Lys, and Leu-Val-Arg. The IC_{50} values of these peptides for ACE from rabbit lung were 13, 4.5, and 14 µM, respectively.	Maruyama et al. 1989
carica	Proteolytic activity	Of the *Ficus* latices examined, latex of *F. stenocarpa* had the highest specific proteolytic activity followed closely by the latices of *F. carica* and *F. glabrata*. The latices were found to contain multiple proteolytic enzymes. The latices of 16 varieties of *F. carica* were all different as determined by free boundary electrophoresis, although the specific proteolytic activity of the latices was reasonably constant.	Williams et al. 1968

TABLE 5.4 (continued)
Pharmacological Studies of *Ficus* spp. Latex

Ficus Species	Pharmacological Action	Details	Reference
elastica	Rubber; *cis*-prenyl transferase (rubber transferase)	Rubber biosynthesis in *F. elastica* involves a rubber particle-bound *F. elastica cis*-prenyl transferase (rubber transferase) associating with a buoyant fraction of latex rubber particles.	Cornish and Siler 1996
elastica "*variegata*"	Inhibitory action on the growth of sarcoma 180 and adenocarcinoma 755	The latex of *F. elastica variegata* showed an inhibitory action on the growth of mouse sarcoma 180 and adenocarcinoma 755 at a dose of 10 and 5 mL/kg daily.	Irie 1964
glabrata	Anticoagulating; proteolytic action on fibrinogen and fibrin	Addition of 0.1 mL latex of *F. glabrata* 1:80 or less to 0.5 mL oxalated plasma prevented coagulation by thromboplastin and $CaCl_2$. Coagulation occurred if also prothrombin-free plasma was added. Fibrin is dissolved by the latex, which thus has a proteolytic action on fibrinogen and fibrin but does not interfere with prothrombin.	Azevedo 1949
glabrata	Inhibits the coagulation of blood; proteolytic	The latex of *F. glabrata* H.B.K. inhibits the coagulation of blood in vitro. It is suggested that its proteolytic enzyme, ficin, destroys the prothrombase.	Cancado 1944
glabrata	Proteolytic	The latices of only 13 of 46 species of *Ficus* examined contained appreciable proteolytic activity. The latex of *F. stenocarpa* had the highest specific activity followed closely by the latices of *F. carica* and *F. glabrata*. A stable, peroxidase-free, proteolytic enzyme is obtained from fig sap latex from *F. glabrata* by clarifying it by filtration, $(NH_4)_2SO_4$ at 40°F added to saturation, the enzyme precipitated, the precipitate filtered off, suspended in a buffer solution at pH 5, dialyzed through cellophane against H_2O, filtered, and lyophilized.	Williams et al. 1968, Douglas and Gaughran 1960

(continued on next page)

TABLE 5.4 (continued)

Pharmacological Studies of *Ficus* spp. Latex

Ficus Species	Pharmacological Action	Details	Reference
glabrata	Peroxidases	*F. glabrata* peroxidases were purified by a procedure comprising CM-cellulose adsorption, $(NH_4)_2SO_4$ fractionation, and chromatography on DEAE-cellulose. Three components were obtained, identical to horseradish peroxidase II with respect to molecular weights, adsorption spectra, and activity–pH relations. They differed from the horseradish enzyme in isoelectric point (pH 4.25 to 4.45 versus 7.2 in acetate buffer) and in having greater heat stability. One of the peroxidases had Vmax. Values similar to those of the horseradish peroxidases on H_2O_2 and guaiacol, while the other two had different values.	Kon and Whitaker 1965
glomerata	Protease activity	Of 24 plants screened, *F. glomerata* (990 units) latex had the highest protease activity.	Chary and Reddy 1983
glomerata	Inhibits glucose-6-phosphatase from rat liver	Extracts from latex of *F. glomerata* Roxb. did not have any significant effect on blood sugar level of diabetic rats. They inhibited glucose-6-phosphatase but not arginase from rat liver.	Rahman et al. 1994
hispida	Proteolytic	A proteolytic enzyme was extracted from the latex of the plant *F. hispida* L.	Chetia et al. 1999
laurifolia	Proteolytic	A stable, peroxidase-free, proteolytic enzyme is obtained from fig sap latex from *F. laurifolia* by clarifying it by filtration, $(NH_4)_2SO_4$ at 40°F added to saturation, the enzyme precipitated, the precipitate filtered off, suspended in a buffer solution at pH 5, dialyzed through cellophane against H_2O, filtered, and lyophilized.	Douglas and Gaughran 1960

TABLE 5.4 (continued)
Pharmacological Studies of *Ficus* spp. Latex

Ficus Species	Pharmacological Action	Details	Reference
microcarpa	Three chitinases: gazyumaru latex chitinase (GLx Chi)-A, -B, and -C; antifungal; chitin-binding	Three chitinases were purified from the latex of *F. microcarpa*. GLx Chi-A, -B, and -C are an acidic class III (33 kDa, pI 4.0), a basic class I (32 kDa, pI 9.3), and a basic class II chitinase (27 kDa, pI > 10), respectively. At low ionic strength, GLx Chi-C exhibited strong antifungal activity, to a similar extent as GLx Chi-B. The antifungal activity of GLx Chi-C became weaker with increasing ionic strength, whereas that of GLx Chi-B became slightly stronger. The results suggest that the chitin-binding domain of basic class I chitinase binds to the chitin in fungal cell walls by hydrophobic interaction and assists the antifungal action of the chitinase.	Taira et al. 2005
platyphylla	The precipitated latex of the fruit had proteolytic activity on egg albumin. The fresh fruit extract and the latex had anthelmintic activity against *Ascardia galli* in chickens.	Fruits contained tartrate, ascorbate, citrate, malate, kojate, maleate, and malonate. The prepared latex of the fruit contained aspartate, hydroxyproline, tryptophan, tyrosine, arginine, lysine, alanine, histidine, and two unidentified acids.	El-Sayyad et al. 1986
pumila	Proteolytic on casein; caseinolytic	Crude extract of fruits of *F. pumila* L. was obtained by clarification of the latex through centrifugation at 16,000 g for 30 min. Subsequently, the supernatant was ultracentrifuged at 100,000 g for 1 h. The crude enzyme preparation showed high proteolytic activity on casein in the presence of 12 mM cysteine, but the activity was inhibited by thiol-specific inhibitors such as $HgCl_2$ and E-64, suggesting the enzymes belonged to the cysteine protease family. This crude enzyme extract showed maximum caseinolytic activity within an alkaline range of pH (7.0–9.0) and marked thermal stability.	Perello et al. 2000

(continued on next page)

TABLE 5.4 (continued)
Pharmacological Studies of *Ficus* spp. Latex

Ficus Species	Pharmacological Action	Details	Reference
stenocarpa	Proteolytic	The latices of only 13 of 46 species of *Ficus* examined exhibited appreciable proteolytic activity. The latex of *F. stenocarpa* had the highest specific activity.	Williams et al. 1968
virgata	Strong toxicity and growth inhibition against lepidopteran larvae; cysteine-protease activity	Leaves of *F. virgata* showed strong toxicity and growth inhibition against lepidopteran larvae, though no apparent toxic factors from the species have been reported. When the latex was washed off, the leaves lost toxicity. Latex was rich in cysteine-protease activity. E-64, a cysteine protease-specific inhibitor, completely deprived the leaves of toxicity when painted on their surface. Also, the cysteine protease ficin showed toxicity.	Konno et al. 2004
Ficus spp.	Allergen	Allergy to latex proteins, in which the responsible antigens are proteins present in the natural rubber latex (NRL), extracted from *Hevea brasiliensis*, has become an important problem, especially in the medical profession. Sensitization can occur via different routes (aerogen, skin, parenteral, etc.) but aerogenic sensitization seems to be very important. Immunological cross-reactions are very frequent with other plant or fruit allergens, but cross-allergy is less frequent. There is also a cross-reactivity with *Ficus*.	Stevens 2000
Ficus spp.	Anthelmintic	Cysteine proteinases from the fruit and latex of papaya, pineapple, and fig cause digestion of the cuticle and decreased the activity of the studied sedentary plant parasitic nematodes of the genera *Meloidogyne* and *Globodera*. The specific inhibitor of cysteine proteinases, E-64, blocked this activity completely, indicating that it was essentially mediated by cysteine proteinases.	Stepek et al. 2005, 2006, 2007a, 2007b

TABLE 5.4 (continued)
Pharmacological Studies of *Ficus* spp. Latex

Ficus Species	Pharmacological Action	Details	Reference
		They also cause rapid loss of motility and digestion of the cuticle of the rodent nematode, *Protospirura muricola* (located in the stomach), leading to death of the nematode in vitro.	
		In vitro, they are active against second-stage juveniles of *Meloidogyne incognita* and *M. javanica*.	
		They are also effective at killing the rodent nematode of the large intestine, *Trichuris muris*. The mechanism of action of these enzymes involved an attack on the structural proteins of the nematode cuticle. Cysteine proteinases from fig also cause marked damage to the cuticle of the rodent gastrointestinal nematode *Heligmosomoides* adult male and female worms, reflected in the loss of surface cuticular layers. The anthelmintic efficacy was comparable for both sexes of worms, dependent on the presence of cysteine, and was completely inhibited by the cysteine proteinase inhibitor E-64. LD_{50} values indicated that the purified ficin was more efficacious than the proteinases in the crude latex. The mechanism of action of these plant enzymes (i.e., an attack on the protective cuticle of the worm) suggests that resistance would be slow to develop in the field.	

FOR THE PLANT

Latex in plants, however we may find ways to make it serve us, also evolved, and most likely evolved primarily, to serve the plant. The latex serves as a vascular fluid, homologous to the blood of animals. It carries nutritive factors, self-healing mechanism, toxicity, and also houses the plant's immune system. In response to stresses, the latex may convey changes, including production of rubber, the actual purpose of which is still not well understood (Kim et al. 2003). Similarly, latex of *F. virgata* shows toxicity against lepidopteran larvae such as oligophagous *Samia ricini*

(Saturniidae), a defense it shares in common with papaya trees (Konno et al. 2004). Translation of such capabilities to benefit human concerns is still, undoubtedly, in a very early stage.

Latex from *F. carica* also inhibits binding of the carcinogen 3H-benzo[a]pyrene to rat liver microsomes, though this chemopreventive effect was much retarded by heating the latex in solvents to 100°C or even by centrifugation (Alwan and Al-Bayati 1988), further highlighting the need to adopt technologies to promote latex stability.

Allergies to *Ficus* that may be cross-reactive between different *Ficus* species have been described. Skin prick tests can be utilized (Antico et al. 2003; Chen et al. 2000). Eleven different allergenic compounds were identified in the latex of *F. benjamina* (weeping fig), and three of them were major IgE-binding allergens with molecular weight in the 25,000–28,000 Da range. The allergenic compounds were denatured from 60°C to 90°C (Axelsson et al. 1990). Allergy to *Ficus* latex should be ruled out as an inclusion criterion for clinical trials with *Ficus* products.

REFERENCES

Abdel-Wahab, S.M., S.F. El-Tohamy, A.A. Seida, and O.A. Rashwan. 1989. Isolation and identification of coumarins of certain *Ficus* species growing in Egypt. *Bull Fac Pharm (Cairo Univ.)* 27: 99–100.

Al-Bayati, Z.A., and A.H. Alwan. 1990. Effects of fig latex on lipid peroxidation and CCl4-induced lipid peroxidation in rat liver. *J Ethnopharmacol.* 30: 215–21.

Alwan, A.H., and Z.A.F. Al-Bayati. 1988. Effects of milk latex of fig (*Ficus carica*) on 3H-benzo[a]pyrene binding to rat liver microsomal protein. *Int J Crude Drug Res.* 26: 209–13.

Amara, A.A., M.H. El-Masry, and H.H. Bogdady. 2008. Plant crude extracts could be the solution: Extracts showing in vivo antitumorigenic activity. *Pak J Pharm Sci.* 21: 159–71.

Antico, A., G. Zoccatelli, C. Marcotulli, and A. Curioni. 2003. Oral allergy syndrome to fig. *Int Arch Allergy Immunol.* 131: 138–142. Comment in *Ibid.* 2004, 133: 316.

Axelsson, I.G., S.G. Johansson, P.H. Larsson, and O. Zetterström. 1990. Characterization of allergenic components in sap extract from the weeping fig (*Ficus benjamina*). *Int Arch Allergy Appl Immunol.* 91: 130–5.

Azevedo, M.P. 1949. [Mechanism of anti-coagulant action of the latex of the *Ficus glabrata* H.B.K.] *Mem Inst Butantan* 22: 25–30.

Bada, K., and M. Kato. 1987. [Controlled release meat tenderizer containing latex enzyme]. Japanese Patent JP 62104575 A 19870515. *Showa Jpn. Kokai Tokkyo Koho.*

Bettolo, G.B.M., P.U. Angeletti, M.L. Salvi, L. Tentori, and G. Vivaldi. 1963. Ficin. I. *Gazz Chim Ital.* 93: 1239–51.

Bircher, A.J., B. Wüthrich, S. Langauer, and P. Schmid. 1993. [*Ficus benjamina*, a perennial inhalation allergen of increasing importance]. *Schweiz Med Wochenschr.* 123: 1153–9.

Bock, H. 1964. *Kreütterbuch darin underscheidt Nammen und Würckung der Kreütter, standen.* [Herbal in which are the Different Names and Properties of Herbs]. Josiam Rihel, Strassburg, 1577. Reprint Konrad Kölbl, München.

Bohlooli, S., A. Mohebipoor, S. Mohammadi, M. Kouhnavard, and S. Pashapoor. 2007. Comparative study of fig tree efficacy in the treatment of common warts (*Verruca vulgaris*) vs. cryotherapy. *Int J Dermatol.* 46: 524–6.

Borensztajn, K.S., and C.A. Spek. 2008. Protease-activated receptors, apoptosis and tumor growth. *Pathophysiol Haemost Thromb.* 36: 137–47.

Brehler, R., and U. Theissen. 1996. [*Ficus benjamina* allergy]. *Hautarzt* 47: 780–2.

Brehler, R., U. Theissen, C. Mohr, and T. Luger. 1997. "Latex-fruit syndrome": Frequency of cross-reacting IgE antibodies. *Allergy* 52: 404–10.

Briggs, D.E.G. 1999. Molecular taphonomy of animal and plant cuticles: Selective preservation and diagenesis. *Philos Trans R Soc Lond B* 354: 7–17.

Brovetto, J., and G. Porcelli. 1967. [Characterization of proteins of *Ficus* latex]. *Ric Sci.* 37: 988–9.

Bykov, V.A., N.B. Demina, N.N. Kataeva, V.A. Kemenova, and V.L. Bagirova. 2000. Enzyme preparations used for the treatment of digestion insufficiency (a review). *Pharm Chem J.* 34: 105–9.

Calabria, L.M., S. Piacente, I. Kapusta et al. 2008. Triterpene saponins from *Silphium radula*. *Phytochemistry* 69: 961–72.

Cancado, J.R. 1944. Ficin, a new anticoagulant. *Rev Bras Biol.* 4: 349–54.

Charpin, D., and D. Vervloet. 1997. [New aero-allergens. Interaction between allergens and the environment]. *Bull Acad Natl Med.* 181: 1551–61.

Chary, M.P., and S.M. Reddy. 1983. Protease activity of some latex bearing plants. *Natl Acad Sci Lett (India)* 6: 183–4.

Chełmińska, M. 2004. [Latex allergy—Part I]. *Pneumonol Alergol Pol.* 72: 143–9.

Chen, Z., M. Düser, A. Flagge et al. 2000. Identification and characterization of cross-reactive natural rubber latex and *Ficus benjamina* allergens. *Int Arch Allergy Immunol.* 123: 291–298.

Chetia, D., L.K. Nath, and S.K. Dutta. 1999. Extraction, purification and physico-chemical properties of a proteolytic enzyme from the latex of *Ficus hispida* Linn. *Indian J Pharm Sci.* 61: 29–33.

Chung, I.M., M.Y. Kim, S.D. Park, W.H. Park, and H.I. Moon. 2009. In vitro evaluation of the antiplasmodial activity of *Dendropanax morbifera* against chloroquine-sensitive strains of *Plasmodium falciparum*. *Phytother Res.* April 15, 2009 [Epub ahead of print].

Cornish, K., and J.L. Brichta. 2002. Some rheological properties of latex from *Parthenium argentatum* Gray compared with latex from *Hevea brasiliensis* and *Ficus elastica*. *J Polym Environ.* 10: 13–18.

Cornish, K., M.D. Myers, and S.S. Kelley. 2004. Latex quantification in homogenate and purified latex samples from various plant species using near infrared reflectance spectroscopy. *Ind Crops Prod.* 19: 283–96.

Cornish, K., and D.J. Silcr. 1996. Characterization of *cis*-prenyl transferase activity localized in a buoyant fraction of rubber particles from *Ficus elastica* latex. *Plant Physiol Biochem. (Paris)* 34: 377–84.

Daniel, F.K., H. Freundlich, and K. Sollner. 1938. Stability of *Ficus elastica* latex. *India Rubber World* 98: 31–3, 38.

De Abreu, H.A., I. Aparecida Dos S Lago, G.P. Souza, D. Piló-Veloso, H.A. Duarte, A.F. de C Alcântara. 2008. Antioxidant activity of (+)-bergenin: A phytoconstituent isolated from the bark of *Sacoglottis uchi* Huber (Humireaceae). *Org Biomol Chem.* 65: 2713–8.

De Cutinella, M.R., A. Guevara, and M. Ruiz. 1964. Proteolytic action of *Ficus carica* latex. *Arch Soc Biol Montevideo* 26: 139–43.

Deeb, D., X. Gao, H. Jiang, S.A. Dulchavsky, and S.C. Gautam. 2009. Oleanane triterpenoid CDDO-Me inhibits growth and induces apoptosis in prostate cancer cells by independently targeting pro-survival Akt and mTOR. *Prostate* 69: 851–60.

De Greef, J.M., F. Lieutier-Colas, J.C. Bessot et al. 2001. Urticaria and rhinitis to shrubs of *Ficus benjamina* and breadfruit in a banana-allergic road worker: Evidence for a cross-sensitization between Moracea, banana and latex. *Int Arch Allergy Immunol.* 125: 182–4.

Devaraj, K.B., L.R. Gowda, and V. Prakash. 2008b. An unusual thermostable aspartic protease from the latex of *Ficus racemosa* (L.). *Phytochemistry* 69: 647–55.

Devaraj, K.B., P.R. Kumar, and V. Prakash. 2008a. Purification, characterization, and solvent-induced thermal stabilization of ficin from *Ficus carica*. *J Agric Food Chem*. 56: 11417–23.

De Wildeman, E. 1949. [The milklike and mucilaginous liquids of the Moraceae (Urticaceae)] *Acad Roy Belg Cl Sci Mem*. 24.

Díez-Gómez, M.L., S. Quirce, E. Aragoneses, and M. Cuevas. 1998. Asthma caused by *Ficus benjamina* latex: Evidence of cross-reactivity with fig fruit and papain. *Ann Allergy Asthma Immunol*. 80: 24–30.

Dioscorides, P. 1902. *Des Pedanios Dioscurides aus Anazarbos Arzneimittellehre in fünf Büchern*, trans. and comm. J. Berendes. Stuttgart: Ferdinand Enke.

Douglas, J.F., E.R.L. Gaughran. 1960. Proteolytic enzyme from fig sap latex. US 2956928 19601018.

Ebo, D.G., C.H. Bridts, M.M. Hagendorens, L.S. De Clerck, and W.J. Stevens. 2003. The prevalence and diagnostic value of specific IgE antibodies to inhalant, animal and plant food, and ficus allergens in patients with natural rubber latex allergy. *Acta Clin Belg*. 58: 183–9.

Echave, D. 1954. In vitro anticoagulant action on blood of the latex of *Ficus domestica*. *Sem Med*. 1: 351–2.

El-Fekih, M., and D. Kertesz. 1968. [The peroxidase from fig latex. I. Purification and properties]. *Bull Soc Chim Biol (Paris)* 50: 547–68.

Erdmann, S.M., U.C. Hipler, H.F. Merk, and M. Raulf-Heimsoth. 2004. Sensitization to fig with cross-sensitization to weeping fig and natural rubber latex. *Int Arch Allergy Immunol*. 133: 316.

Eristavi, K.D., M.G. Gachechiladze, S.G. Gonashvili, and M.S. Machabeli. 1963. The fibrinolytic action of ficin. *Soobshch Akad Nauk Gruz Ssr* 30: 667–70.

Espy, S.C., J.D. Keasling, J. Castillón, and K. Cornish. 2006. Initiator-independent and initiator-dependent rubber biosynthesis in *Ficus elastica*. *Arch Biochem Biophys*. 448: 13–22.

Fadýloğlu, S. 2001. Immobilization and characterization of ficin. *Nahrung* 45: 143–6.

Feleke, S., and A. Brehane. 2005. Triterpene compounds from the latex of *Ficus sur* I. *Bull Chem Soc Ethiop*. 19: 307–310.

Focke, M., W. Hemmer, S. Wöhrl, M. Götz, and R. Jarisch. 2003. Cross-reactivity between *Ficus benjamina* latex and fig fruit in patients with clinical fig allergy. *Clin Exp Allergy* 33: 971–7.

Gaughran, E.R.L. 1976. Ficin: History and present status. *Q J Crude Drug Res*. 14: 1–21.

Gauthier, C., J. Legault, K. Girard-Lalancette, V. Mshvildadze, and A. Pichette. 2009. Haemolytic activity, cytotoxicity and membrane cell permeabilization of semi-synthetic and natural lupane- and oleanane-type saponins. *Bioorg Med Chem*. 17: 2002–8.

Gerard, J. 1633. *The herball or generall historie of plantes*. London: Adam Islip Ioice Norton and Richard Whitakers.

Gerber, C. 1912. Ferments of the latex of the fig (*Ficus carica* L.) compared with those of the paper mulberry (*Broussonetia papyrifera* L.). *Bull soc bot France* 4: 2.

Gerber, C. 1913. Latex of *Ficus coronata*. *C R* 156: 1917–9.

Gonashvili, S.H.G. 1964. [Proteolytic properties of the latex from the fig tree (*Ficus carica* Z.)]. *Vopr Pitan*. 23: 26–30.

Gonashvili, S.G., and M.S. Gonashvili. 1968. Proteolytic enzymes of some Georgian plants. *Rastitel'nye Resursy* 4: 356–65.

Guerrini, A., I. Lampronti, N. Bianchi et al. 2009. Bergamot (*Citrus bergamia* Risso) fruit extracts as gamma-globin gene expression inducers: Phytochemical and functional perspectives. *J Agric Food Chem*. April 16, 2009 [Epub ahead of print].

Guillet, M.H., and G. Guillet. 2004. [Contact urticaria to natural rubber latex in childhood and associated atopic symptoms: A study of 27 patients aged under 15 years]. *Ann Dermatol Venereol.* 131: 35–7.

Hai, M.A., R.K. Sutradhar, and M.U. Ahmad. 1991. Chemical constituents of *Ficus glomerata* Roxb (Moraceae). *J Bangladesh Chem Soc.* 4: 247–50.

Hansen, A. 1886. Ferments. *Bot. Ztg.* 1886: 137.

Hansson, A., G. Veliz, C. Naquira, M. Amren, M. Arroyo, and G. Arevalo. 1986. Preclinical and clinical studies with latex from *Ficus glabrata* HBK, a traditional intestinal anthelminthic in the Amazonian area. *J Ethnopharmacol.* 17: 105–38.

Hansson, A., J.C. Zelada, and H.P. Noriega. 2005. Reevaluation of risks with the use of *Ficus insipida* latex as a traditional anthelmintic remedy in the Amazon. *J Ethnopharmacol.* 98: 251–7.

Hemmatzadeh, F., A. Fatemi, and F. Amini. 2003. Therapeutic effects of fig tree latex on bovine papillomatosis. *J Vet Med B Infect Dis Vet Public Health* 50: 473–6.

Hemmer, W., M. Focke, M. Götz, and R. Jarisch. 2004. Sensitization to *Ficus benjamina*: Relationship to natural rubber latex allergy and identification of foods implicated in the *Ficus*-fruit syndrome. *Clin Exp Allergy* 34: 1251–8.

Huang, L., H.Z. Qu, L. Zhang et al. 2008. Purification and characterization of a proteolytic enzyme from fig latex. *Chem Res Chin Univ.* 24: 348–52.

Huang, S.D., H.C. Shu, and T.S. Hsieh. 1972. [Plant proteases]. *Huaxue* 3: 96–120.

Husain, S.S., and G. Lowe. 1970. The amino acid sequence around the active-site cysteine and histidine residues, and the buried cysteine residue in ficin. *Biochem J.* 117: 333–40.

Irie, S. 1964. Anticancer activity of some latexes and luciferin. *Gann* 55: 263–6.

Jaffe, W.G. 1945. The activation of papain and related plant enzymes with sodium thiosulfate. *Arch Biochem.* 8: 385–93.

Jones, I.K., and A.N. Glazer. 1970. Comparative studies on four sulfhydryl endopeptidases ("ficins") of *Ficus glabrata* latex. *J Biol Chem.* 245: 2765–72.

Kang, H., Y.S. Kim, and G.C. Chung. 2000. Characterization of natural rubber biosynthesis in *Ficus benghalensis*. *Plant Physiol Biochem (Paris)* 38: 979–87.

Kawaii, S., and E.P. Lansky. 2004. Differentiation-promoting activity of pomegranate (*Punica granatum*) fruit extracts in HL-60 human promyelocytic leukemia cells. *J Med Food* 7: 13–8.

Kim, J.S., Y.O. Kim, H.J. Ryu, Y.S. Kwak, J.Y. Lee, and H. Kang. 2003. Isolation of stress-related genes of rubber particles and latex in fig tree (*Ficus carica*) and their expressions by abiotic stress or plant hormone treatments. *Plant Cell Physiol.* 44: 412–9.

Kim, J.P., J.S. Suh, and J.S. Kim. 1986. [Isolation and purification of ficin from fig latex]. *Han'guk Sikp'um Kwahakhoechi* 18: 270–7.

Kon, S., and J.R. Whitaker. 1965. Separation and partial characterization of the peroxidases of *Ficus glabrata* latex. *J Food Sci.* 30: 977–85.

Konno, K., C. Hirayama, M. Nakamura et al. 2004. Papain protects papaya trees from herbivorous insects: Role of cysteine protease in latex. *Plant J.* 37: 370–8.

Kortt, A.A., S. Hamilton, E.C. Webb, and B. Zerner. 1974. Ficins (E.C. 3.4.22.3). Purification and characterization of the enzymic components of the latex of *Ficus glabrata*. *Biochemistry* 13: 2023–8.

Kramer, D.E., and J.R. Whitaker. 1969b. Ficin-catalyzed reactions, hydrolysis of alpha-*n*-benzoyl-l-arginine ethyl ester and alpha *n* benzoyl-l-argininamide. *Plant Physiol.* 44: 609–14.

Kramer, D.E., and J.R. Whitaker. 1964. *Ficus* enzymes. II. Properties of the proteolytic enzymes from the latex of *Ficus carica* variety *Kadota*. *J Biol Chem.* 239: 2178–83.

Kramer, D.E., and J.R. Whitaker. 1969a. Multiple molecular forms of ficin—evidence against autolysis as explanation. *Plant Physiol.* 44: 1560–1565.

Kramer, D.E., and J.R. Whitaker. 1969c. Nature of the conversion of *Ficus carica* variety *Kadota* ficin component D to component C. some physicochemical properties of components C and D. *Plant Physiol.* 44: 1566–73.

Krishnaswamy, M. A., D.S. Johar, and V. Subrahmanyam. 1961. Vegetable rennet for Cheddar and processed cheese. *Res Ind. (New Delhi)* 6: 43–4.

Lansky, E.P., and R.A. Newman. 2007. *Punica granatum* (pomegranate) and its potential for prevention and treatment of inflammation and cancer. *J Ethnopharmacol.* 109: 177–206.

Lansky, E.P., H.M. Paavilainen, A.D. Pawlus, and R.A. Newman. 2008. *Ficus* spp. (fig): Ethnobotany and potential as anticancer and anti-inflammatory agents. *J Ethnopharmacol.* 119: 195–213.

Li, S.C., M. Asakawa, Y. Hirabayashi, and Y. Li. 1981. Isolation of two endo-beta-*N*-acetylglucosaminidases from fig latex. *Biochim Biophys Acta* 660: 278–83.

Liener, I.E., and B. Friedenson. 1970. Ficin. *Methods Enzymol.* 19: 261–73.

Luna, L.E. 1984b. The concept of plants as teachers among four Mestizo shamans of Iquitos, northeastern Peru. *J Ethnopharmacol.* 11: 134–56.

Luna, L.E. 1984a. The healing practices of a Peruvian shaman. *J Ethnopharmacol.* 11: 123–33.

Lynn, K.R., and N.A. Clevette-Radford. 1986. Ficin E, a serine-centred protease from *Ficus elastica. Phytochemistry* 25: 1559–61.

Major, R.T., and A. Walti. 1939. Proteolytic enzyme material from *Ficus* latex. US Patent 2162737 19390620, 1939.

Malthouse, J.P., and K. Brocklehurst. 1976. Preparation of fully active ficin from *Ficus glabrata* by covalent chromatography and characterization of its active centre by using 2,2′-depyridyl disulphide as a reactivity probe. *Biochem J.* 159: 221–34.

Maruyama, S., S. Miyoshi, and H. Tanaka. 1989. Angiotensin I-converting enzyme inhibitors derived from *Ficus carica. Agric Biol Chem.* 53: 2763–7.

Mavlonov, G.T., K.A. Ubaidullaeva, M.I. Rakhmanov, I.Y. Abdurakhmonov, and A. Abdukarimov. 2008. Chitin-binding antifungal protein from *Ficus carica* latex. *Chem Nat Compd.* 44: 216–9.

Mavromoustakos, T., C. Petrou, E. Kokkalou et al. 2008. *Ficus sycomorus* sap: A psoralene source with potential for the treatment of psoriasis. In Abstracts of the 7th Joint Meeting of the Association Francophone pour l'Enseignement de la Recherche en Pharmacognosie (AFERP), American Society of Pharmacognosy (ASP), Society for Medicinal Plant Research (GA), Phytochemical Society of Europe (PSE), and Societa Italiana di Fitochimica (SIF), August 3–8, 2008, Athens, Greece, published in *Planta Med.* 74, 116, 2008.

Mechoulam, R., S. Rubnov, Y. Kashman, R. Rabinowitz, and M. Schlesinger. 2002. 6-*O*-acyl-β-d-glucosyl-β-sitosterols from fig (*Ficus carica*) resin: Isolation, structure elucidation, synthesis and their use as antiproliferative agents. PCT Int. Patent Application WO 2002-IB918 20020326, 31 pp, 2002.

Murachi, T., and N. Takahashi. 1970. Structure and function of stem bromelain. In *Struct Funct Relat Proteolytic Enzymes Proc Int Symp.* School of Medicine, Nagoya City University, Nagoya, Japan. Meeting Date 1969, ed. P. Desnuelle, 298–309.

Nagaty, H.F., M.A. Rifaat, and T.A. Morsy. 1959. Trials of the effect on dog Ascaris in vivo produced by the latex of *Ficus carica* and *Papaya carica* growing in Cairo gardens. *Ann Trop Med Parasitol.* 53: 215–9.

Nassar, A.H., and H.J. Newbury. 1987. Ficin production by callus cultures of *Ficus carica. J Plant Physiol.* 31: 171–9.

Nazir, N., S. Koul, M.A. Qurishi et al. 2007. Immunomodulatory effect of bergenin and norbergenin against adjuvant-induced arthritis—a flow cytometric study. *J Ethnopharmacol.* 112: 401–5.

NIST Chemistry WebBook. 1996. [Online Database]. NIST Standard Reference Database Number 69. The National Institute of Standards and Technology (NIST). URL: http://webbook.nist.gov/chemistry/contact.html [accessed August 14, 2009].

Orlacchio, A., C. Maffei, C. Emiliani, and J.A. Reinosa. 1985. On the active site of β-hexosaminidase from latex of *Ficus glabrata*. *Phytochemistry* 24: 659–62.

Pant, R., and S.C. Srivastava. 1966. Proteolytic activity of some plant latex: Effect of time variation. *Curr Sci*. 35: 42–3.

Park, W.G. 2008. [Method for preparing products containing refined latex of *Ficus carica* for re-growth and avian influenza prevention of birds]. Korean Patent Application KR 2008-70894, 20080718. *Repub. Korean Kongkae Taeho Kongbo*.

Pawlus, A.D., C.A. Cartwright, M. Vijjeswarapu, Z. Liu, E. Woltering, and R.A. Newman. 2008. Antiangiogenic activity from the fruit latex of *Ficus carica* (Fig). In Abstracts of the 7th Joint Meeting of the Association Francophone pour l'Enseignement de la Recherche en Pharmacognosie (AFERP), American Society of Pharmacognosy (ASP), Society for Medicinal Plant Research (GA), Phytochemical Society of Europe (PSE), and Societa Italiana di Fitochimica (SIF), August 3–8, 2008, Athens, Greece, published in *Planta Med*. 74, 72, 2008.

Perello, M., M.C. Arribere, N.O. Caffini, and N.S. Priolo. 2000. Proteolytic enzymes from the latex of *Ficus pumila* L. (Moraceae). *Acta Farm Bonaer*. 19: 257–62.

Petrov, V., R. Fagard, and P. Lijnen. 2000. Effect of protease inhibitors on angiotensin-converting enzyme activity in human T-lymphocytes. *Am J Hypertens*. 13: 535–39.

Plinius, C., Pliny (the Elder). 1967–1970. *C. Plini Secundi naturalis historiae, libri XXXVII*. [The Natural History]. Ed. K.F.T. Mayhoff. Stuttgart: B.C. Teubner.

Porcelli, G. 1967. Ficin. III. Purification of ficin by gel filtration and the characterization of other protein fractions of *Ficus* latex. *J Chromatogr*. 28: 44–8.

Pradalier, A., E. Leriche, C. Trinh, and J.L. Molitor. 2004. [The return of the prodigal child or allergy to ficus]. *Eur Ann Allergy Clin Immunol*. 36: 326–9.

PubChem. The PubChem Project. 2001. [Online Database]. National Center for Biotechnology Information. U.S. National Library of Medicine. 8600 Rockville Pike, Bethesda, MD 20894. URL: http://pubchem.ncbi.nlm.nih.gov/ [accessed August 16, 2009].Rahman, N.N., M. Khan, and R. Hasan. 1994. Bioactive components from Ficus glomerata. Pure Appl Chem. 66: 2287–90.

Rajabi, O., B.J. Danaie, A.R. Varasateh, L. Jahangiri, and A. Baratian. 2006. [Quantitative analysis of amount and activity of ficin in Khorasan's fig tree latex retrieved from different organs of trees in different seasons]. *Faslnamah-i Giyahan-i Daruyi* 5: 11–20, 98.

Richter, G., H.P. Schwarz, F. Dorner, and P.L. Turecek. 2002. Activation and inactivation of human factor X by proteases derived from *Ficus carica*. *Br J Haematol*. 119: 1042–51.

Rubnov, S., Y. Kashman, R. Rabinowitz, M. Schlesinger, and R. Mechoulam. 2001. Suppressors of cancer cell proliferation from fig (*Ficus carica*) resin: Isolation and structure elucidation. *J Nat Prod*. 64: 993–6.

Rudack, C., F. Sachse, and S. Jörg. 2003. [Aeroallergens becoming more significant for allergic rhinitis]. *HNO* 51: 694–703.

Santana, C.A.R. 2008. [Topical pharmaceutical composition]. Brazilian Patent Application BR 2006-2051 *Braz. Pedido PI.*.

Santhakumari, T.N., and P.P. Pillay. 1959. Chemical examination of the latex of *Ficus bengalensis*. *Bull Res Inst Univ Kerala Trivandrum A Phys Sci*. 6: 6–10.

Sapozhnikova, E.V. 1940. [The chemical composition of the fruits and latex of *Ficus carica* L]. *Biokhim Kul'tur Rastenii* 7: 485–8.

Schenkelberger, V., M. Freitag, and P. Altmeyer. 1998. [*Ficus benjamina*—the hidden allergen in the house]. *Hautarzt* 49: 2–5.

Sgarbieri, V.C. 1965. [Proteolytic enzymes in latex from several varieties of *Ficus carica*]. *Bragantia* 24: 109–24.

Sgarbieri, V.C., S.M. Gupte, D.E. Kramer, and J.R. Whitaker. 1964. *Ficus* enzymes. I. Separation of the proteolytic enzymes of *Ficus carica* and *Ficus glabrata* latices. *J Biol Chem.* 239: 2170–7.

Shukla, O.P., and C.R. Krishna Murti. 1961. Bacteriolytic activity of plant latexes. *J Sci Ind Res.* 20C: 225–6.

Shukla, O.P., and C.R. Krishna Murti. 1971. Biochemistry of plant latex. *J Sci Ind Res.* 30: 640–62.

Stepek, G., D.J. Buttle, I.R. Duce, A. Lowe, and J.M. Behnke. 2005. Assessment of the anthelmintic effect of natural plant cysteine proteinases against the gastrointestinal nematode, *Heligmosomoides polygyrus*, in vitro. *Parasitology* 130: 203–211.

Stepek, G., R.H. Curtis, B.R. Kerry et al. 2007b. Nematicidal effects of cysteine proteinases against sedentary plant parasitic nematodes. *Parasitology* 134: 1831–8.

Stepek, G., A.E. Lowe, D.J. Buttle, I.R. Duce, and J.M. Behnke. 2007a. Anthelmintic action of plant cysteine proteinases against the rodent stomach nematode, *Protospirura muricola*, in vitro and in vivo. *Parasitology* 134: 103–12, Erratum 134: 607.

Stepek, G., A.E. Lowe, D.J. Buttle, I.R. Duce, and J.M. Behnke. 2006. In vitro and in vivo anthelmintic efficacy of plant cysteine proteinases against the rodent gastrointestinal nematode, *Trichuris muris*. *Parasitology* 132: 681–9.

Stevens, H.P. 1939. *Ficus elastica* and *Castilloa* latexes. *India Rubber World* 100: 27–30, 42.

Stevens, W. 2000. [Latex allergy and cross-reactions: A new threat?] *Verh K Acad Geneeskd Belg.* 62: 155–70.

Sugiura, M., and M. Sasaki. 1973. [Studies on proteinases from *Ficus carica* var. *Horaishi*. II. Physicochemical properties of ficin A, B, C, and D.] *Yakugaku Zasshi* 93: 63–7.

Sugiura, M., and M. Sasaki. 1974. Studies on proteinases from *Ficus carica* var *Horaishi*. V. Purification and properties of a sugar-containing proteinase (Ficin S). *Biochim Biophys Acta* 350: 38–47.

Taira, T., A. Ohdomari, N. Nakama, M. Shimoji, and M. Ishihara. 2005. Characterization and antifungal activity of gazyumaru (*Ficus microcarpa*) latex chitinases: Both the chitin-binding and the antifungal activities of class I chitinase are reinforced with increasing ionic strength. *Biosci Biotechnol Biochem.* 69: 811–8.

Tata, S.J., A.N. Boyce, B.L. Archer, and B.G. Audley. 1976. Lysozymes: Major components of the sedimentable phase of *Hevea brasiliensis* latex. *J Rubber Res Inst Malays.* 24: 233–6.

Ten Cate, H., and A. Falanga. 2008. Overview of the postulated mechanisms linking cancer and thrombosis. *Pathophysiol Haemost Thromb.* 36: 122–30.

Ullman, S.B. 1952. The inhibitory and necrosis-inducing effects of the latex of *Ficus carica* L. on transplanted and spontaneous tumours. *Exp Med Surg.* 10: 26–49.

Ullman, S.B., G.M. Clark, and K.M. Roan. 1952. The effects of the fraction R3 of the latex of *Ficus carica* L. on the tissues of mice bearing spontaneous mammary tumors. *Exp Med Surg.* 10: 287–305.

Ullman, S.B., L. Halberstaedter, and J. Leibowitz. 1945. Some pharmacological and biological effects of the latex of *Ficus carica* L. *Exp Med Surg.* 3: 11–23.

Ultee, A.J. 1949. Components of samples of latex [from a number of tropical trees]. *Pharm Weekbl.* 84: 65–70.

Ultee, A.J. 1922a. Stearic acid in the latex of *Ficus fulva*, Reinw. *Bull Jard Bot Buitenzorg Ser. 3* 5: 105–6.

Ultee, A.J. 1922b. A variety of wax in the latex of *Ficus alba* Reiner. *Bull Jard Bot Buitenzorg* 5: 241–3.

Urech, K., J.M. Scher, K. Hostanska, and H. Becker. 2005. Apoptosis inducing activity of viscin, a lipophilic extract from *Viscum album* L. *J Pharm Pharmacol.* 57: 101–9.

van der Bie, G.J. 1946. *Ficus elastica* latex and rubber. *Rev Gen Caoutchouc* 23: 285–90.

Visco, S. 1924. Enzymes present in the latex of *Ficus carica*. I. Esterase. *Arch Farmacol Sper Sci Affin.* 38: 243–50.

Walti, A. 1954. Isolating proteolytic enzyme and suspended matter in *Ficus* latex. US Patent 2694032 19541109.

Wang, K.W. 2007. A new fatty acid ester of triterpenoid from *Celastrus rosthornianus* with anti-tumor activities. *Nat Prod Res.* 21: 669–74.

Wang, J., X. Wang, S. Jiang et al. 2008. Cytotoxicity of fig fruit latex against human cancer cells. *Food Chem Toxicol.* 46: 1025–33.

Wang, J., B.J. Wang, C.M. Wei et al. 2009. Determination of bergenin in human plasma after oral administration by HPLC-MS/MS method and its pharmacokinetic study. *Biomed Chromatogr.* 23: 199–203.

Werfel, S., F. Ruëff, and B. Przybilla. 2001. [Anaphylactic reaction to *Ficus benjamina* (weeping fig)]. *Hautarzt* 52: 935–7.

Whitaker, J.R. 1958. The ficin content of the latex from different varieties of *Ficus carica* and a comparison of several micro methods of protein determination. *J Food Sci.* 23: 364–70.

Williams, D.C., V.C. Sgarbieri, and J.R. Whitaker. 1968. Proteolytic activity in the genus *Ficus*. *Plant Physiol.* 43: 1083–8.

Yu, W., Y. Wang, Y. Zhang et al. 2009. Quantitation of bergenin in human plasma by liquid chromatography/tandem mass spectrometry. *J Chromatogr B Analyt Technol Biomed Life Sci.* 877: 33–6.

6 Bark, Wood, and Stems

When considering the wood and bark of the tree, we are likely to find ourselves out on a limb, suffering from the ambiguity of trying to figure out exactly what is being referred to by the word *stem*. Ethnomedical assessments of *Ficus* frequently use the term "bark of stem." In fact, stem may refer first to the trunk of the tree. It may also refer to processes originating from branches and going outward, and may even refer to the stem of a leaf, or of a fruit. Usually, however, we must assume it is woody, since there is reference to its "bark." Many of the most provocative pharmacognostical (the pharmacology of crude drugs) investigations of *Ficus* in the past few months have been done in stem bark, and for the most part, this may often refer to bark of the trunk.

These investigations fall into several partially overlapping groups: 1. fibrin generation, 2. anticonvulsant, 3. antioxidant, 4. hepato- and radioprotective, 5. antimicrobial, including antibacterial and antifungal, 6. antiviral, 7. mosquito larvicidal, 8. antiallergic, and 9. antidrepanocytic. The studies from 2007 to the present writing (May 31, 2009) are summarized in Table 6.1.

Earlier work in the past decade with stem bark of various *Ficus* spp. highlighted some familiar themes, such as protection of the liver from carbon tetrahydrochloride, antidiabetic, and also antiatherosclerotic, muscle-relaxing activity, antidiuretic, anti-inflammatory, antiplasmodial (antimalarial), antioxidant, and anticancer effects. These studies, from 2000–2006, are summarized in Table 6.2.

FIGURE 6.1   Small branchlet of *Ficus microcarpa* (pen and ink by Zipora Lansky).

FIGURE 6.2 *Ficus carica*, trunk (27.7.2009, Seven Species Garden, Binyamina, Israel, by Helena Paavilainen).

Having reviewed Tables 6.1 to 6.4, we are struck by how this field of *Ficus* stem bark pharmacology has grown up and started to walk over really just the past year. From a single pioneering paper by Professor Augusti from India in 1975 on antidiabetic effects of an ethanolic extract of *F. benghalensis* bark, the field has proliferated within the past decade with the good Professor Augusti remaining at the forefront throughout the period. There were no further papers from 1975 to 1993 when Cherian and Augusti (1993) published their work focusing on the in vivo antidiabetic effect of a single glycoside of leucopelargonidin found within the original *F. benghalensis* ethanolic extract. In 1994, Rahman et al., working with a petroleum ether extract of *F. glomerata* bark, showed specific effects on enzymes related to glucose metabolism, thus delving more deeply into the antidiabetic effects previously reported. In 1998, Daniel et al. isolated two flavonoids from *F. benghalensis* bark and found them to possess potent antioxidant properties in vivo.

From 2000 to 2006, an additional ten papers were published on the physiological actions of *Ficus* stem bark preparations from *F. benghalensis* and five other species

FIGURE 6.3 *Ficus carica*, branch (27.7.2009, Seven Species Garden, Binyamina, Israel, by Helena Paavilainen).

not previously investigated: *F. citrifolia, F. fistulosa, F. racemosa, F. thonningii,* and *F. sycomorus.* In addition to ongoing work in diabetes in vivo with extracts of *F. racemosa* by Bhaskara et al. (2002) and with *F. thonningii* by Musabayane et al. (2006), new studies with the extracts of these barks and their derivatives entered into the realms of hepatoprotection from carbon tetrachloride *in vivo* (*F. benghalensis*; Augusti et al. 2005), anti-inflammatory (*F. racemosa*; Li et al. 2003), antipyretic (Rao et al. 2002), antidiuretic (*F. racemosa*; Ratnasooriya et al. 2003), muscle relaxant (*F. sycomorus*; Sandabe et al. 2006), as a potential adjunct to cancer chemotherapy to overcome cancer multidrug resistance (*F. citrifolia*; Simon et al. 2001), and as an antimalarial (*F. fistulosa*; Zhang et al. 2002). The work of Daniel et al. (2003) took the work against diabetes with a *F. benghalensis* leucopelargonin derivative two steps further, first by moving to alloxan-induced diabetic dogs, and second, by noting effects in vivo on cholesterol biosynthesis and probable protection against atherogenesis.

Since 2007, until the present—it is now June 3, 2009—14 more papers have been published on *Ficus* spp. stem bark pharmacology. Investigations into antioxidant effects (*F. benghalensis, F. racemosa*; Manian et al. 2008) have also extended this concept further into chemoprotection against hepatic damage from carbon tetrachloride (*F. glomerata*; Channabasavaraj et al. 2008) and against radiation damage in Chinese hamster fibroblasts (Veerapur et al. 2009). Musabayane et al. (2006) showed that their extract of *F. thonningii* not only lowered blood sugar in vivo, but also stimulated health and vitality of kidney and heart preparations, signaling likely nascent cardioprotective and renal protective effects in situ. Several workers in the last couple years have plunged into antimicrobial studies with very promising findings, often against difficult and highly resistant organisms. The antimicrobial compounds and complexes, including one case with antiviral properties (*F. sycomorus*; Maregesi et al. 2008), were also derived from *F. asperifolia* (Annan and Houghton 2008),

FIGURE 6.4   *Ficus binnendyijkii*, trunk (19.7.2009, University of Turku, Botanical Garden, Ruissalo, Turku, Finland, by Kaarina Paavilainen).

F. microcarpa (Ao et al. 2008), *F. chlamydocarpa*, *F. cordata* (Kuete et al. 2008), and *F. ovata* (Kuete et al. 2009). Chindo et al. (2009) have demonstrated an intriguing pharmaceutical lead for an anticonvulsive drug from *F. platyphylla*. A drug for wound healing, to promote fibroblast growth as well as for affording broad-spectrum antimicrobial protection is suggested from the work of Annan and Houghton (2008) with *F. asperifolia*. Taur et al. (2007) elucidated antiallergic mechanisms underlying the action of *F. benghalensis*, and Nongonierma et al. (2007) showed antidrepanocytic effect from an extract of the bark of *F. gnaphalocarpa*, suggesting a putative drug for use in sickle cell anemia. Rahuman et al. (2008) elaborated the potential for an extract of *F. racemosa* to kill mosquito larvae and its consequent wide-scale potential public health applications.

Tables 6.5 and 6.6 give an overview of the historical medicinal uses of the stem, wood, and bark of different *Ficus* spp.

For references used in Table 6.6, see Chapter 3, Table 3.6. The scoring system is based on the number of times a reference discusses the use of *Ficus* stems, bark, or

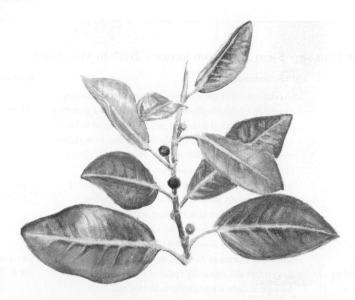

FIGURE 6.5 **(See color insert after page 128.)** *Ficus microcarpa*, stem (18.8.2009, Haifa, Israel, watercolor by Zipora Lansky).

wood for the purpose indicated. If the number of citations is regular, the reference is given in regular type and scored as "1." If the number of citations is strong, the reference is given in *italic* and scored as "2." If the number of citations is strongest, the reference is given in **bold** and scored as "3." The final score in the last column is the sum of these scores for the individual references. The particular parts of the plant used are starred, but this information does not enter into the calculation of the total score for the function. The indications are shown in the order of importance based on their score values.

For the historical uses of ashes and lye prepared from the stems and wood of *Ficus* spp., see Chapter 3, Tables 3.9 and 3.10.

CHEMISTRY

The bark is rich in polyphenols, including flavonoids and their glycosides. Triterpenoids and furocoumarins, sterols, and other complex phenolic acids, alcohols, and ketones have all been isolated and reported from the bark of various species. This work is summarized in Table 6.7.

The challenge and also the trend in plant chemical analysis is evermore focused on awareness of minor compounds, which, by themselves, are often not physiologically active in animals, at least not physiologically active in a way that is easily measured or noticed. The challenge is to measure the presence of all compounds in the matrix, and this idea is best understood as relating to metabolomics. Still, some specific techniques have been proved for *F. carica* stem bark for the measurement of polyphenolics and furocoumarins using reversed-phase HPLC (high-performance liquid chromatography) (Govindarajan et al. 2008).

TABLE 6.1
Studies on *Ficus* spp. Stem Bark from January 2007 to May 2009

Actions	Study Summary	Reference
Antioxidant, antibacterial, fibroblast stimulation	Aqueous extract of the traditional Ghana wound-healing *Ficus asperifolia* Miq bark stimulated fibroblast growth at 50 µg/mL, inhibited hydrogen peroxide-induced oxidative damage and growth of *Staphylococcus aureus* (including resistant strains), *Bacillus subtilis, Micrococcus flavus, Escherichia coli,* and *Pseudomonas aeruginosa*	Annan and Houghton 2008
Anticonvulsant	*Ficus platyphylla* stem bark, a saponin-rich fraction of methanol extract protected mice against pentylene-tetrazole- and strychnine-induced seizures, *and* blocked sustained repetitive firing (SRF) and spontaneous action potential firing in neonatal rat brain slice model	Chindo et al. 2009
Antioxidant, hepatoprotective	Methanolic extract of *Ficus glomerata* was potently antioxidant and markedly reduced biochemical hepatic damage in carbon tetrachloride-treated rats	Channabasavaraj et al. 2008
Antioxidant, radioprotective	Ethanolic and water extracts of *Ficus racemosa* stem bark dose-dependently improved free radical scavenging in silico and reduced percentage of micronucleated binuclear cells, an index of genetic damage, in irradiated Chinese hamster lung fibroblasts	Veerapur et al. 2009
Antimicrobial	Methanolic extracts of the stem bark of *Ficus chlamydocarpa, Ficus cordata,* a mixture of the two plants, and pure isolated stem bark flavonoids alpinumisoflavone, genistein, laburnetin, luteolin (from *F. chlamydocarpa*), and catechin and epiafzelechin (from *F. cordata*) were tested and found active against mycobacteria, fungi, Gram-positive and Gram-negative bacteria	Kuete et al. 2008
Antimicrobial	Minimal inhibition concentrations of *Ficus ovata* stem bark methanolic extract, selected fractions and pure isolated compounds were found for Gram-positive and Gram-negative bacteria and fungal pathogens, especially *Streptococcus faecalis, Candida albicans* and *Microsporum audouinii,* and *Staphylococcus aureus*	Kuete et al. 2009
Antiviral	The stem bark of *Ficus sycomorus* was one of four medicinal plant parts from which an extract exhibited notable activity against human simplex virus in a panel of 50 medicinal Tanzanian plant parts from 39 plants tested against the virus and selected microbes	Maregesi et al. 2008
Mosquito larvicidal	An acetone fraction of *Ficus racemosa* exerted potent toxic activity against early fourth-instar larvae of *Culex quinquefasciatus* (Diptera: Culicidae) and other mosquito species and yielded a unique tetracyclic triterpene, gluanol acetate, which was also potently larvicidal against several mosquito species	Rahuman et al. 2008

TABLE 6.1 (continued)
Studies on *Ficus* spp. Stem Bark from January 2007 to May 2009

Actions	Study Summary	Reference
Antioxidant, antibacterial	Effects of extract of *Ficus microcarpa*	Ao et al. 2008
Antioxidant	Antioxidant action for extracts of stem bark of *Ficus benghalensis* and *Ficus racemosa*	Manian et al. 2008
Antiallergic, antistress	Aqueous, ethanolic, and ethyl acetate extracts of the stem bark of *Ficus benghalensis* reduced milk-induced leukocytosis and eosinophilia	Taur et al. 2007
Antidrepanocytic (sickle cell)	Drepanocytes in the presence of acetone extract of stem barks of *Ficus gnaphalocarpa* take again a normal configuration	Nongonierma et al. 2007
Hypoglycemic, cardioprotective and renal protective, benefiting diabetes	Similar to metformin, ethanolic extract of *Ficus thonningii* (Blume) decreased blood glucose levels in nondiabetic and STZ-diabetic rats, and improved metabolism in Madin-Darby bovine kidney cells (a model of renal function) with negative chronotropic and inotropic actions in isolated rat atria (modeling cardiac health)	Musabayane et al. 2007

One small curiosity may lead to epochal ideas, especially when concerning fungi and fermentation. Recall the examples of penicillin, artificial rubber, or LSD from ergot. In another case, a new alkaloid was discovered in a "fermentation broth of an unidentified endophytic fungus obtained from the bark of *Ficus microcarpa* L." The alkaloid, dubbed "nomofungin" (Ratnayake et al. 2001), possesses a unique and novel natural chemical structure and the actions of disrupting microfilaments in cultured mammalian cells and moderate cytotoxicity with minimum inhibitory concentrations (MICs) of 2 and 4.5 μg/mL against LoVo (human colon cancer) and KB (human oral squamous cell cancer) cells, respectively. The possibility of creating new hybrid medicinal agents based on overall metabolomic patterning between species, or "interkingdom metabolomes" of associations between particular medicinal fungi and higher plant species, should not be overlooked (Lansky and Nevo 2009). This is a topic to which we shall return before closing.

TABLE 6.2

Studies on *Ficus* spp. Stem Bark between 2000 and 2006

Actions	Study Summary	Reference
Hepatoprotective	Leucopelargonidin derivative isolated from *F. benghalensis* prevented derangement to alkaline phosphatase, alanine transaminase, aspartate transaminase, glutathionze reductase, HMGCoA reductase, catalase, gluc.6.PDH, and malic enzyme in rats treated with CCl₄ (carbon tetrachloride)	Augusti et al. 2005
Antidiabetic	Significantly reduced blood sugar in both normal and alloxan-induced diabetic rats by *F. racemosa* stem bark extract 200 and 400 mg/kg p.o. (similar to standard hypoglycemic agent)	Bhaskara et al. 2002
Antiatherogenic	Treatment of alloxan diabetic dogs with leucopelargonin derivative (100 mg/kg/day) from the bark of *F. benghalensis* decreased fasting blood sugar and glycosylated hemoglobin by 34% and 28%, respectively, while decreasing cholesterol synthesis in rats	Daniel et al. 2003
Anti-inflammatory	With lower numbers denoting greater potency, aspirin with IC(50) Mg/mL inhibition of 241, and indomethacin with a score of 1.2, the ethanolic *F. racemosa* stem bark extract has 100, the second-best score among 33 extracts of 24 medicinal plant species tested for inhibition of cyclooxygenase type 1 (COX-1)	Li et al. 2003
Antidiabetic, appetite suppressant	Extract of *Ficus thonningii* reduced blood sugar and food consumption in rats over 5 weeks, with dosing comparable to classical hypoglycemic drugs	Musabayane et al. 2006
Antipyretic	Subcutaneous injection of methanol extract of *Ficus racemosa* reduced yeast-induced hyperpyrexia in rats; antipyretic dose equivalent of paracetamol	Rao et al. 2002
Antidiuretic	50% antidiuretic effect in rats of antidiuretic hormone (ADH) with *F. racemosa* aqueous decoction per os as per reduced urinary Na+ level and Na+/K+ ratio, and an increase in urinary osmolarity	Ratnasooriya et al. 2003
Muscle relaxant	Extract of *F. sycomorus* reduced acetylcholine contractile responses of guinea pig duodena and frog recti abdominis muscles significantly, thus inhibiting both striated and smooth muscle contraction	Sandabe et al. 2006
Anticancer	Dichloromethane extract of *F. citrifolia* stem bark bound to purified C-terminal cytosolic domain of P-glycoprotein induced accumulation and cytotoxicity of vinblastine and daunomycin in two model cell lines overexpressing P-glycoprotein, comparable to cyclosporin A in increasing intracellular accumulation of daunomycin in K562/R7 leukemic cells and enhanced cytotoxic effect of vinblastine on the growth of MESSA/Dx5 cells	Simon et al. 2001
Antimalarial	*F. fistulosa* extract inhibited *Plasmodium falciparum* growth with IC₅₀ values <1 ng/mL	Zhang et al. 2002

TABLE 6.3
Studies on *Ficus* spp. Stem Bark prior to 2000

Actions	Study Summary	Reference
Antidiabetic	Ethanolic extract of *F. benghalensis* bark and a glucoside within it are hypoglycemic both in normal and alloxan diabetic rabbits, with the glucoside being more active than the crude extract and half as potent as tolbutamide.	Augusti 1975
Antidiabetic	*F. benghalensis* leucopelargonidin glycoside exerted significant hypoglycemic, hypolipidemic, and serum insulin raising effects in moderately diabetic rats similar to minimal dose of glibenclamide	Cherian and Augusti 1993
Antioxidant	Two *F. benghalensis* flavonoids (Table 6.7) significantly inhibited oxidation in hyperlipidemic rats.	Daniel et al. 1998
Antidiabetic	Petroleum ether extract of *Ficus glomerata* Roxb completely inhibited enzymes glucose-6-phosphatase and arginase and activated glucose-6-phosphate dehydrogenase from rat liver while significantly reducing serum glucose.	Rahman et al. 1994

FIGURE 6.6 *Ficus carica*, trunk (27.7.2009, Seven Species Garden, Binyamina, Israel, by Helena Paavilainen).

TABLE 6.4
Studies on *Ficus* spp. Stem Bark According to Species (total studied = 14)

Ficus species	Actions	References
asperifolia	Antioxidant, antimicrobial, fibroblast stimulation	Annan and Houghton 2008
benghalensis	Antiallergic, antiatheromatous, antidiabetic, antioxidant, hepatoprotective	Augusti 1975, Augusti et al. 2005, Cherian and Augusti 1993, Daniel et al. 1998, 2003, Manian et al. 2008, Taur et al. 2007
chlamydocarpa	Antimicrobial	Kuete et al. 2008
citrifolia	Anticancer	Simon et al. 2001
cordata	Antimicrobial	Kuete et al. 2008
fistulosa	Antimalarial	Zhang et al. 2002
gnaphalocarpa	Antidrepanocytic	Nongonierma et al. 2007
glomerata	Antidiabetic, hepatoprotective	Rahman et al. 1994, Channabasavaraj et al. 2008
microcarpa	Antioxidant, antimicrobial	Ao et al. 2008
ovata	Antimicrobial	Kuete et al. 2009
platyphylla	Anticonvulsant	Chindo et al. 2009
racemosa	Antioxidant, antidiuretic, anti-inflammatory, antipyretic, mosquito larvicidal, radioprotective	Bhaskara et al. 2002, Li et al. 2003, Manian et al. 2008, Rao et al. 2002, Rahuman et al. 2008, Ratnasooriya et al. 2003, Veerapur et al. 2009
sycomorus	Antiviral, muscle relaxant	Maregesi et al. 2008, Sandabe et al. 2006
thonningii	Antidiabetic, appetite suppressant, cardioprotective, and renal protective	Musabayane et al. 2006, 2007

TABLE 6.5

Historical Uses of *Ficus* spp. Stem, Bark, and Wood: Formulas

How to Prepare and Store Stems, Bark, and Wood from *Ficus* for Internal and External Applications

Variations	Notes
Shoots as a poultice	Externally for dog bites; on warts
Juice of shoots taken before the leaves appear	Locally on a tooth with caries
Shoots mixed with wax	Externally for furuncles, boils, and bites of shrew mouse
Shoots taken in wine	Internally for scorpion stings and shrew mouse bites. The wound is additionally treated with latex and leaves of fig tree
Shoots and leaves mixed with chick-pea flour and wine	Antidote to the poison of marine animals
Twigs for mixing milk when preparing whey	Whey internally as a purgative
Branches as a cooking aid	Branches are put to boil with hard beef in order to make it tender and soft
Scrapings of branches	For impetigo: branches are first barked and then scraped; the scrapings, which are as fine as sawdust, being applied topically to the parts affected
Bark beaten up with oil	Internally for ulcerations in the abdomen

TABLE 6.6

Historical Uses of *Ficus* spp. Stem, Bark, and Wood

Indication	Branches/ Twigs	Juice of Branches/ Twigs	Ashes of Branches/ Twigs	Tree	Juice of Tree	Ashes of Tree	Wood	Ashes of Wood	Ashes (general)	Bark	Juice of Bark	Sap of Bark	References	Score
Warming	*	*		*	*	*							1, 3, 6, 11, 14, 15, 20	15
Tenderizes meat (in cooking)	*						*						1, 3, 6, 7, 10, 11, 13, 16, 17, 20	12
Rarefying	*			*									3, 6, 7, 10, 11, 14, 15, 20	10
Against injuries due to falls from high, particularly internal bleeding						*			*				1, 6, 13, 14, 15	9
Against wounds						*		*	*	*		*	1, 6, 14, 18, 19	9
Closing [wounds, etc.]						*		*	*	*		*	1, 6, 14, 18	8
Against abdominal, intestinal ulcers						*		*	*	*			1, 6, 7, 14	7
Against dysentery				*		*		*	*	*			1, 6, 14	6
Cleaning, absterging		*		*		*			*				1, 6, 14, 20	6
Laxative	*			*									2, 6, 11, 15	6
Spicy, sharp		*		*		*				*			1, 6, 15, 20	6
Against the sun (as a sheltering thicket)											*		15, 18, 20	5
Against warts	*	*		*	*								6, 11, 20	5
Burning, stinging		*		*	*								6, 11, 15, 20	5

Property								References	Count
Causes ulcers				*		*	*	6, 11, 20	5
Dissolving coagulated blood					*	*	*	1, 6, 13, 15	5
Against cramps, convulsions				*	*			1, 14	4
Against ordinary dog bite	*		*	*		*		1, 2, 6, 14	4
Analgesic					*	*	*	1, 6, 14, 18	4
Opens mouths of veins (causing bleeding)	*		*	*	*		*	6, 15, 20	4
Against colic				*			*	1, 14, 18	3
Against injuries due to falls from height				*	*	*		1, 13, 14	3
Against honeycomb ulcers	*							1, 2, 14	3
Against shrew mouse bite	*a							1, 14	3
Against weakness of sight	*			*		*		1, 6, 14	3
Benefits skin	*	*b		*	*			1, 14	3
Against bruises				*	*			1, 14	2
Against marks in eyes	*					*		6	2
Against proud flesh						*		6	2
Against ruptures	*c			*				1, 14	2
Against skin inflammations					*	*		1, 14	2
Against spoiled putrid materials								6, 17	2
Against tetanus				*	*		*	1, 14	2
Against tooth disease				*	*		*	1, 14	2
Against ulcers of eyes		*e				*		1, 14	2
Helpful in chronic diseases								6	2

(continued on next page)

TABLE 6.6 (continued)
Historical Uses of *Ficus* spp. Stem, Bark, and Wood

Indication	Branches/ Twigs	Juice of Branches/ Twigs	Ashes of Branches/ Twigs	Tree	Juice of Tree	Ashes of Tree	Wood	Ashes of Wood	Ashes (general)	Bark	Juice of Bark	Sap of Bark	References	Score
Bitter										*	*		15	2
Cleans ulcers									*				17	2
Coagulating milk					*								2, 14	2
Corroding					*					*			15	2
Dissolving coagulated milk					*								2, 14	2
For woodwork							*						15, 20	2
Healing burns						*			*				1, 14	2
Opening hemorrhoids				*	*								6	2
Removes splinters of bones after accident	*												1, 14	2
Against all swellings/tumors of spleen		*e										*	18	1
Against all swellings/tumors under ears (mumps?)				*									7	1
Against diseases of the buttocks						*			*				17	1
Against chilblains								*					1	1
Against diarrhea	*												6	1
Against epilepsy										*			17	1
Against ergotism								*					17	1
Against fistulas								*					6	1
Against fluxes								*					6	1

Against gangrene	*					17	1
Against hard swelling, thickening of spleen	*				*	18	1
Against hemiplegia	*	*				6	1
Against hemorrhoids		*				6	1
Against leprosy			*			6	1
Against looseness of uvula						1	1
Against madness	*	*				17	1
Against malignant ulcers		*				6	1
Against marks on skin				*		17	1
Against nerve pain	*	*				6	1
Against pimples, pustules	*					1	1
Against poison of marine animals	*					1	1
Against putrid ulcers	*	*				6	1
Against scorpion sting	*					1	1
Against snakebite		*				18	1
Against spreading ulcers					*	17	1
Against thick humors					*	18	1
Against toothache	*f					7	1
Against torsion of eye	*					6	1
Against vitiligo	*					2	1
Against worry, obsession	*			*	*	17	1
Astringent					*	15	1
Benefits body	*			*		10	1
Benefits gums						17	1
Blackens hair						17	1

(continued on next page)

TABLE 6.6 (continued)
Historical Uses of *Ficus* spp. Stem, Bark, and Wood

Indication	Branches/ Twigs	Juice of Branches/ Twigs	Ashes of Branches/ Twigs	Tree	Juice of Tree	Ashes of Tree	Wood	Ashes of Wood	Ashes (general)	Bark	Juice of Bark	Sap of Bark	References	Score
Causes intestinal ulcers		*											20	1
Caustic		*g											11	1
Cooling, calming heat				*									10	1
Prevents cataract	*												6	1
Softening (in general)												*	18	1
Sudorific								*					6	1
Whitens teeth									*				17	1

a Also: white stalks/shoots.
b Ashes of stalks springing from roots.
c White stalks/shoots.
d Ashes of stalks springing from roots.
e Juice of branches that have no leaves yet.
f Branches before leaves come.
g Juice of branches before leaves come.

TABLE 6.7
Compounds Found in Stem Bark of Various *Ficus* spp.

Ficus Species	Compounds Found	Technical Details	Reference
benghalensis	Delphinidin-3-*O*-α-l-rhamnoside, pelargonidin-3-*O*-α-l-rhamnoside, and leucocyanidin-3-*O*-β-d-galactosylcellobioside	Spectral data and degradation studies	Subramanian and Misra 1977
benghalensis	20-Tetratriacontene-2-one, 6-heptatriacontene-10-one, pentatriacontan-5-one, β-sitosterol-α-d-glucose, meso-inositol	Spectral data and degradation studies	Subramanian and Misra 1978
benghalensis	5,7-Dimethyl ether of leucopelargonidin 3-*O*-α-l-rhamnoside, 5,3′-dimethyl ether of leucocyanidin 3-*O*-α-d-galactosyl cellobioside		Daniel et al. 1998
benghalensis	Leucopelargonidin and leucocyanidin derivatives; quercetin		Daniel et al. 2003
benghalensis	Leucopelargonidin glycoside		Cherian and Augusti 1993
benghalensis	Bengalenoside		Augusti 1975
exasperata	Unusual fatty acid, new acyl-glucosylsterol (unnamed structures)	1-D and 2-D NMR spectroscopic techniques (DEPT, COSY, HSQC, and NOESY) and mass spectral analysis	Dongfack et al. 2008
fistulosa Reinw. ex Blume	Verrucarin L acetate, 3α-hydroxyisohop-22(29)-en-24-oic acid, 3β-gluco-sitosterol, 3,4-dihydro-6,7-dimethoxyisocarbostyril, 3,4,5-trimethoxybenzyl alcohol, α-methyl-3,4,5-trimethoxybenzyl alcohol, indole-3-carboxaldehyde, palmanine, aurantiamide acetate	Bioassay-directed fractionation of extract of both leaves and stem barks	Zhang et al. 2002
glomerata	Lupeol, β-sitosterol, stigmasterol	Petroleum ether extract	Singhal and Saharia 1980
hispida	*n*-Triacontanyl acetate, β-amyrin acetate, gluanol acetate	Petroleum ether extract	Acharya and Kumar 1984
infectoria Roxb.	Methyl-ricinolate, β-sitosterol, lanosterol, caffeic acid, bergenin		Swami et al. 1989a

(continued on next page)

TABLE 6.7 (continued)
Compounds Found in Stem Bark of Various *Ficus* spp.

Ficus Species	Compounds Found	Technical Details	Reference
nymphaefolia Mill.	5,7-Dihydroxy-4′-methoxy-3′-(2,3-dihydroxy-3-methylbutyl)isoflavone, genistein, erycibenin A, cajanin, 5,7,2′-trihydroxy-4′-methoxyisoflavone, erythrinin C, alpinumisoflavone, derrone, 3′-(3-methylbut-2-enyl)biochanin A	Spectral analysis, including 1-D and 2-D NMR	Darbour et al. 2007
platyphylla Del.	7-Methoxycoumarin, α- and β-amyrin, ceryl alc., β-sitosterol, ursolic acid, an unknown flavonoid aglycon, 2 unknown flavonoids, rutin	Extd. with 70% EtOH then extracts. successively fractionated with petroleum ether, CHCl$_3$, and EtOAc	El-Sayyad et al. 1986
platyphylla	Saponins	Saponin-rich fraction (SFG) obtained from the methanol extract of *F. platyphylla* stem bark	Chindo et al. 2009
racemosa	β-Sitosterol	Ethanolic extract	Bhatt and Agrawal 1973
racemosa	Leucocyanidin 3-*O*-β-d-glucopyranoside, leucopelargonidin 3-*O*-α-l-rhamnopyranoside	Ethanolic extract	Agrawal and Misra 1977
racemosa	Ceryl behenate, lupeol, lupeol acetate, α-amyrin acetate	Petroleum ether extract	Shrivastava et al. 1977
racemosa	Gluanol acetate, β-sitosterol	Bioassay-guided fractionation of acetone extract	Joshi et al. 1977, Rahuman et al. 2008
religiosa	Vitamin K1, *n*-octacosanol, methyl-oleanolate, lanosterol, β-sitosterol, stigmasterol, lupen-3-one		Swami et al. 1989b

FIGURE 6.7 *Ficus benghalensis*, trunks (9.7.2009, in a park at Lahaina, Maui, by Forest & Kim Starr; Image 090709-2561; *Plants of Hawaii*).

FIGURE 6.8 *Ficus carica*, branches (22.11.2005, Haleakala Ranch, Maui, by Forest & Kim Starr; Image 051122-5356; *Plants of Hawaii*).

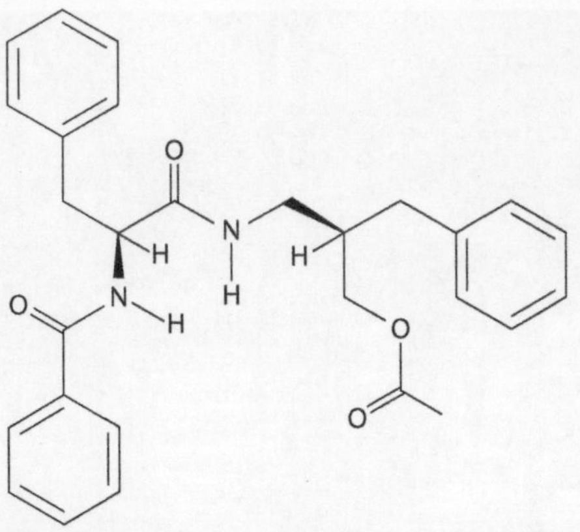

282

Figs: The genus *Ficus*

FIGURE 6.9 Aurantiamide acetate. (*PubChem*. The PubChem Project. 2001. [Online Database]. National Center for Biotechnology Information. U.S. National Library of Medicine. 8600 Rockville Pike, Bethesda, MD 20894. URL: http://pubchem.ncbi.nlm.nih.gov/ [accessed August 16, 2009].)

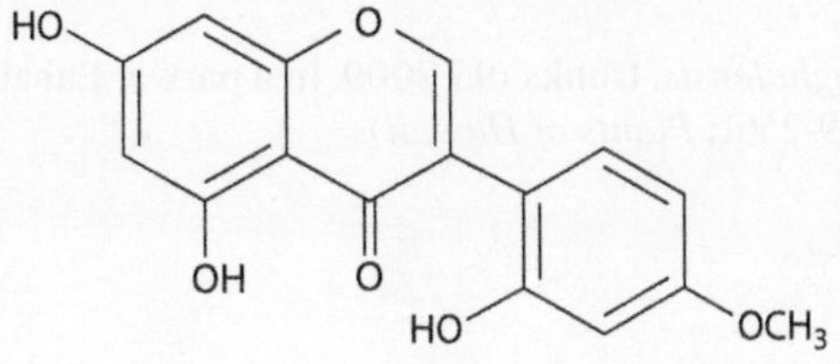

FIGURE 6.10 5,7,2′-Trihydroxy-4′-methoxyisoflavone. (Reprinted with permission from Lansky et al. 2008. *Ficus* spp. (fig): Ethnobotany and potential as anticancer and anti-inflammatory agents. *J Ethnopharmacol.* 119: 195–213. Reprinted with permission of Elsevier.)

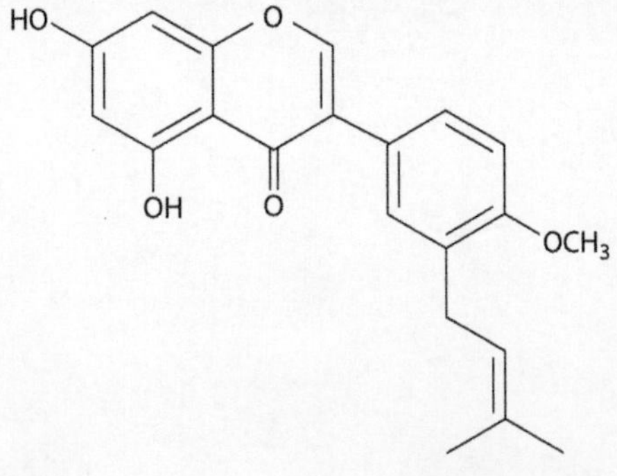

FIGURE 6.11 3′-(3-Methylbut-2-enyl)biochanin A. (Reprinted with permission from Lansky et al. 2008. *Ficus* spp. (fig): Ethnobotany and potential as anticancer and anti-inflammatory agents. *J Ethnopharmacol.* 119: 195–213. Reprinted with permission of Elsevier.)

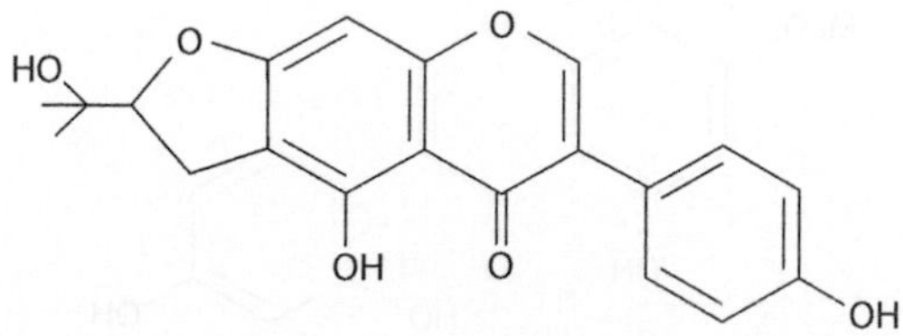

FIGURE 6.12 Erythrinin C. (Reprinted with permission from Lansky et al. 2008. *Ficus* spp. (fig): Ethnobotany and potential as anticancer and anti-inflammatory agents. *J Ethnopharmacol.* 119: 195–213. Reprinted with permission of Elsevier.)

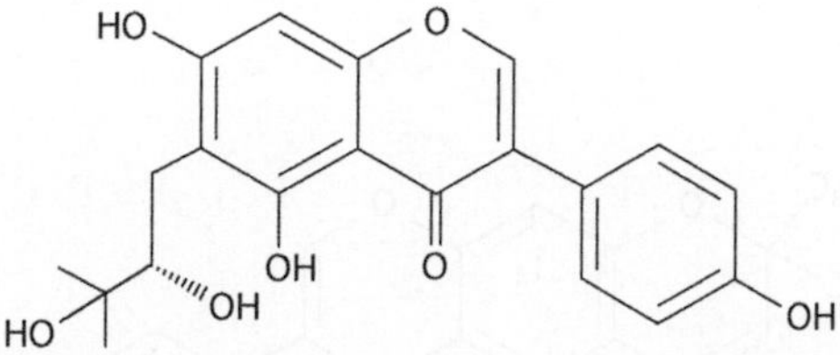

FIGURE 6.13 Erycibenin A. (Reprinted with permission from Lansky et al. 2008. *Ficus* spp. (fig): Ethnobotany and potential as anticancer and anti-inflammatory agents. *J Ethnopharmacol.* 119: 195–213. Reprinted with permission of Elsevier.)

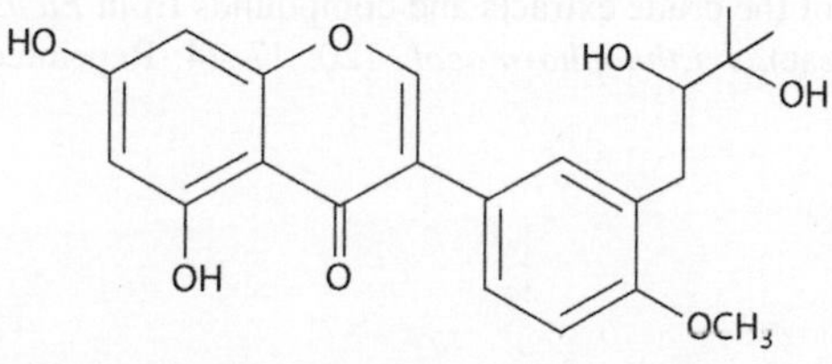

FIGURE 6.14 5,7-Dihydroxy-4-methoxy-3′-(2,3-dihydroxy-3-methylbutyl)isoflavone. (Reprinted with permission from Lansky et al. 2008. *Ficus* spp. (fig): Ethnobotany and potential as anticancer and anti-inflammatory agents. *J Ethnopharmacol.* 119: 195–213. Reprinted with permission of Elsevier.)

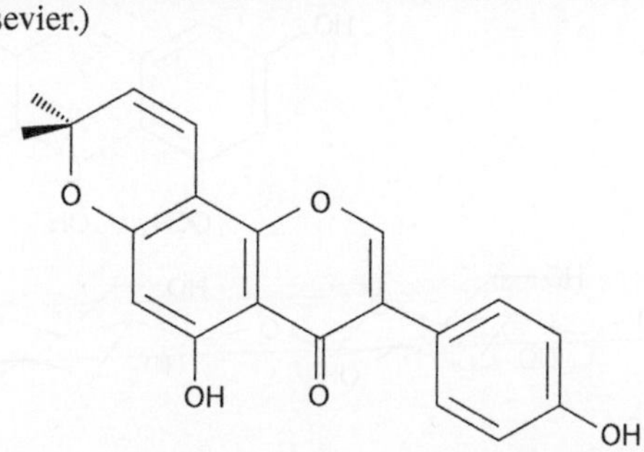

FIGURE 6.15 Derrone. (Reprinted with permission from Lansky et al. 2008. *Ficus* spp. (fig): Ethnobotany and potential as anticancer and anti-inflammatory agents. *J Ethnopharmacol.* 119: 195–213. Reprinted with permission of Elsevier.)

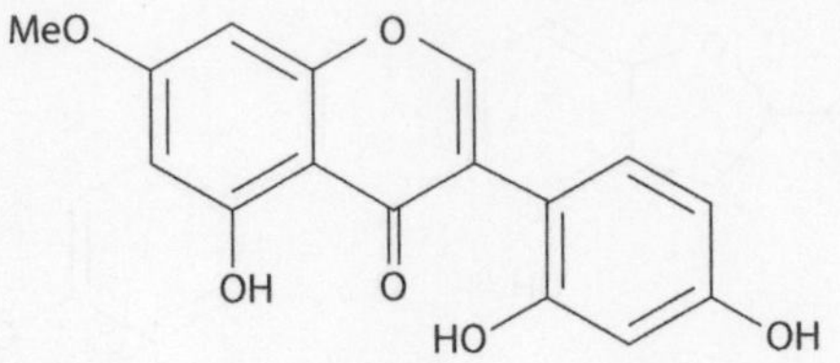

FIGURE 6.16 Cajanin. (Reprinted with permission from Kuete et al. 2009. Antimicrobial activity of the crude extract, fractions and compounds from stem bark of *Ficus ovata* (Moraceae). *J Ethnopharmacol.* 124: 556–61. Reprinted with permission of Elsevier.)

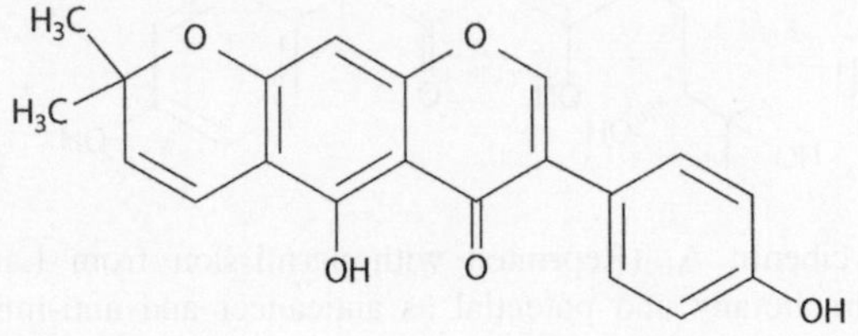

FIGURE 6.17 Alpinum isoflavone. (Reprinted with permission from Kuete et al. 2008. Antimicrobial activity of the crude extracts and compounds from *Ficus chlamydocarpa* and *Ficus cordata* (Moraceae). *J Ethnopharmacol.* 120: 17–24. Reprinted with permission of Elsevier.)

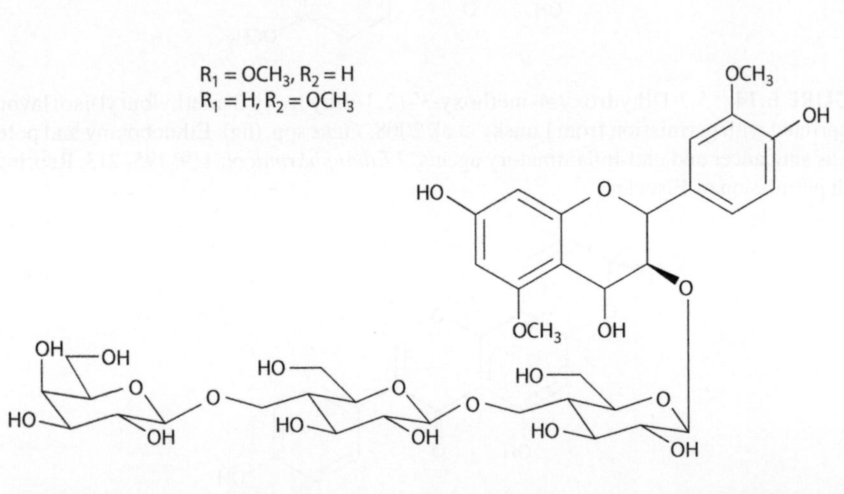

FIGURE 6.18 5,3′-Dimethyl leucocyanidin-3-*O*-β-galactosyl cellobioside. (Reprinted with permission from Lansky et al. 2008. *Ficus* spp. (fig): Ethnobotany and potential as anticancer and anti-inflammatory agents. *J Ethnopharmacol.* 119: 195–213. Reprinted with permission of Elsevier.)

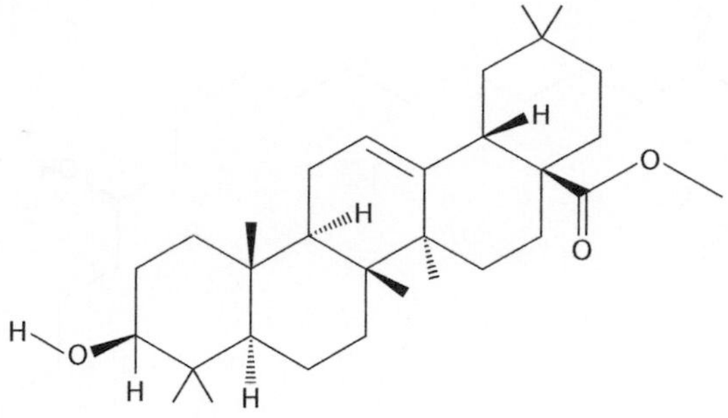

FIGURE 6.19 Methyl oleanolate. (*PubChem*. The PubChem Project. 2001. [Online Database]. National Center for Biotechnology Information. U.S. National Library of Medicine. 8600 Rockville Pike, Bethesda, MD 20894. URL: http://pubchem.ncbi.nlm.nih. gov/ [accessed August 16, 2009].)

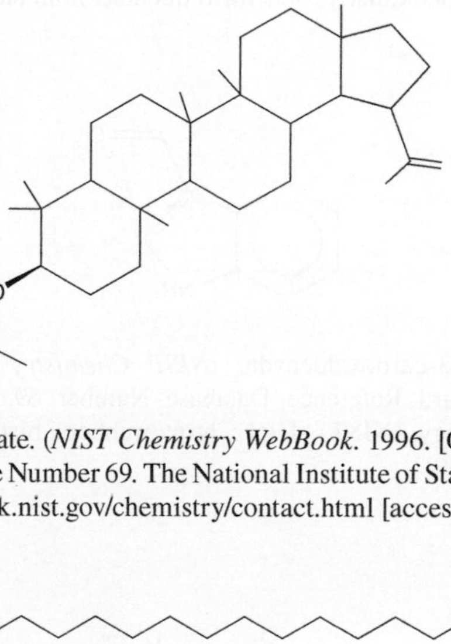

FIGURE 6.20 Lupeol acetate. (*NIST Chemistry WebBook*. 1996. [Online Database]. NIST Standard Reference Database Number 69. The National Institute of Standards and Technology (NIST). URL: http://webbook.nist.gov/chemistry/contact.html [accessed August 14, 2009].)

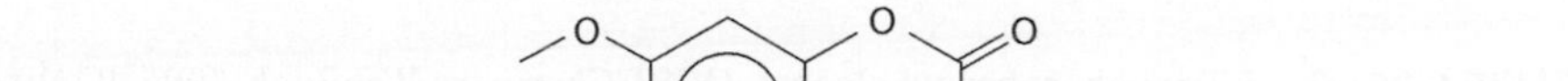

FIGURE 6.21 Ceryl alcohol. (*NIST Chemistry WebBook*. 1996. [Online Database]. NIST Standard Reference Database Number 69. The National Institute of Standards and Technology (NIST). URL: http://webbook.nist.gov/chemistry/contact.html [accessed August 14, 2009].)

FIGURE 6.22 7-Methoxycoumarin. (*NIST Chemistry WebBook*. 1996. [Online Database]. NIST Standard Reference Database Number 69. The National Institute of Standards and Technology (NIST). URL: http://webbook.nist.gov/chemistry/contact.html [accessed August 14, 2009].)

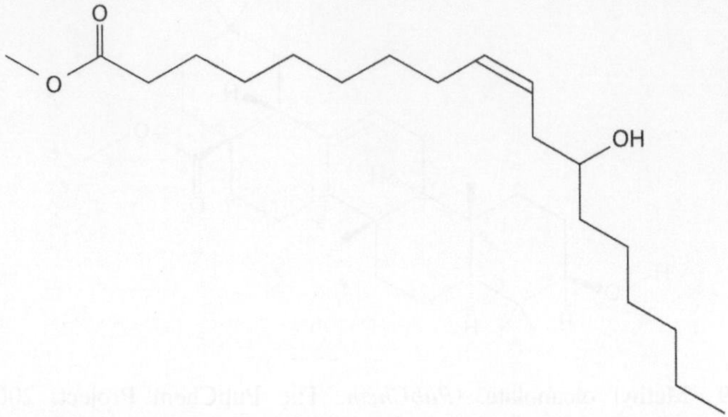

FIGURE 6.23 Methyl ricinolate. (*NIST Chemistry WebBook*. 1996. [Online Database]. NIST Standard Reference Database Number 69. The National Institute of Standards and Technology (NIST). URL: http://webbook.nist.gov/chemistry/contact.html [accessed August 14, 2009].)

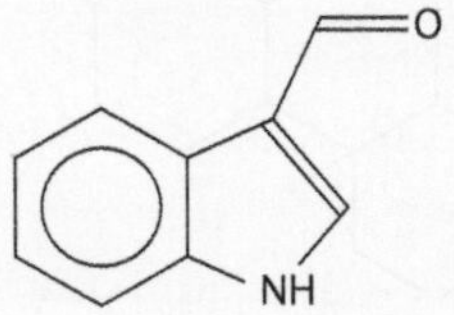

FIGURE 6.24 Indole-3-carboxaldehyde. (*NIST Chemistry WebBook*. 1996. [Online Database]. NIST Standard Reference Database Number 69. The National Institute of Standards and Technology (NIST). URL: http://webbook.nist.gov/chemistry/contact.html [accessed August 14, 2009].)

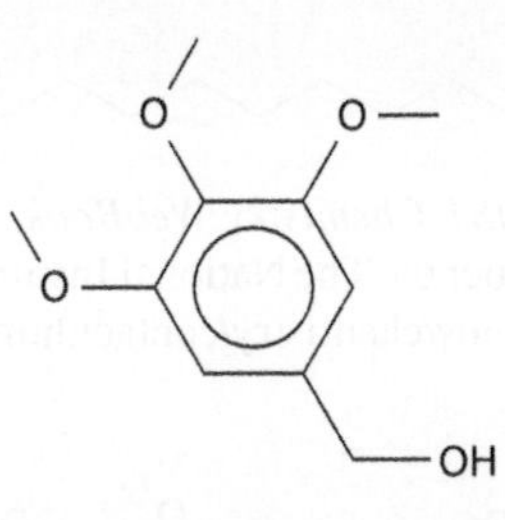

FIGURE 6.25 3,4,5-Trimethoxybenzyl alcohol. (*NIST Chemistry WebBook*. 1996. [Online Database]. NIST Standard Reference Database Number 69. The National Institute of Standards and Technology (NIST). URL: http://webbook.nist.gov/chemistry/contact.html [accessed August 14, 2009].)

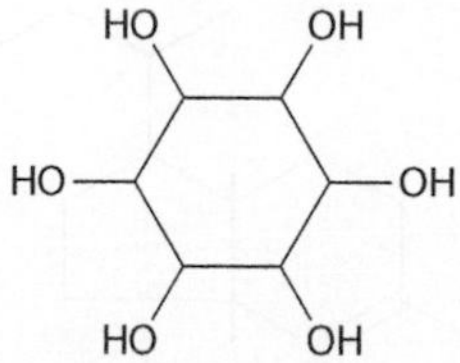

FIGURE 6.26 Meso-inositol. (*NIST Chemistry WebBook*. 1996. [Online Database]. NIST Standard Reference Database Number 69. The National Institute of Standards and Technology (NIST). URL: http://webbook.nist.gov/chemistry/contact.html [accessed August 14, 2009].)

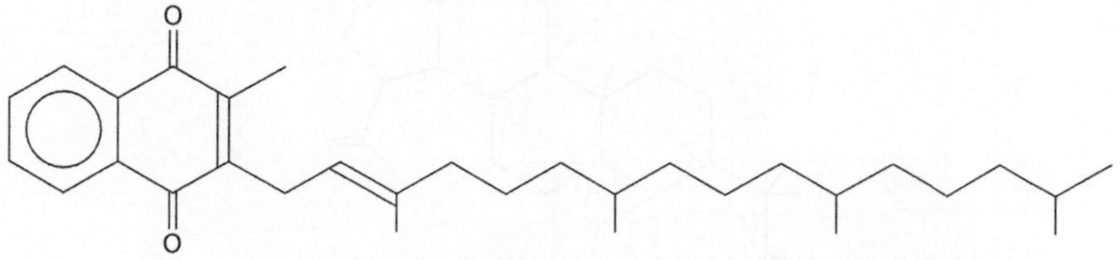

FIGURE 6.27 Vitamin K1. (*NIST Chemistry WebBook*. 1996. [Online Database]. NIST Standard Reference Database Number 69. The National Institute of Standards and Technology (NIST). URL: http://webbook.nist.gov/chemistry/contact.html [accessed August 14, 2009].)

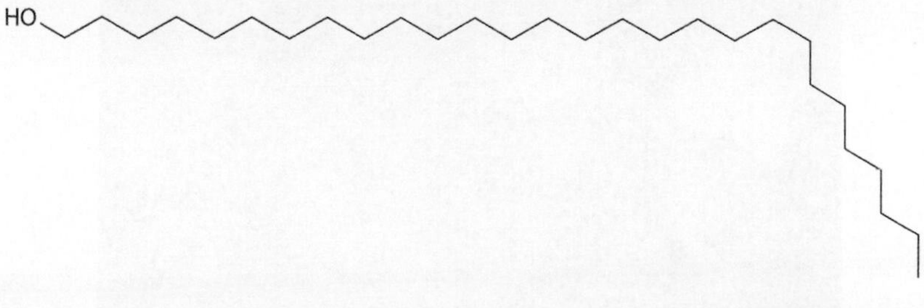

FIGURE 6.28 *n*-Octacosanol. (*NIST Chemistry WebBook*. 1996. [Online Database]. NIST Standard Reference Database Number 69. The National Institute of Standards and Technology (NIST). URL: http://webbook.nist.gov/chemistry/contact.html [accessed August 14, 2009].)

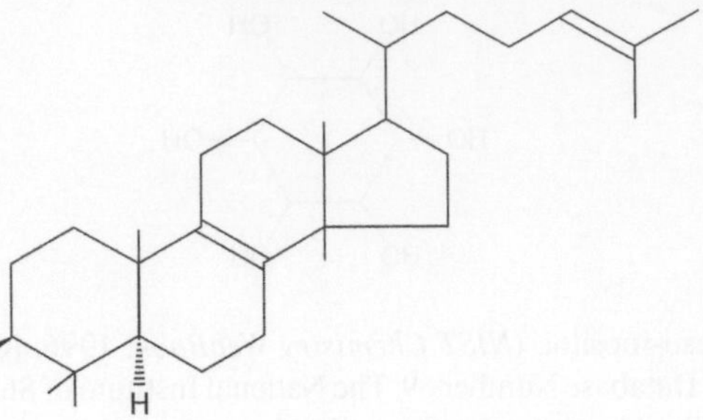

FIGURE 6.29 Lanosterol. (*NIST Chemistry WebBook*. 1996. [Online Database]. NIST Standard Reference Database Number 69. The National Institute of Standards and Technology (NIST). URL: http://webbook.nist.gov/chemistry/contact.html [accessed August 14, 2009].)

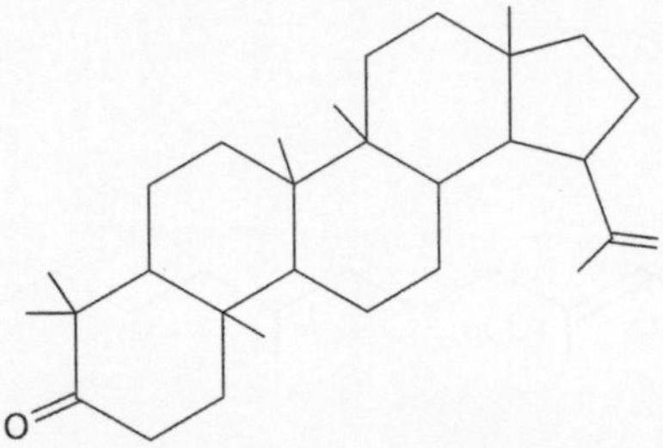

FIGURE 6.30 Lupen-3-one. (*NIST Chemistry WebBook*. 1996. [Online Database]. NIST Standard Reference Database Number 69. The National Institute of Standards and Technology (NIST). URL: http://webbook.nist.gov/chemistry/contact.html [accessed August 14, 2009].)

FIGURE 6.31 *Ficus carica*, trunk (27.7.2009, Seven Species Garden, Binyamina, Israel, by Helena Paavilainen).

FIGURE 6.32   *Ficus carica* from the ground up (29.7.2009, Hebrew University Botanical Garden, Mount Scopus, Jerusalem, Israel, by Krina Brandt-Doekes).

REFERENCES

Acharya, B.M., and K.A. Kumar. 1984. Chemical examination of the bark of *Ficus hispida* Linn. *Curr Sci.* 53: 1034–5.

Agrawal, S., and K. Misra. 1977. Leucoanthocyanins from *Ficus racemosa* bark. *Chem Scr.* 12: 37–9.

Annan, K., and P.J. Houghton. 2008. Antibacterial, antioxidant and fibroblast growth stimulation of aqueous extracts of *Ficus asperifolia* Miq. and *Gossypium arboreum* L., wound-healing plants of Ghana. *J Ethnopharmacol.* 119: 141–4.

Ao, C., A. Li, A.A. Elzaawely, T.D. Xuan, and S. Tawata. 2008. Evaluation of antioxidant and antibacterial activities of *Ficus microcarpa* L. fil. *Food Control* 19: 940–8.

Augusti, K.T. 1975. Hypoglycaemic action of bengalenoside, a glucoside isolated from *Ficus bengalensis* Linn, in normal and alloxan diabetic rabbits. *Indian J Physiol Pharmacol.* 19: 218–20.

Augusti, K.T., P.S.P. Anuradha, K.B. Smitha, M. Sudheesh, A. George, and M.C. Joseph. 2005. Nutraceutical effects of garlic oil, its nonpolar fraction and a *Ficus* flavonoid as compared to vitamin E in CCl4 induced liver damage in rats. *Indian J Exp Biol.* 43: 437–44.

Bhaskara, R.R., T. Murugesan, S. Sinha, B.P. Saha, M. Pal, and S.C. Mandal. 2002. Glucose lowering efficacy of *Ficus racemosa* bark extract in normal and alloxan diabetic rats. *Phytother Res.* 16: 590–2.

Bhatt, K., and Y.K. Agrawal. 1973. Chemical investigation of the trunk-bark from *Ficus racemosa. J Indian Chem Soc.* 50: 611.

Channabasavaraj, K.P., S. Badami, and S. Bhojraj. 2008. Hepatoprotective and antioxidant activity of methanol extract of *Ficus glomerata. Nat Med (Tokyo)* 62: 379–83.

Cherian, S., and K.T. Augusti. 1993. Antidiabetic effects of a glycoside of leucopelargonidin isolated from *Ficus bengalensis* Linn. *Indian J Exp Biol.* 31: 26–9.

Chindo, B.A., J.A. Anuka, L. McNeil et al. 2009. Anticonvulsant properties of saponins from *Ficus platyphylla* stem bark. *Brain Res Bull.* 78: 276–82.

Daniel, R.S., B.C. Mathew, K.S. Devi, and K.T. Augusti. 1998. Antioxidant effect of two flavonoids from the bark of *Ficus bengalensis* Linn in hyperlipidemic rats. *Indian J Exp Biol.* 36: 902–6.

Daniel, R.S., K.S. Devi, K.T. Augusti, C.R. Sudhakaran Nair. 2003. Mechanism of action of antiatherogenic and related effects of *Ficus bengalensis* Linn. flavonoids in experimental animals. *Indian J Exp Biol.* 41: 296–303.

Darbour, N., C. Bayet, S. Rodin-Bercion et al. 2007. Isoflavones from *Ficus nymphaefolia. Nat Prod Res.* 21: 461–4.

El-Sayyad, S.M., H.M. Sayed, and S.A. Mousa. 1986. Chemical constituents and preliminary anthelmintic activity of *Ficus platyphylla* (Del). *Bull Pharm Sci Assiut Univ.* 9: 164–77.

Govindarajan, R., D.P. Singh, and A.K. Rawat. 2008. Validated reversed-phase column high-performance liquid chromatographic method for separation and quantification of polyphenolics and furocoumarins in herbal drugs. *J AOAC Int.* 91: 1020–4.

Joshi, K.C., L. Prakash, and R.K. Shah. 1977. Chemical constituents of *Clerodendron infortunatum* Linn. and *Ficus racemosa* Linn. *J Indian Chem Soc.* 54: 1104.

Kuete, V., F. Nana, B. Ngameni, A.T. Mbaveng, F. Keumedjio, and B.T. Ngadjui. 2009. Antimicrobial activity of the crude extract, fractions and compounds from stem bark of *Ficus ovata* (Moraceae). *J Ethnopharmacol.* 124: 556–61.

Kuete, V., B. Ngameni, C.C. Simo et al. 2008. Antimicrobial activity of the crude extracts and compounds from *Ficus chlamydocarpa* and *Ficus cordata* (Moraceae). *J Ethnopharmacol.* 120: 17–24.

Lansky, E.P., and E. Nevo. 2009. Plant Immunity May Benefit Human Medicine. *Open Systems Biology Journal* 2: 18–19.

Lansky, E.P., H.M. Paavilainen, A.D. Pawlus, and R.A. Newman. 2008. *Ficus* spp. (fig): Ethnobotany and potential as anticancer and anti-inflammatory agents. *J Ethnopharmacol.* 119: 195–213.

Li, R.W., S.P. Myers, D.N. Leach, G.D. Lin, and G. Leach. 2003. A cross-cultural study: anti-inflammatory activity of Australian and Chinese plants. *J Ethnopharmacol.* 85: 25–32.

Manian, R., N. Anusuya, P. Siddhuraju, and S. Manian. 2008. The antioxidant activity and free radical scavenging potential of two different solvent extracts of *Camellia sinensis* (L.) O. Kuntz, *Ficus bengalensis* L. and *Ficus racemosa* L. *Food Chem.* 107: 1000–1007.

Maregesi, S.M., L. Pieters, O.D. Ngassapa et al. 2008. Screening of some Tanzanian medicinal plants from Bunda district for antibacterial, antifungal and antiviral activities. *J Ethnopharmacol.* 119: 58–66.

Musabayane, C.T., P.T. Bwititi, and J.A. Ojewole. 2006. Effects of oral administration of some herbal extracts on food consumption and blood glucose levels in normal and streptozotocin-treated diabetic rats. *Methods Find Exp Clin Pharmacol.* 28: 223–8.

Musabayane, C.T., M. Gondwe, D.R. Kamadyaapa, A.A. Chuturgoon, and J.A. Ojewole. 2007. Effects of *Ficus thonningii* (Blume) [Morarceae] stem-bark ethanolic extract on blood glucose, cardiovascular and kidney functions of rats, and on kidney cell lines of the proximal (LLC-PK1) and distal tubules (MDBK). *Ren Fail.* 29: 389–97.

NIST Chemistry WebBook. 1996. [Online Database]. NIST Standard Reference Database Number 69. The National Institute of Standards and Technology (NIST). URL: http://webbook.nist.gov/chemistry/contact.html [accessed August 14, 2009].

Nongonierma, B.R., L. Ndiaye, D. Thiam, I. Samb, and A. Samb. 2007. [Antidrepanocytic action of the acetone extract of stem barks of *Ficus gnaphalocarpa* (MIQ) Stend]. *J Soc Ouest-Africaine Chim.* 12: 95–101.

Plants of Hawaii. 2009. [Online Database]. Starr, F., and K. Starr. 149 Hawea Place, Makawao, Maui, Hawaii, 96768. URL: http://www.hear.org/starr/plants/images/species/ [accessed October 10, 2009].

PubChem. The PubChem Project. 2001. [Online Database]. National Center for Biotechnology Information. U.S. National Library of Medicine. 8600 Rockville Pike, Bethesda, MD 20894. URL: http://pubchem.ncbi.nlm.nih.gov/ [accessed August 16, 2009].

Rahman, N.N., M. Khan, and R. Hasan. 1994. Bioactive components from *Ficus glomerata*. *Pure Appl Chem.* 66: 2287–90.

Rahuman, A.A., P. Venkatesan, K. Geetha, G. Gopalakrishnan, A. Bagavan, and C. Kamaraj. 2008. Mosquito larvicidal activity of gluanol acetate, a tetracyclic triterpene derived from *Ficus racemosa* Linn. *Parasitol Res.* 103: 333–9.

Rao, R.B., K. Anuparna, K.R. Swaroop, T. Murugesan, M. Pal, and S.C. Mandal. 2002. Evaluation of anti-pyretic potential of *Ficus racemosa* bark. *Phytomedicine* 9: 731–3.

Ratnasooriya, W.D., J.R. Jayakody, and T. Nadarajah. 2003. Antidiuretic activity of aqueous bark extract of Sri Lankan *Ficus racemosa* in rats. *Acta Biol Hung.* 54: 357–63.

Ratnayake, A.S., W.Y. Yoshida, S.L. Mooberry, and T.K. Hemscheidt. 2001. Nomofungin: a new microfilament disrupting agent. *J Org Chem.* 66: 8717–21.

Sandabe, U.K., P.A. Onyeyili, and G.A. Chibuzo. 2006. Phytochemical screening and effect of aqueous extract of *Ficus sycomorus* L. (Moraceae) stem bark on muscular activity in laboratory animals. *J Ethnopharmacol.* 104: 283–5.

Shrivastava, P.N., G.S. Mishra, and Y.N. Sukla. 1977. Chemical constituents of *Ficus racemosa* Linn. *Proc Natl Acad Sci India Sect A Phys Sci.* 47: 1–3.

Simon, P.N., A. Chaboud, N. Darbour et al. 2001. Modulation of cancer cell multidrug resistance by an extract of *Ficus citrifolia*. *Anticancer Res.* 21: 1023–28.

Singhal, R.K., and H.S. Saharia. 1980. Chemical examination of *Ficus glomerata* Roxb. *Herba Hung* 19: 17–20.

Subramanian, P.M., and G.S. Misra. 1977. Chemical constituents of *Ficus bengalensis*. *Indian J Chem Sect B Org Chem Incl Med Chem.* 15: 762–3.

Subramanian, P.M., and G.S. Misra. 1978. Chemical constituents of *Ficus bengalensis* (part II). *Pol J Pharmacol Pharm.* 30: 559–62.

Swami, K.D., G.S. Malik, and N.P.S. Bisht. 1989a. Chemical examination of stem bark of *Ficus infectoria* Roxb. *J Indian Chem Soc.* 66: 141–2.

Swami, K.D., G.S. Malik, and N.P.S. Bisht. 1989b. Chemical investigation of stem bark of *Ficus religiosa* and *Prosopis spicigera*. *J Indian Chem Soc.* 66: 288–9.

Taur, D.J., S.A. Nirmal, R.Y. Patil, and M.D. Kharya. 2007. Antistress and antiallergic effects of *Ficus bengalensis* bark in asthma. *Nat Prod Res.* 21: 1266–70.

Veerapur, V.P., K.R. Prabhakar, V.K. Parihar et al. 2009. *Ficus racemosa* stem bark extract: a potent antioxidant and a probable natural radioprotector. *Evid Based Complement Alternat Med.* 6: 317–24.

Zhang, H.J., P.A. Tamez, Z. Aydogmus et al. 2002. Antimalarial agents from plants. III. Trichothecenes from *Ficus fistulosa* and *Rhaphidophora decursiva*. *Planta Med.* 68: 1088–91.

7 Roots (Including Aerial Roots and Root Bark)

If the fruits of fig trees are the (most) edible parts, the roots may be the least edible, but may also possess some of the most potent medicinal powers. Two new studies from 2009 demonstrate, for example, the powerful antidiabetic properties of aqueous extracts of banyan tree (*Ficus benghalensis*) roots, which may be largely due to the presence of pentacyclic triterpenes or their esters, such as α-amyrin acetate (Singh A.B. et al. 2009). Further, this action may involve mechanisms related to transient increased levels of divalent cations that may affect transport pumps or other cellular regulatory mechanisms dependent on calcium and magnesium balance (Singh R.K. et al. 2009). Thus, the first main potential use for *Ficus* roots may be considered to be antidiabetic, at least for the roots of *F. benghalensis*.

FIGURE 7.1 Aerial roots of the curtain tree in Tiberias, Israel (pen and wash by Zipora Lansky).

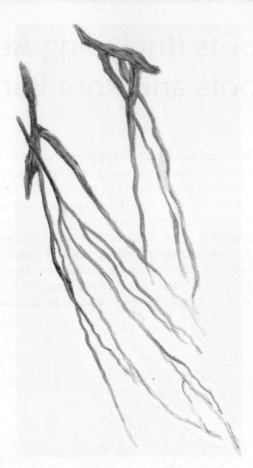

FIGURE 7.2   (See color insert after page 128.) Aerial roots of *Ficus microcarpa* (18.8.2009, Haifa, Israel, watercolor by Zipora Lansky).

An aqueous decoction of the roots of the widespread *F. elastica* exhibited potent anti-inflammatory actions in vivo in vigorous pharmacological tests (Sackeyfio and Lugeleka 1986). Given the central role of inflammation in so many diseases, not only acute forms such as carrageenan-induced ear edema or adjunct-induced arthritis but also chronic states including heart disease, cancer, and metabolic syndrome, it would be wise to explore this potential further. There are many problems with conventional anti-inflammatory drugs such as indomethacin, relating both to gastric irritation and, in some cases, even to cardiopathy, so reasonable botanic alternatives that likely operate in a more redundant and gentler manner are attractive. To this large and lucrative pharmaceutical megamarket, *Ficus* roots such as those of *F. elastica* may offer attractive complex pharmaceutical candidates.

Protection against hepatic damage is an important pharmacological target since it reflects the possibility of minimizing or reversing completely compromise to the liver following various types of toxic exposures. Since hepatic metabolism is required for all life functions, and particularly for the utilization of medicines, correcting

FIGURE 7.3 (**See color insert after page 128.**) *Ficus carica* roots over the edge (29.7.2009, Hebrew University Botanical Garden, Mount Scopus, Jerusalem, Israel, by Krina Brandt-Doekes).

and/or preventing disturbances of the liver is a key and pivotal objective that underlies internal medicine. ("Is life worth living? It depends on the liver." Alan Ginsburg, Swarthmore Pennsylvania, c. 1980.) Accordingly, studies that have shown extracts of the roots of *Ficus glomerata* (syn. *F. racemosa*) (Channabasavaraj et al. 2008) and *Ficus sycomorus* (Garba et al. 2007) to prevent hepatic damage following oral dosing with the known hepatotoxic solvent carbon tetrachloride, or of an aqueous extract from the roots of the East Asian *F. hirta* to prevent hepatotoxicity following injection of cocaine (Cai et al. 2007), open new possibilities for hepatoprotective drugs, such as the silymarin derived from the milk thistle. It is possible, and even likely, that a combination of such agents as silymarin with *Ficus* root extracts may provide a higher and/or synergistic level of protection in cases where a chemical or drug induced suppression of hepatic function is a major concern.

    Ficus root extracts have also shown activity against various cancer cell lines. The activity has been linked to specific compounds within the roots, and these compounds occur in more than one chemical class. The two most important classes are

FIGURE 7.4   *Ficus binnendyijkii*, aerial roots (19.7.2009, University of Turku, Botanical Garden, Ruissalo, Turku, Finland, by Kaarina Paavilainen).

terpenoids, found in *F. microcarpa* (Chiang et al. 2005), and phenanthroindolizidine alkaloids from *F. septica* (Damu et al. 2009). It would certainly be interesting and possibly fruitful to more carefully elucidate the mechanisms underlying the anticancer mechanisms of these two chemical classes of compounds, on the one hand, and possible synergistic effects accruing from combining them in natural complex agents, on the other. In that anticancer drugs must often be used in conjunction with each other to achieve maximum efficacy, the possibility that these two different types of agents from roots of the same genus might interact in an additive or supraadditive manner should be explored, also ruling out subtractive or negative interactions.

Just as in many other pharmacological settings, the search for novel effective and safe agents is paramount, so also in the field of malaria treatment. A methanolic extract of the root of the African species, *Ficus sur*, was found to potently and significantly suppress parasitemia caused by a *Plasmodium* species resistant to the conventional quinine agent, chloroquine. Although extracts of leaf and stem bark of *F. sur*

FIGURE 7.5 *Ficus microcarpa* with aerial roots hanging from branch and wrapping around the trunk (Haifa, Israel, in early September, Zipora Lansky).

also produced, to a degree, the antimalarial effect, only the extract taken from the roots of *F. sur* acted synergistically with chloroquine (Muregi et al. 2003, 2007).

In short, the actions attributable to *Ficus* spp. root extracts on the basis of pharmacological studies may be remembered as "CDMH."

- Anticancer
- Antidiabetic
- Antimalarial
- Hepatoprotective

Further details of the studies underlying this conclusion are summarized in Table 7.1.

Fortunately, chemical investigation into *Ficus* spp. roots has followed, and even preceded, the aforementioned pharmacological explorations. Sharma et al. (1963), working with a petroleum ether extract of *F. benghalensis* roots, elaborated two

TABLE 7.1
Pharmacological Studies of *Ficus* spp. Roots (including aerial roots and root bark)

Ficus Species	Pharmacological Action	Details	Reference
benghalensis	Improved diabetic condition in streptozotocin-induced diabetic rats and in diabetic db/db mice at 50 mg/kg	Orally, α-amyrin acetate (α-AA) isolated from aerial roots, db/db mice are a type II diabetes model	Singh, A.B. et al. 2009
benghalensis	Lowered blood sugar of about 40%–45% in normal rats by glucose tolerance test (GTT) and about 50%–55% in mildly diabetic rats	Correlated with high concentrations of Ca^{++} and Mg^{++} in serum at p.o. dose of aqueous extract at 300 mg kg(-1)	Singh, R.K. et al. 2009
elastica	Inhibited carrageenin-induced edema and adjuvant-induced arthritis in rat, i.e., potent anti-inflammatory	Aqueous extract inhibition of secondary arthritic changes, comparable to indomethacin	Sackeyfio and Lugeleka 1986
glomerata (syn. *racemosa*)	Antioxidant activity and protection against CCl_4-induced hepatic damage	In rats at 250–500 mg methanolic extract p.o.	Channabasavaraj et al. 2008
hirta	Protected mouse from hepatic damage secondary to subcutaneous injection of cocaine hydrochloride	Aqueous extract at 100–300 g/kg dosages intragastric preceding cocaine	Cai et al. 2007
microcarpa	Cytotoxic to HONE-1 nasopharyngeal carcinoma, KB oral epidermoid carcinoma, and HT29 colorectal carcinoma cells	3β-Acetoxy-25-hydroxylanosta-8,23-diene, acetylbetulinic acid, betulonic acid, acetylursolic acid, ursonic acid, ursolic acid, 3-oxofriedelan-28-oic acid all with C-28 carboxylic acid functionality, active IC_{50}'s 4.0–9.4 µM (more details: Table 7.2)	Chiang et al. 2005
septica	Phenanthroindolizidine alkaloids toxic to adenocarcinoma (NUGC) and nasopharyngeal carcinoma (HONE-1) cell lines	Twenty compounds, including 10 new, with 13a R-antofine with ED (50) < 0.1 µg/mL against L-1210, P-388, A-549, and HCT-8 cell lines	Damu et al. 2009
sur	Significant suppression of chloroquine-resistant plasmodium, alone (like *F. sur* stems, leaves) and (*synergistically*) with chloroquine, potent antimalarial, suppression of parasitemia	Methanolic extract of roots, strong potential antimalarial agent	Muregi et al. 2003, 2007

TABLE 7.1 (continued)

Pharmacological Studies of *Ficus* spp. Roots (including aerial roots and root bark)

Ficus Species	Pharmacological Action	Details	Reference
sycomorus	Protective trends against CCl_4 hepatic damage in rats	Per activity of enzymes aspartate transaminase (AST), alanine-transaminase (alt), alkaline phosphatase (ALP), and bilirubin), effect on liver parenchymal architecture, fibroplasias, and cirrhosis	Garba et al. 2007

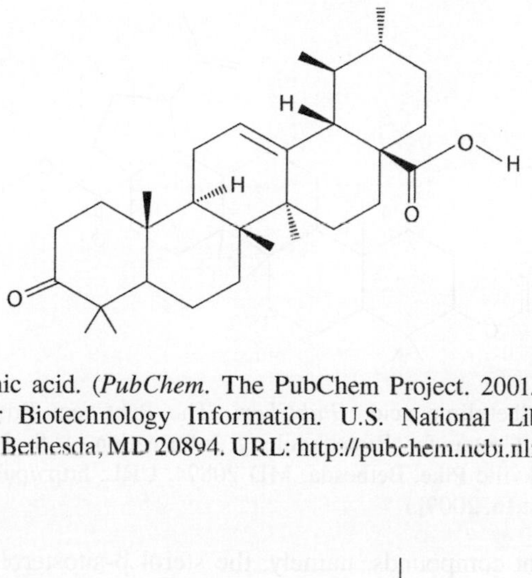

FIGURE 7.6 Ursonic acid. (*PubChem*. The PubChem Project. 2001. [Online Database]. National Center for Biotechnology Information. U.S. National Library of Medicine. 8600 Rockville Pike, Bethesda, MD 20894. URL: http://pubchem.ncbi.nlm.nih.gov/ [accessed August 16, 2009].)

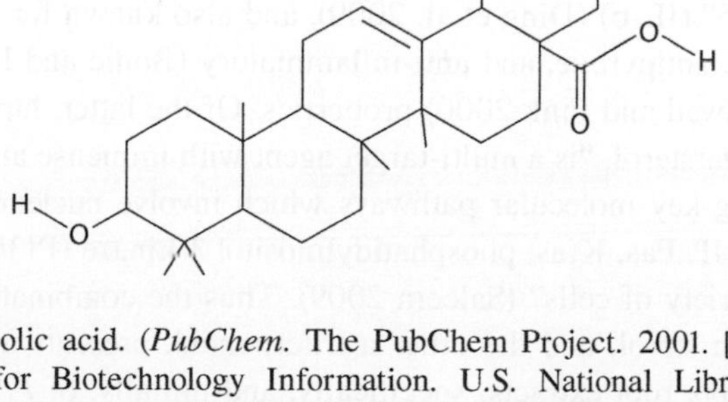

FIGURE 7.7 Ursolic acid. (*PubChem*. The PubChem Project. 2001. [Online Database]. National Center for Biotechnology Information. U.S. National Library of Medicine. 8600 Rockville Pike, Bethesda, MD 20894. URL: http://pubchem.ncbi.nlm.nih.gov/ [accessed August 16, 2009].)

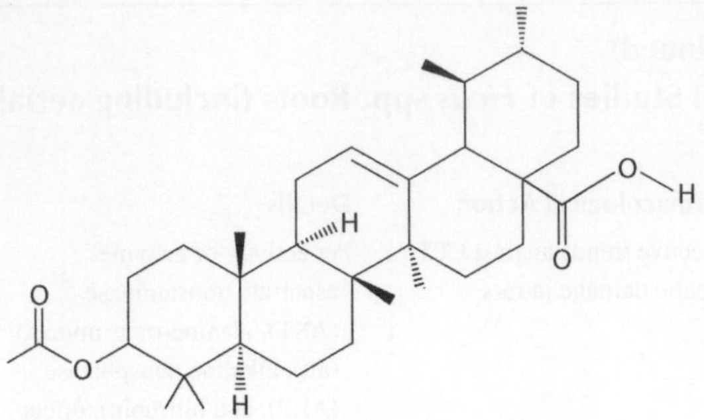

FIGURE 7.8 Acetylursolic acid. (*PubChem*. The PubChem Project. 2001. [Online Database]. National Center for Biotechnology Information. U.S. National Library of Medicine. 8600 Rockville Pike, Bethesda, MD 20894. URL: http://pubchem.ncbi.nlm.nih. gov/ [accessed August 16, 2009].)

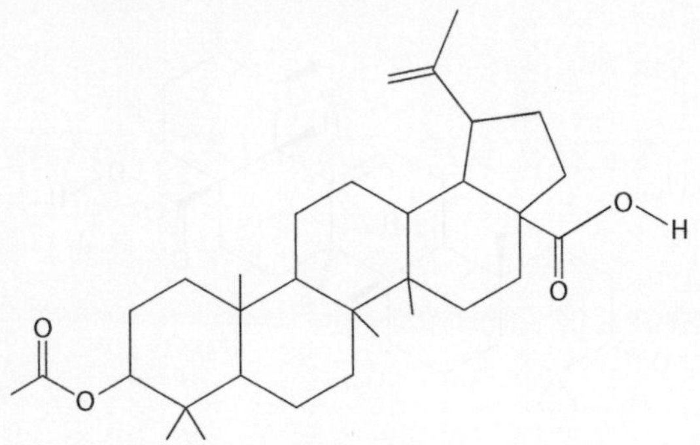

FIGURE 7.9 Acetylbetulinic acid. (*PubChem*. The PubChem Project. 2001. [Online Database]. National Center for Biotechnology Information. U.S. National Library of Medicine. 8600 Rockville Pike, Bethesda, MD 20894. URL: http://pubchem.ncbi.nlm.nih. gov/ [accessed August 16, 2009].)

extremely important compounds, namely, the sterol β-sitosterol, and the triterpenoid lupeol. Of the former, β-sitosterol is a potent modulator of lipopolysaccharide-induced secretion of inflammatory cytokines "tumor necrosis factor α" (TNF-α) and "interleukin 6" (IL-6) (Ding et al. 2009), and also known for its more general immunopotentive, antipyretic, and anti-inflammatory (Bouic and Lamprecht 1999) and anticancer (Awad and Fink 2000) properties. Of the latter, lupeol, a triterpene also known as fagarsterol, "is a multi-target agent with immense anti-inflammatory potential targeting key molecular pathways which involve nuclear factor kappa B (NFkappaB), cFLIP, Fas, Kras, phosphatidylinositol-3-kinase (PI3K)/Akt and Wnt/β-catenin in a variety of cells" (Saleem 2009). Thus the combination of these two compounds, one a sterol, and the other a triterpenoid, presents an excellent base by which *Ficus* spp. root extracts, specifically, and initially, of *F. glomerata*, may produce its complex pharmacological effects, including the antidiabetic one, since insulin resistance is a function of chronic inflammation, and β-sitosterol and lupeol

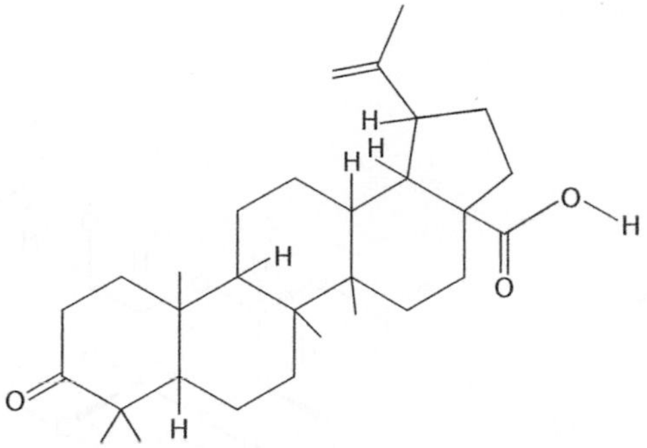

FIGURE 7.10 Betulonic acid. (*PubChem*. The PubChem Project. 2001. [Online Database]. National Center for Biotechnology Information. U.S. National Library of Medicine. 8600 Rockville Pike, Bethesda, MD 20894. URL: http://pubchem.ncbi.nlm.nih.gov/ [accessed August 16, 2009].)

provide a nontoxic and potent means of mediating inflammatory pathways at multiple and redundant targets. Further, the possibility that these two compounds may interact with synergistic or supraadditive power must also be considered. The combination, therefore, provides a basis for complex antidiabetic, anti-inflammatory, immune system-enhancing and antineoplastic therapies.

Subsequent work on roots of *F. hirta* revealed additional sterols (stigmasterol, daucosterol), triterpenoids (amyrin and derivatives), furocoumarins (psoralen), and flavonoids (kaempferol, hesperidin, naringenin) (Li et al. 2006, Ya et al. 2008). The multiplicity of compounds, such as additional sterols, could lend synergistic enhancement when combined, but the opposite could also be true. Detailed studies comparing the complex combinations to individual compounds will be needed to sort this out. Psoralen has been cited as the basis for the hepatoprotective effect of *F. hirta* root extract against cocaine exposure (Cai et al. 2007). Pertinent to its high absorbance of ultraviolet rays, psoralen has been effectively used in photoreactive treatments of psoriasis known as "psoralen baths with long wave ultraviolet radiation" (PUVA) (Rodriguez-Granados et al. 2009). Psoralen is also, though, a mutagen, and recently the carcinogenic potential of PUVA treatments has been reviewed and a caution issued regarding its excessive use (Patel et al. 2009).

Flavonoids found in *Ficus* spp. roots and root bark, such as sterols and triterpenoids, exert numerous signal transduction modulating effects on animal cells. Kaempferol and its derivatives, for example, may modulate pathways through reduction of cellular oxidative tension and thus help prevent neurodegenerative processes, including Parkinson's disease (Qu et al. 2009). Hesperidin plays a key role in reducing serum lipids, cholesterol, and blood pressure, as well as regulating body temperature and cutaneous blood flow (Shen et al. 2009), and it is radioprotective in human peripheral blood lymphocytes (Kalpana et al. 2009). Naringenin is another flavonoid found in *Ficus* spp. roots, which inhibits neuroinflammatory signaling in neuroglial cells and may subsequently help to prevent neuroinflammatory disease (Vafeiadou et al. 2009). Naringenin also helps activate the aryl hydrocarbon receptor (AhR), which is

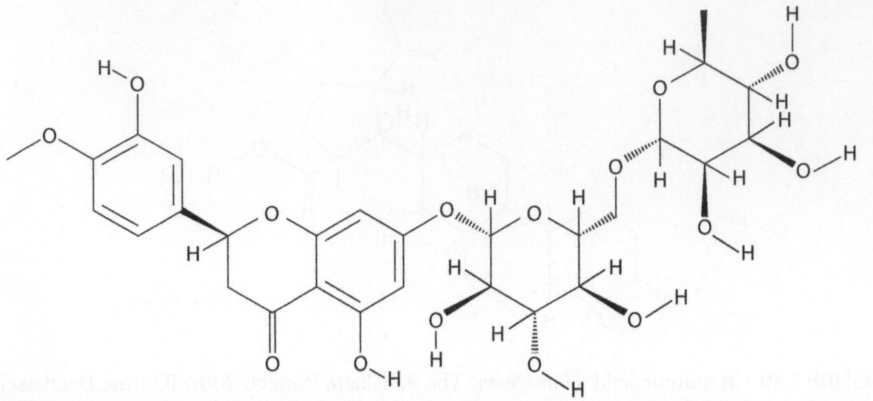

FIGURE 7.11 Hesperidin. (*PubChem*. The PubChem Project. 2001. [Online Database]. National Center for Biotechnology Information. U.S. National Library of Medicine. 8600 Rockville Pike, Bethesda, MD 20894. URL: http://pubchem.ncbi.nlm.nih.gov/ [accessed August 16, 2009].)

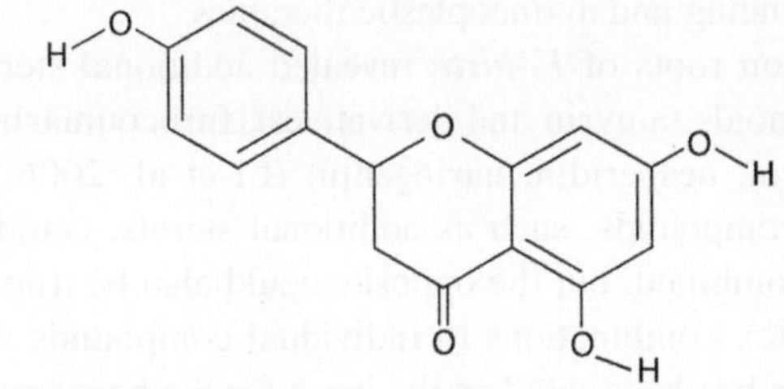

FIGURE 7.12 Naringenin. (*PubChem*. The PubChem Project. 2001. [Online Database]. National Center for Biotechnology Information. U.S. National Library of Medicine. 8600 Rockville Pike, Bethesda, MD 20894. URL: http://pubchem.ncbi.nlm.nih.gov/ [accessed August 16, 2009]. Reprinted with permission of Elsevier.)

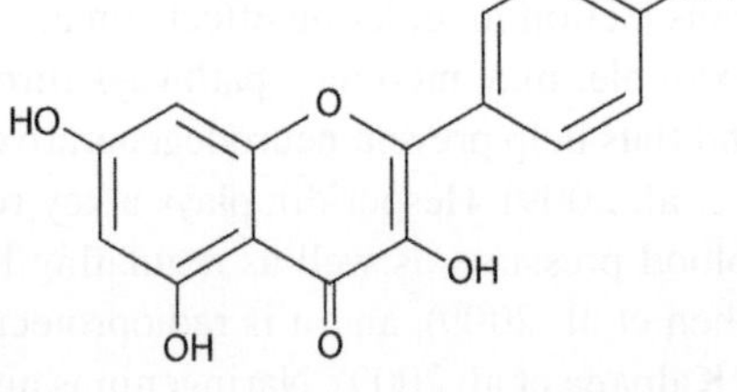

FIGURE 7.13 Kaempferol. (Reprinted with permission from Lansky and Newman 2007. *Punica granatum* (pomegranate) and its potential for prevention and treatment of inflammation and cancer. *J Ethnopharmacol.* 109: 177–206. Reprinted with permission of Elsevier.)

involved in neutralization of many environmental hydrocarbon pollutants (Amakura et al. 2008).

Additional terpenoids were discovered in *F. microcarpa* (Chiang et al. 2005, Wang et al. 2009) along with sterols. Sterols have anticancer and cholesterol-lowering properties (Jones and AbuMweis 2009), while terpenoids provide potential drug leads for malaria (Kaur et al. 2009), hypotensive drugs (Loizzo et al. 2008), muscle relaxants, antiallergenics, and more (Asakawa 2008), and also may be integrated into pharmaceutical products to facilitate and enhance natural and safe transdermal penetration (Sapra et al. 2008).

Work in *F. stenophylla* roots revealed additional coumarins/psoralen derivatives, including bergapten, terpenoids, and flavonoids (Jiang et al. 2007, Zhang et al. 2008). One flavonoid found, apigenin, possesses anxiolytic and progestinic activities (Lansky and Newman 2007). The full range of chemical investigations into *Ficus* spp. roots is summarized in Table 7.2.

HISTORICAL USES

Tables 7.3 and 7.4 give an overview of the historical medicinal uses of the roots of various *Ficus* spp.

For references used in Table 7.4, see Chapter 3, Table 3.6. The scoring system is based on the number of times a reference discusses the use of *Ficus* roots for the purpose indicated. If the number of citations is regular, the reference is given in regular type and scored as "1." If the number of citations is strong, the reference is given in *italic* and scored as "2." If the number of citations is strongest, the reference is given in **bold** and scored as "3." The final score in the last column is the sum of these scores for the individual references.

TABLE 7.2
Chemicals Identified and/or Isolated from Roots of Various *Ficus* spp.

Ficus Species	Compounds	Details	Reference
carica	Root bark: α-amyrin, β-sitosterol, β-sitosterol-β-d-glucoside; known coumarins: psoralen, bergapten, xanthotoxin; a new coumarin: $C_{13}H_{12}O_3$, m.p. 190°C identified as 6-(2-methoxyvinyl)-7-methylcoumarin	Petroleum ether and benzene fractions from the alcohol extract of the root bark on column chromatography over silica gel	Jain et al. 2007
glomerata (syn. racemosa)	β-Sitosterol, m.p. and mixed m.p. 137–138°C; acetyl and benzoyl derivs., m.p. and mixed m.p. 126–127°C and 145–146°C, respectively; lupeol, m.p. and mixed m.p. 212–213°C; acetyl and benzoyl derivatives, m.p. and mixed m.p. 214–215°C and 269–270°C	Petr. ether extract of the root bark isolated by chromatography of the concentrated extract over acid-treated Al_2O_3 and elution with petroleum ether and C_6H_6–petroleum ether mixture, respectively	Sharma et al. 1963
hirta	β-Sitosterol, stigmasterol, psoralen, 3-β-hydroxy-stigmast-5-en-7-one, 5-hydroxy-4', 6, 7, 8-tetramethoxy flavone, 4', 5, 6, 7, 8-pentamethoxy flavone, 4', 5, 7-trihydroxy-flavone, 3-β-acetoxy-β-amyrin, 3-β-acetoxy-α-amyrin, hesperidin	Isolated by chromatography on silica gel and HP-20 resin columns, structures elucidated by chemical and spectroscopic methods	Li et al. 2006
hirta	Psoralen, umbelliferon, 5,3',4'-trihydroxy-3,7-dimethoxyflavone, norartocarpetin, 5-hydroxy-3,7,4'-trimethoxyflavone, kaempferol, astragalin, acacetin 7-*O*-β-d-glucopyranoside, luteolin 7-*O*-β-d-glucopyranoside, naringenin, daucosterol	5,3',4'-Trihydroxy-3,7-dimethoxyflavone, norartocarpetin, 5-hydroxy-3,7,4'-trimethoxyflavone, acacetin 7-*O*-β-d-glucopyranoside, luteolin 7-*O*-β-d-glucopyranoside isolated for the first time from *Ficus* genus	Ya et al. 2008
microcarpa	Six *new triterpenes*: 3β-acetoxy-12,19-dioxo-13(18)-oleanene, 3β-acetoxy-19(29)-taraxasten-20α-ol, 3β-acetoxy-21α,22α-epoxytaraxastan-20α-ol, 3,22-dioxo-20-taraxastene, 3β-acetoxy-11α,12α-epoxy-16-oxo-14-taraxerene, 3β-acetoxy-25-methoxylanosta-8,23-diene; nine known triterpenes: 3β-acetoxy-11α,12α-epoxy-14-taraxerene,	Structures elucidated by spectroscopic methods, details of cytotoxic function in Table 7.1	Chiang et al. 2005

TABLE 7.2 (continued)
Chemicals Identified and/or Isolated from Roots of Various *Ficus* spp.

Ficus Species	Compounds	Details	Reference
microcarpa (continued)	3β-acetoxy-25-hydroxylanosta-8,23-diene, oleanonic acid, acetylbetulinic acid, betulonic acid, acetylursolic acid, ursonic acid, ursolic acid, 3-oxofriedelan-28-oic acid		
microcarpa	(2S, 3S, 4R)-2-[(2′R)-2′-Hydroxypentracosanoylamino]-heptadecane-1, 3, 4-triol (named microcarpaceramide A), 12, 20 (30)-ursa-dien-3α-ol, epifriedelanol, α-amyrin acetate, β-sitosterol, β-daucosterol, hexacosanoic acid, heneicosanoic acid	Solvent extraction, silica gel and Sephadex LH-20 column chromatography to isolate and purify the constituents, structures elucidated by physicochemical properties, analysis of spectroscopic data	Wang et al. 2009
stenophylla	Methyl 3-(6-hydroxy-4-methoxybenzofuran-5-yl) propanoate, kaempferol, kaempferol 3-*O*-β-d-glucoside, quercetin, tricin	Isolated and purified by repeated column chromatography on silica gel and Sephadex LH-20, structures based on physicochemical and spectral data	Zhang et al. 2008
stenophylla	3,4-Dihydropsoralen (new natural product), 7-hydroxycoumarin, bergapten, psoralen, (+)-catechin, apigenin, sucrose, vanillic acid, daucosterol, stigmasterol	Isolated from the roots by silica gel and Sephadex LH-20 column chromatography	Jiang et al. 2007

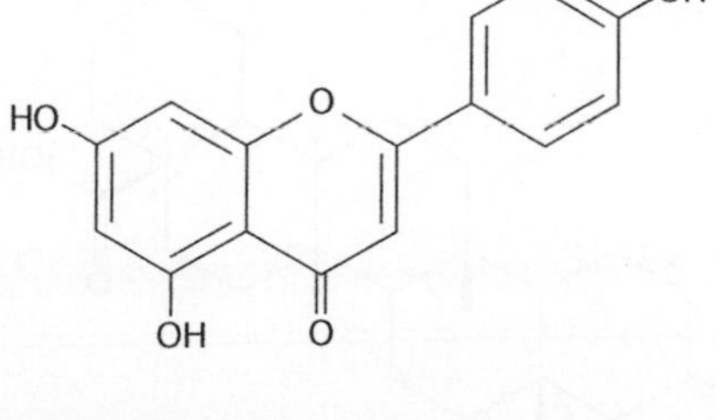

FIGURE 7.14 Apigenin. (Reprinted with permission from Lansky and Newman 2007. *Punica granatum* (pomegranate) and its potential for prevention and treatment of inflammation and cancer. *J Ethnopharmacol.* 109: 177–206.)

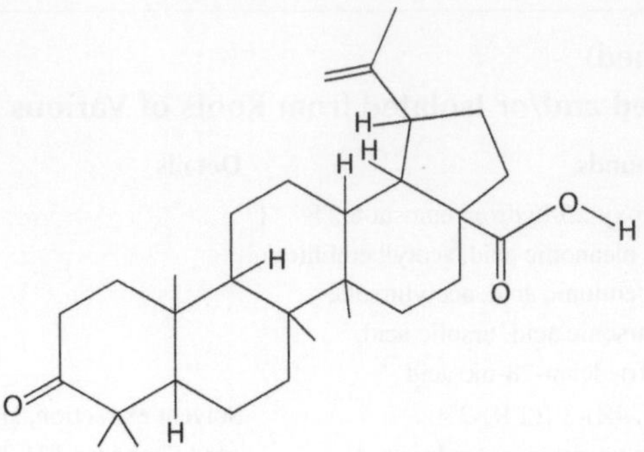

FIGURE 7.15 Astragalin. (*PubChem*. The PubChem Project. 2001. [Online Database]. National Center for Biotechnology Information. U.S. National Library of Medicine. 8600 Rockville Pike, Bethesda, MD 20894. URL: http://pubchem.ncbi.nlm.nih.gov/ [accessed August 16, 2009].)

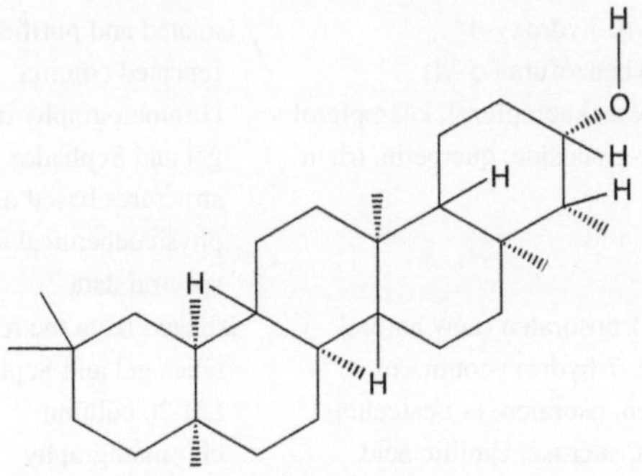

FIGURE 7.16 Epifriedelanol. (*PubChem*. The PubChem Project. 2001. [Online Database]. National Center for Biotechnology Information. U.S. National Library of Medicine. 8600 Rockville Pike, Bethesda, MD 20894. URL: http://pubchem.ncbi.nlm.nih.gov/ [accessed August 16, 2009]. Reprinted with permission of Elsevier.)

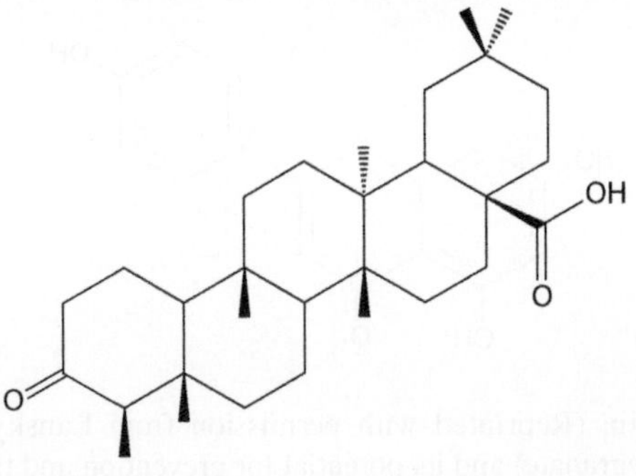

FIGURE 7.17 3-Oxofriedelan-28-oic acid. (Reprinted with permission from Lansky et al. 2008. *Ficus* spp. (fig): Ethnobotany and potential as anticancer and anti-inflammatory agents. *J Ethnopharmacol.* 119: 195–213.)

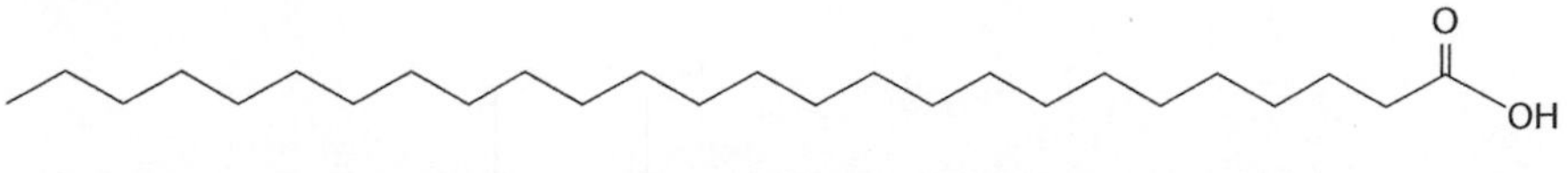

FIGURE 7.18 Hexacosanoic acid. (*NIST Chemistry WebBook*. 1996. [Online Database]. NIST Standard Reference Database Number 69. The National Institute of Standards and Technology (NIST). URL: http://webbook.nist.gov/chemistry/contact.html [accessed August 14, 2009].)

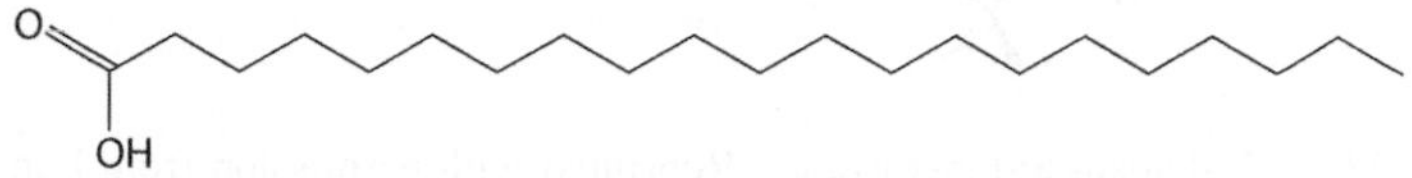

FIGURE 7.19 Heneicosanoic acid. (*NIST Chemistry WebBook*. 1996. [Online Database]. NIST Standard Reference Database Number 69. The National Institute of Standards and Technology (NIST). URL: http://webbook.nist.gov/chemistry/contact.html [accessed August 14, 2009].)

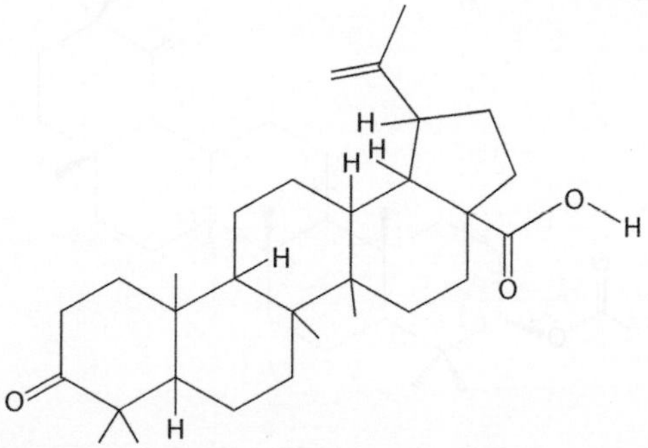

FIGURE 7.20 Oleanonic acid. (*PubChem*. The PubChem Project. 2001. [Online Database]. National Center for Biotechnology Information. U.S. National Library of Medicine. 8600 Rockville Pike, Bethesda, MD 20894. URL: http://pubchem.ncbi.nlm.nih.gov/ [accessed August 16, 2009]. Reprinted with permission of Elsevier.)

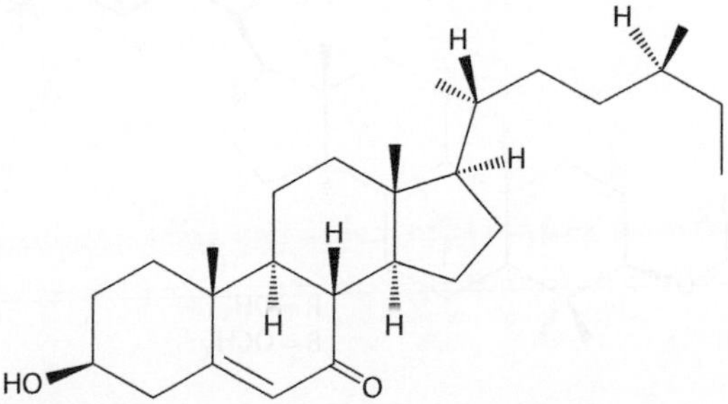

FIGURE 7.21 3β-Hydroxy-stigmast-5-en-7-one. (Reprinted with permission from Lansky et al. 2008. *Ficus* spp. (fig): Ethnobotany and potential as anticancer and anti-inflammatory agents. *J Ethnopharmacol*. 119: 195–213. Reprinted with permission of Elsevier.)

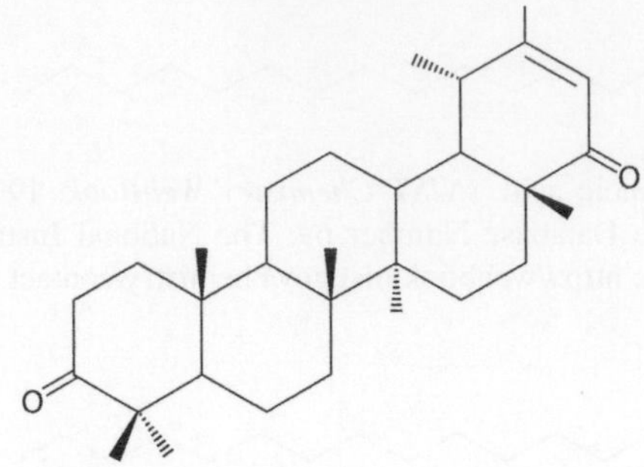

FIGURE 7.22 3,22-Dioxo-20-taraxastene. (Reprinted with permission from Lansky et al. 2008. *Ficus* spp. (fig): Ethnobotany and potential as anticancer and anti-inflammatory agents. *J Ethnopharmacol.* 119: 195–213. Reprinted with permission of Elsevier.)

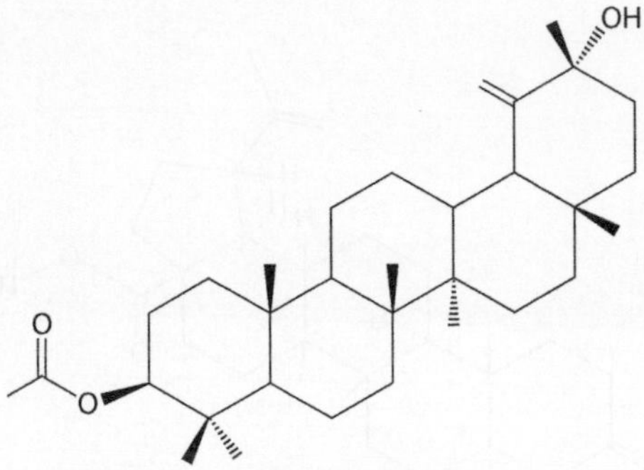

FIGURE 7.23 3β-Acetoxy-19(29)-taraxasten-20α-ol. (Reprinted with permission from Lansky et al. 2008. *Ficus* spp. (fig): Ethnobotany and potential as anticancer and anti-inflammatory agents. *J Ethnopharmacol.* 119: 195–213. Reprinted with permission of Elsevier.)

R = OH
R = OCH₃

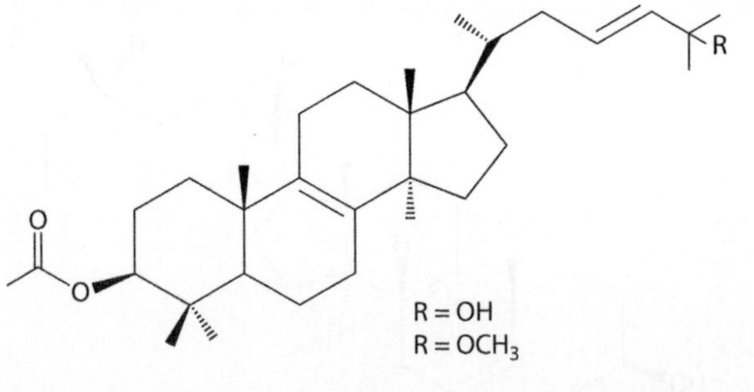

FIGURE 7.24 3β-Acetoxy-25-hydroxylanosta-8,23-diene; 3β-acetoxy-25-methoxylanosta-8,23-diene. (Adapted from Lansky et al. 2008. *Ficus* spp. (fig): Ethnobotany and potential as anticancer and anti-inflammatory agents. *J Ethnopharmacol.* 119: 195–213. Reprinted with permission of Elsevier.)

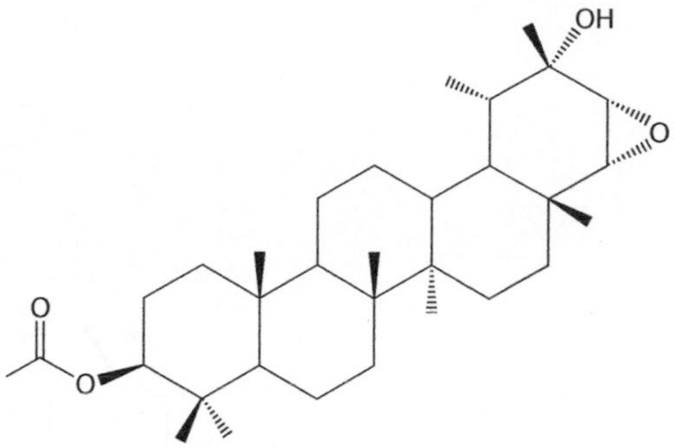

FIGURE 7.25 3β-Acetoxy-21α,22α-epoxytaraxastan-20α-ol. (Reprinted with permission from Lansky et al. 2008. *Ficus* spp. (fig): Ethnobotany and potential as anticancer and anti-inflammatory agents. *J Ethnopharmacol*. 119: 195–213. Reprinted with permission of Elsevier.)

R = O
R = H₂

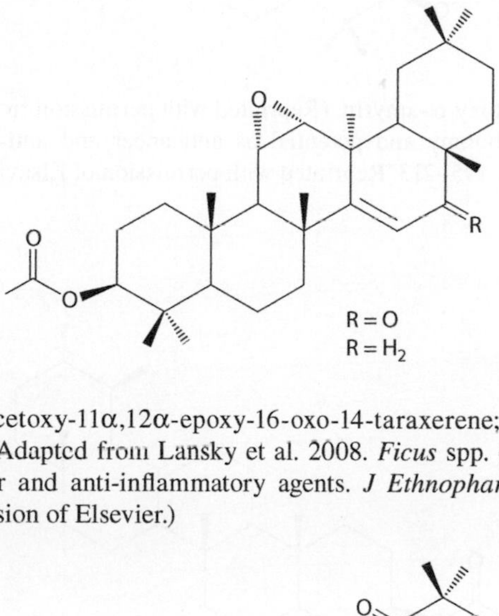

FIGURE 7.26 3β-Acetoxy-11α,12α-epoxy-16-oxo-14-taraxerene; 3β-acetoxy-11α,12α-epoxy-14-taraxerene. (Adapted from Lansky et al. 2008. *Ficus* spp. (fig): Ethnobotany and potential as anticancer and anti-inflammatory agents. *J Ethnopharmacol*. 119: 195–213. Reprinted with permission of Elsevier.)

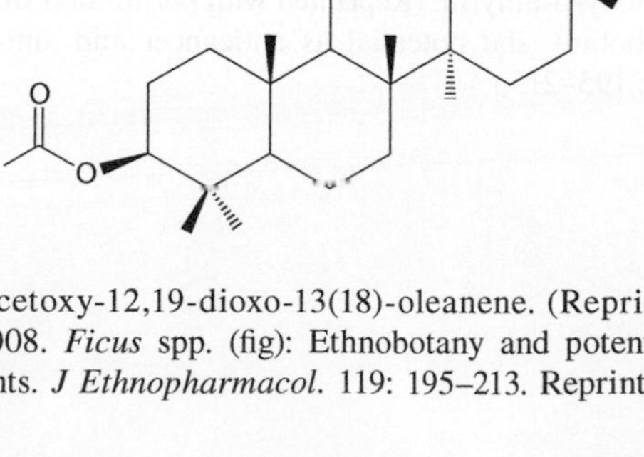

FIGURE 7.27 3β-Acetoxy-12,19-dioxo-13(18)-oleanene. (Reprinted with permission from Lansky et al. 2008. *Ficus* spp. (fig): Ethnobotany and potential as anticancer and anti-inflammatory agents. *J Ethnopharmacol*. 119: 195–213. Reprinted with permission of Elsevier.)

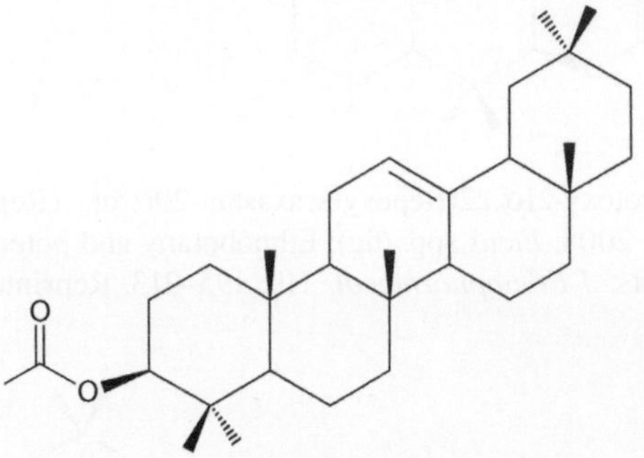

FIGURE 7.28 3β-Acetoxy-α-amyrin. (Reprinted with permission from Lansky et al. 2008. *Ficus* spp. (fig): Ethnobotany and potential as anticancer and anti-inflammatory agents. *J Ethnopharmacol.* 119: 195–213. Reprinted with permission of Elsevier.)

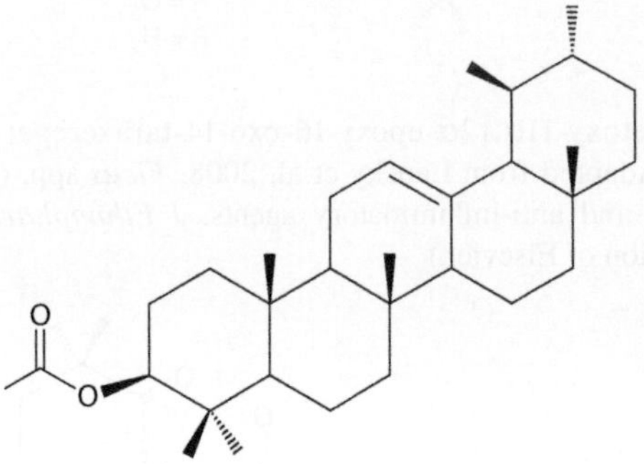

FIGURE 7.29 3β-Acetoxy-β-amyrin. (Reprinted with permission from Lansky et al. 2008. *Ficus* spp. (fig): Ethnobotany and potential as anticancer and anti-inflammatory agents. *J Ethnopharmacol.* 119: 195–213.)

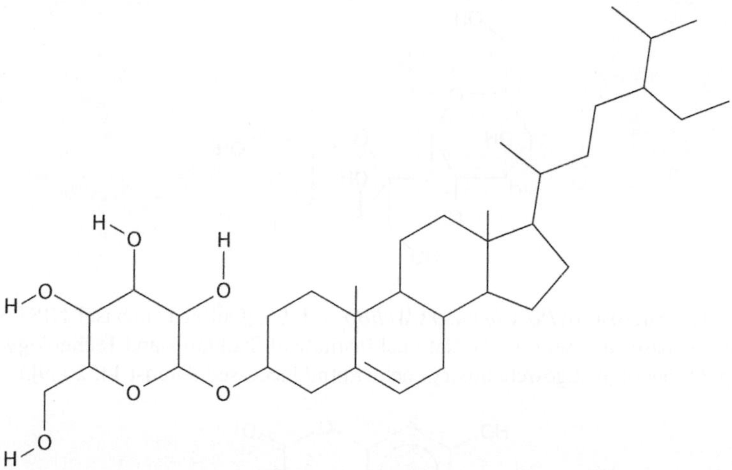

FIGURE 7.30 Daucosterol. (*PubChem*. The PubChem Project. 2001. [Online Database]. National Center for Biotechnology Information. U.S. National Library of Medicine. 8600 Rockville Pike, Bethesda, MD 20894. URL: http://pubchem.ncbi.nlm.nih.gov/ [accessed August 16, 2009].)

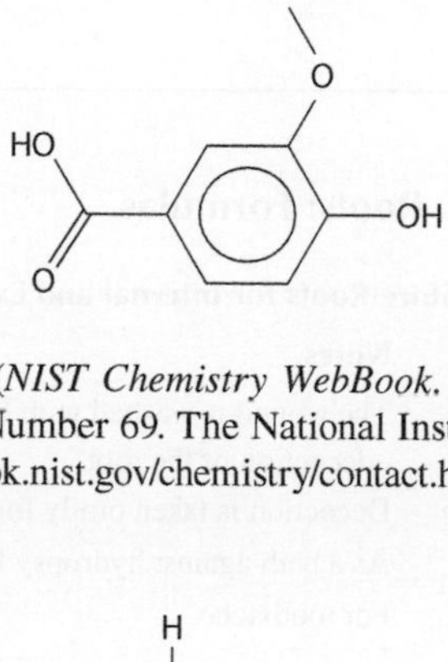

FIGURE 7.31 Vanillic acid. (*NIST Chemistry WebBook*. 1996. [Online Database]. NIST Standard Reference Database Number 69. The National Institute of Standards and Technology (NIST). URL: http://webbook.nist.gov/chemistry/contact.html [accessed August 14, 2009].)

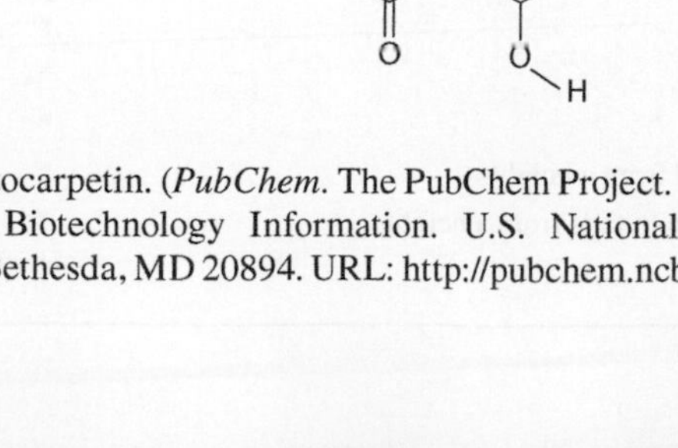

FIGURE 7.32 Norartocarpetin. (*PubChem*. The PubChem Project. 2001. [Online Database]. National Center for Biotechnology Information. U.S. National Library of Medicine. 8600 Rockville Pike, Bethesda, MD 20894. URL: http://pubchem.ncbi.nlm.nih.gov/ [accessed August 16, 2009].)

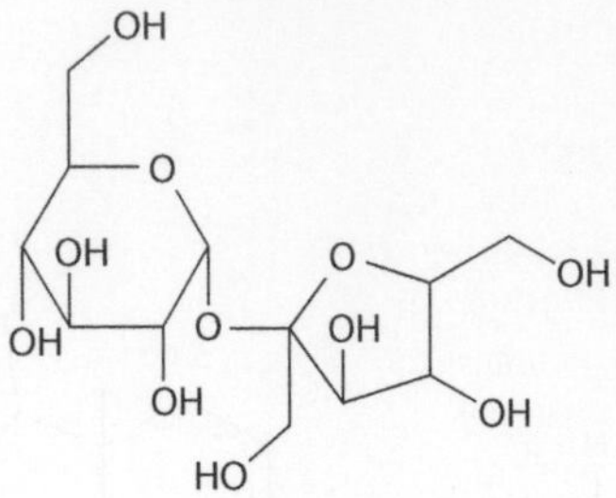

FIGURE 7.33 Sucrose. (*NIST Chemistry WebBook*. 1996. [Online Database]. NIST Standard Reference Database Number 69. The National Institute of Standards and Technology (NIST). URL: http://webbook.nist.gov/chemistry/contact.html [accessed August 14, 2009].)

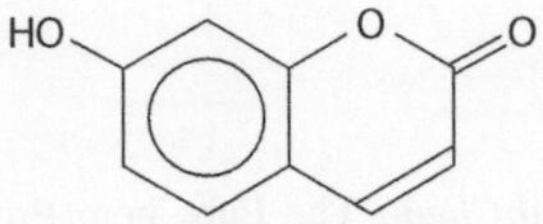

FIGURE 7.34 7-Hydroxycoumarin. (*NIST Chemistry WebBook*. 1996. [Online Database]. NIST Standard Reference Database Number 69. The National Institute of Standards and Technology (NIST). URL: http://webbook.nist.gov/chemistry/contact.html [accessed August 14, 2009].)

TABLE 7.3
Historical Uses of *Ficus* spp. Roots: Formulas

How to Prepare and Store Roots for Internal and External Applications

Variations	Notes
Decoction of the root	The wound is washed with hot wine and treated with a decoction of the root.
Decoction of the root of *F. racemifera*	Decoction is taken orally for recovery after a fall from a height. As a bath against hydropsy for children.
Decoction of the root in wine	For toothache.

TABLE 7.4
Historical Uses of *Ficus* spp. Roots

Indication	Root	References	Score
Analgesic	*	1	1
Against wounds	*	19	1
Against toothache	*	1	1
Against damages due to falls from a height	*	19	1
Against internal bleeding due to falls from a height	*	19	1
Against hydropsy	*	19	1

REFERENCES

Amakura, Y., T. Tsutsumi, K. Sasaki, M. Nakamura, T. Yoshida, and T. Maitani. 2008. Influence of food polyphenols on aryl hydrocarbon receptor-signaling pathway estimated by in vitro bioassay. *Phytochemistry* 69: 3117–30.

Asakawa, Y. 2008. Liverworts—potential source of medicinal compounds. *Curr Pharm Des.* 14: 3067–88.

Awad, A.B., and C.S. Fink. 2000. Phytosterols as anticancer dietary components: Evidence and mechanism of action. *J Nutr.* 130: 2127–30.

Bouic, P.J., and J.H. Lamprecht. 1999. Plant sterols and sterolins: A review of their immune-modulating properties. *Altern Med Rev.* 4: 170–7.

Cai, Q.Y., H.B. Chen, S.Q. Cai et al. 2007. [Effect of roots of *Ficus hirta* on cocaine-induced hepatotoxicity and active components]. *Zhongguo Zhong Yao Za Zhi* 32: 1190–3.

Channabasavaraj, K.P., S. Badami, and S. Bhojraj. 2008. Hepatoprotective and antioxidant activity of methanol extract of *Ficus glomerata*. *Nat Med (Tokyo)* 62: 379–83.

Chiang, Y.M., J.Y. Chang, C.C. Kuo, C.Y. Chang, and Y.H. Kuo. 2005. Cytotoxic triterpenes from the aerial roots of *Ficus microcarpa*. *Phytochemistry* 66: 495–501.

Damu, A.G., P.C. Kuo, L.S. Shi, C.Y. Li, C.R. Su, and T.S. Wu. 2009. Cytotoxic phenanthroin-dolizidine alkaloids from the roots of *Ficus septica*. *Planta Med.* Mar 18 2009 [Epub ahead of print].

Ding, Y., H.T. Nguyen, S.I. Kim, H.W. Kim, and Y.H. Kim. 2009. The regulation of inflammatory cytokine secretion in macrophage cell line by the chemical constituents of *Rhus sylvestris*. *Bioorg Med Chem Lett.* 19: 3607–10.

Garba, S.H., J. Prasad, and U.K. Sandabe. 2007. Hepatoprotective effect of the aqueous root-bark extract of *Ficus sycomorus* (Linn) on Carbon tetrachloride induced hepatotoxicity in rats. *J Biol Sci (Faisalabad)* 7: 276–281.

Jain, R., S. Jain, and S.C. Jain. 2007. Secondary metabolites from *Ficus carica* roots. *Proc Natl Acad Sci India Sect A Phys Sci* 77: 99–100.

Jiang, W.W., X.Q. Zhang, Q. Li, W.C. Ye, and X.S. Yao. 2007. [Chemical constituents of root of *Ficus stenophylla*]. *Tianran Chanwu Yanjiu Yu Kaifa* 19: 588–590.

Jones, P.J., and S.S. AbuMweis. 2009. Phytosterols as functional food ingredients: Linkages to cardiovascular disease and cancer. *Curr Opin Clin Nutr Metab Care* 12: 147–51.

Kalpana, K.B., N. Devipriya, M. Srinivasan, and V.P. Menon. 2009. Investigation of the radio-protective efficacy of hesperidin against gamma-radiation induced cellular damage in cultured human peripheral blood lymphocytes. *Mutat Res.* 676: 54–61.

Kaur, K., M. Jain, T. Kaur, and R. Jain. 2009. Antimalarials from nature. *Bioorg Med Chem.* 17: 3229–56.

Lansky, E.P., and R.A. Newman. 2007. *Punica granatum* (pomegranate) and its potential for prevention and treatment of inflammation and cancer. *J Ethnopharmacol.* 109: 177–206.

Lansky, E.P., H.M. Paavilainen, A.D. Pawlus, and R.A. Newman. 2008. *Ficus* spp. (fig): Ethnobotany and potential as anticancer and anti-inflammatory agents. *J Ethnopharmacol.* 119: 195–213.

Li, C., P.B. Bu, D.K. Yue, and Y.F. Sun. 2006. [Chemical constituents from roots of *Ficus hirta*]. *Zhongguo Zhong Yao Za Zhi* 31: 131–3.

Loizzo, M.R., R. Tundis, F. Menichini, G.A. Statti, and F. Menichini. 2008. Hypotensive natural products: Current status. *Mini Rev Med Chem.* 8: 828–55.

Muregi, F.W., S.C. Chhabra, E.N. Njagi et al. 2003. In vitro antiplasmodial activity of some plants used in Kisii, Kenya against malaria and their chloroquine potentiation effects. *J Ethnopharmacol.* 84: 235–9.

Muregi, F.W., A. Ishih, T. Miyase et al. 2007. Antimalarial activity of methanolic extracts from plants used in Kenyan ethnomedicine and their interactions with chloroquine (CQ) against a CQ-tolerant rodent parasite, in mice. *J Ethnopharmacol.* 111: 190–5.

NIST Chemistry WebBook. 1996. [Online Database]. NIST Standard Reference Database Number 69. The National Institute of Standards and Technology (NIST). URL: http://webbook.nist.gov/chemistry/contact.html [accessed August 14, 2009].

Patel, R.V., L.N. Clark, M. Lebwohl, and J.M. Weinberg. 2009. Treatments for psoriasis and the risk of malignancy. *J Am Acad Dermatol.* 60: 1001–17.

PubChem. The PubChem Project. 2001. [Online Database]. National Center for Biotechnology Information. U.S. National Library of Medicine. 8600 Rockville Pike, Bethesda, MD 20894. URL: http://pubchem.ncbi.nlm.nih.gov/ [accessed August 16, 2009].

Qu, W., L. Fan, Y.C. Kim et al. 2009. Kaempferol derivatives prevent oxidative stress-induced cell death in a DJ-1-dependent manner. *J Pharmacol Sci.* 110: 191–200.

Rodríguez-Granados, M.T., M.J. Pereira-Rodríguez, and F.L. Vázquez-Vizoso. 2009. [Therapeutic effectiveness of psoralen-UV-A bath therapy in psoriasis]. *Actas Dermosifiliogr.* 100: 212–21.

Sackeyfio, A.C., and O.M. Lugeleka. 1986. The anti-inflammatory effect of a crude aqueous extract of the root bark of *"Ficus elastica"* in the rat. *Arch Int Pharmacodyn Ther* 281: 169–76.

Saleem, M. 2009. Lupeol, a novel anti-inflammatory and anti-cancer dietary triterpene. *Cancer Lett.* May 21 2009 [Epub ahead of print].

Sapra, B., S. Jain, and A.K. Tiwary. 2008. Percutaneous permeation enhancement by terpenes: Mechanistic view. *AAPS J.* 10: 120–32.

Sharma, R.C., A. Zaman, and A.R. Kidwai. 1963. Chemical examination of *Ficus racemosa.* *Indian J Chem* 1: 365–6.

Shen, J., H. Nakamura, Y. Fujisaki et al. 2009. Effect of 4G-alpha-glucopyranosyl hesperidin on brown fat adipose tissue- and cutaneous-sympathetic nerve activity and peripheral body temperature. *Neurosci Lett.* 461: 30–5.

Singh, A.B., D.K. Yadav, R. Maurya, and A.K. Srivastava. 2009. Antihyperglycaemic activity of alpha-amyrin acetate in rats and db/db mice. *Nat Prod Res.* 23: 876–82.

Singh, R.K., S. Mehta, D. Jaiswal, P.K. Rai, and G. Watal. 2009. Antidiabetic effect of *Ficus bengalensis* aerial roots in experimental animals. *J Ethnopharmacol.* 123: 110–4.

Vafeiadou, K., D. Vauzour, H.Y. Lee, A. Rodriguez-Mateos, R.J. Williams, and J.P. Spencer. 2009. The citrus flavanone naringenin inhibits inflammatory signalling in glial cells and protects against neuroinflammatory injury. *Arch Biochem Biophys.* 484: 100–9.

Wang, X., K. Liu, and H. Xu. 2009. [Studies on chemical constituents of aerial roots of *Ficus microcarpa*]. *Zhongguo Zhong Yao Za Zhi* 34: 169–71.

Ya, J., X.Q. Zhang, Y. Wang, Y.L. Li, and W.C. Ye. 2008. [Studies on flavonoids and coumarins in the roots of *Ficus hirta* Vahl.] *Linchan Huaxue Yu Gongye* 28: 49–52.

Zhang, X.Q., W.W. Jiang, Y. Wang, Y.L. Li, and W.C. Ye. 2008. [New phenylpropanoic acid derivatives from the roots of *Ficus stenophylla*]. *Yao Xue Xue Bao* 43: 281–3.

8 Fig Wasps

A defining aspect of all trees of the *Ficus* genus is their obligate pollination by a single species of "fig wasp." In the literature, one may frequently encounter the expression fig/fig wasp mutualism, or some similar relationship. Do not be fooled by thinking the fig is related to itself reflexively, or that two figs are involved here. The relationship is between the fig, on the one hand, and the "fig wasp" on the other. It might be more clearly written fig/fig-wasp interaction, instead of fig/fig wasp interaction, but the latter is the convention today because of printing space considerations, even at the expense of a decrease in clarity.

It turns out in looking more closely at the fig/fig-wasp mutualism that many larger issues, including anthropocentric issues, come ever more sharply into view the longer one stares at the tiny wasps, each with its own special fig tree species, and in many cases, by its own particular fig. Indeed, certain species of male fig wasps typically spend their entire lives inside a single syconium, a single fig! The tension of being cooped up so long, especially where the ratio

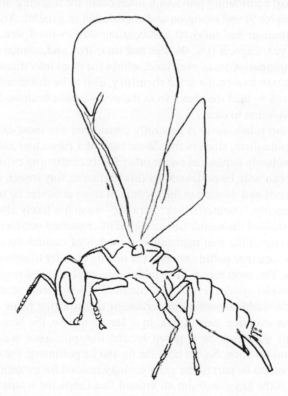

FIGURE 8.1 Fig wasp (pen and ink by Zipora Lansky).

315

of males to females is impossibly high, makes for some very violent little critters. Living under tremendous social and sexual pressures in a dark, cooped-up environment, the male fig wasps of a certain species take to fighting with their brothers. A way out of the mess is simply to get out of Mother Fig, and, the wasp often being wingless by nature, slides over to a nearby leaf until the tensions inside calm down.

Wasp society around a single fig can be incredibly complex. Not only is each fig species required to be pollinated by its own obligate fig-wasp species (and consequently, for its survival as a species), but also additional fig wasps, of at least three other genera. These preferably or obligatively utilize a particular fig species for their mating and birthing grounds (and in some cases, also burial grounds), while not contributing (at least not directly) to the pollination process. Such nonpollinating wasps have been traditionally referred to in the literature as "counterfeit" or "parasitic" wasps because they are not "required" or "utilized" by the fig for its pollination process. The shifting of language is in step with our politically correct times, and also, happily, opens the door to a genuinely revised and expanded and hopefully also more accurate understanding of the role of these counterfeit wasps in the overall scheme of things, from the vantage points of both the *Ficus* host and the wasp pollinator, its cohorts, enemies, and cousins. Further—and here is the really surprising part—the fig wasp community provides a microcosm for thinking about *Ficus* and ourselves (in Chapter 9) and evolution and coevolution in general. And because this (successful) mutualism has survived and continuously evolved since its inception over 60 million years ago, it (i.e., the figs and their true and counterfeit wasps and their all-around mutualism) may even hold, within the deep interstices of a receptive fig, some lessons (not to mention some chemistry, also to be discussed in Chapter 9) for us and our work to heal the planet from the evil we have bestowed upon it in the process of our evolution to date.

The fig/fig-wasp relationship is generally considered the most extreme biological example of mutualism, since in the classic case of a *Ficus* tree and its pollinator wasp, each is absolutely dependent on the other for its continuing existence as a species: the fig tree can only be pollinated by this particular tiny insect, and this insect is only able to breed and develop in the syconia of this particular fig tree.

The nonpollinating, "counterfeit" or "parasitic" wasp has likely always seemed to those who have studied the mutualism as a kind of unwanted outsider, its very existence seeming to upset the neat mathematical models of mutualism and the natural constraints of the one true pollinator upon the one fig in order to actualize the singular goals of each. Yet, most recent work in this ever-evolving and exciting biological and ecological model system reveals that counterfeit wasps may actually—no, they most certainly do—add something important to the mix that helps it maintain its balance in an ever-evolving state of flux. In at least one case, the presence of certain counterfeit wasps may even be required for the true pollinator wasps to exercise their singular vital function. So, for both the fig and its pollinator, the nonpollinating counterfeiters seem to be part of the great ecology needed for evolution to go on.

Competition is the key constraint all around that keeps the wasps bound to their fig. This relates in some degree to the internal topology of the syconia and the ovules it presents to wasps.

Syconia of all figs contain ovules, tubular structures that can hold a fertilized seed of the fig to maturity, or, alternatively, if it is "galled" by a wasp, to house and allow to develop, also to maturity, a wasp. The wasp may be a true pollinator or a counterfeiter, that is, a nonpollinator.

The best ovules from the point of view of a wasp are short ones, because these are the easiest for the female to insert her ovipositor and leave behind an egg nice and snug at the bottom of the tubule. The short ovules the fig concentrates in the center part of the syconium, making this part of the fig the best for the wasps for several reasons. The first, mentioned earlier, is because these ovules are shorter and easier for oviposition, and second, the inner areas are away from the periphery and its vulnerability to attacking insect, fungal, and bacterial invaders.

The outer ovules, therefore, the fig reserves for its own internal uses, that is, for development of seed. In spite of all this tightly engineered control, however, figs will be figs, and wasps will be wasps, meaning that each species has its own individual agenda in all such encounters, be it the pollinator with its central role in enabling reproduction, the counterfeits with their shifty and mysterious synergistic contributions to maintaining the ecological balance, and the fig herself, the great *Ficus* host of the party. Each species "wants" the most it can get "for itself."

Because the space available in those short tubules is very restricted for courtship and mating and later egg development, small (figwasp) females are better, that is, more "fit," reproductively speaking, than large females. However, larger males can more easily fight off competing males and open the tubule to mate with the still-entrapped newly emergent young adult females so that, reproductively speaking, the larger males are the more fit. Overall, this favors male production by fit females (almost all females are fit, i.e., for finding a mate) by the so-called Trivers–Willard effect (Greeff and Ferguson 1999). The latter note that theory and data suggest that a male in "good condition" at the conclusion of the parental investment period will, all other things being equal, "outreproduce" a sister in similar condition, while she can be expected to "outreproduce" him if both are in "poor condition," suggesting that natural selection favors parental ability to adjust the sex ratio of their offspring according to their ability to invest. In mammals, the adult female produces fewer males compared to females as her condition declines (Trivers and Willard 1973).

Between figs and wasps, these phenomena lead to intense competition among males for mating space within the fig. It is hard, perhaps, to say "who" is choreographing the drama, the fig wasps or the fig tree, but the entire interphylum zone works in a seemingly creative and intelligently responsive way to regulate itself. Fig wasps do learn to leave the outer ovules, which are longer and tastier, to the fig and, when the pressure becomes too great, to exit the fig altogether and hang out on an adjacent leaf to wait until conditions may again become favorable to taking on the activity within the syconium. Human beings (see Chapter 9), on the other hand, have learned to select certain varieties, subspecies, or cultivars of *Ficus carica*, the so-called common, edible fig of commerce. Many strains of these edible figs require no pollination to bear fruit. Virtually all edible figs in an edible condition are completely female.

Greeff and Ferguson (1999) observed that apterous (wingless) and seemingly nondispersing male wasps of the species *Otitesella digitata* routinely left their *Ficus ingens* figs to repose on leaves close to their natal fig. Such dispersal can affect both

(1) sex allocation by reducing the degree of local mate competition between brothers and (2) male dimorphism by reducing the mating opportunities of males with a dispersing morphology. An extremely male-biased sex ratio resulted in almost all *O. longicauda* females being mated.

The volatile compounds released by *Ficus* species to attract their wasps always occur in complex blends. The complexes consist primarily of terpenoids, aliphatic compounds, and products from the shikimic acid pathway. Major compounds, typically those particularly common among floral fragrances in general, occur in many of the 20 *Ficus* species examined by Grison-Pigé et al. (2002), but also very rare compounds in small amounts that were more or less specific to individual *Ficus* species. The work, which identified 99 different compounds, provides a preliminary scaffold for better understanding of the chemical basis that binds specific figs to their dedicated pollinating as well as opportunist counterfeit wasps.

Two other groups specifically investigated the dynamics of these volatile compounds in *Ficus hispida* and their attractiveness to this species' wasps. Song et al. (2001) demonstrated that linalool was the major constituent of steam-distilled oil of either male or female receptive figs, while di-butyl phthalate was prominent in the oils of postparasitized and postpollinated figs. In the petroleum ether extracts, palmitic oil and 9,12-octadecadienoic acid were the main constituents of male and female receptive figs, while hexadecanoic acid ethyl ester predominated in postparasitized and postpollinated figs. In dichloromethane extracts, linalool was the major constituent of male and female receptive figs, 1-hydroxylinalool was the major component of male postparasitized figs, and 1-hydroxylinalool and benzyl alcohol were the major constituents of female postpollinated figs. Bioassays with sticky traps showed *Ceratosolen solmsi marchali* to be attracted to the aforementioned dichloromethane extracts of both male and female receptive figs and to petroleum ether extracts of female receptive figs, but not to dichloromethane and petroleum ether extracts of male postparasitized or female postpollinated figs. Figs were attractive to pollinating wasps only at the receptive stage and this was reflected in a difference in the patterns of volatile constituents of receptive figs versus postpollinated or postparasitized figs. From a receptive to a postpollinated state, figs altered their volatile compositions, with linalool, linalool oxide, terpeneol, and 2,6-dimethyl-1,7-octadiene-3,6-diol decreasing. Thus, these may act as the attractants of the wasps. An increase of dibutyl phthalate, 1-hydroxylinalool, and benzyl alcohol repelled the wasps. That dichloromethane extracts of male and female receptive figs both attracted fig wasps indicates that receptive figs of both sexes are similarly attractive to fig wasps. Extracts of leaf of the plant at different stages were also prepared but did not attract the wasps. Further, no repellant effects of the leaves at any stage were described.

The odor of the *F. hispida* figs was studied at two floral stages, namely, the female floral stage, attractive to *C. solmsi marchali* and enabling of pollination, and the interfloral stage (between the female floral and male floral stages). Odors of the receptive and interfloral figs were attractive to pollinators, but the receptive figs were greatly preferred in such a way that the wasps would eschew large concentrations of interfloral volatiles in preference to very small amounts of female floral receptive compounds. Three monoterpenes in the receptive female fig blend predominate: linalool, the main compound, and two minor compounds limonene and β-pinene.

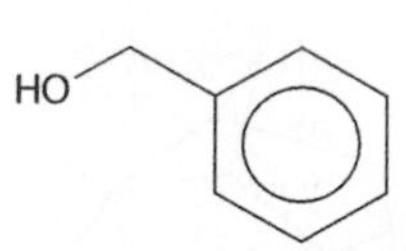

FIGURE 8.2 Benzyl alcohol. (*NIST Chemistry WebBook*. 1996. [Online Database]. NIST Standard Reference Database Number 69. The National Institute of Standards and Technology (NIST). URL: http://webbook.nist.gov/chemistry/contact.html [accessed August 14, 2009].)

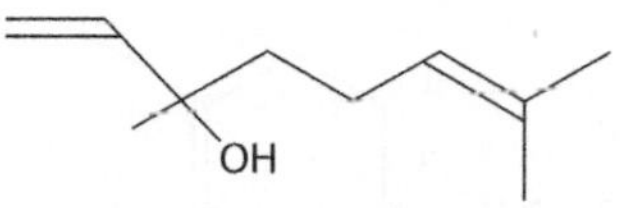

FIGURE 8.3 Linalool. (*NIST Chemistry WebBook*. 1996. [Online Database]. NIST Standard Reference Database Number 69. The National Institute of Standards and Technology (NIST). URL: http://webbook.nist.gov/chemistry/contact.html [accessed August 14, 2009].)

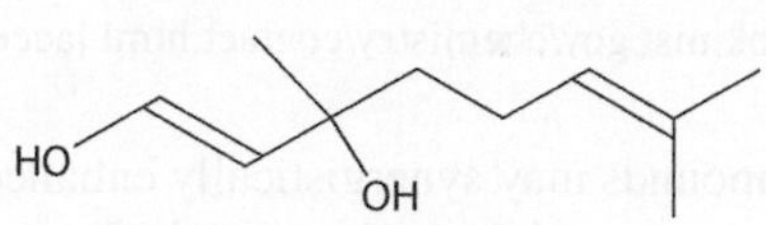

FIGURE 8.4 1-Hydroxylinalool. (*NIST Chemistry WebBook*. 1996. [Online Database]. NIST Standard Reference Database Number 69. The National Institute of Standards and Technology (NIST). URL: http://webbook.nist.gov/chemistry/contact.html [accessed August 14, 2009].)

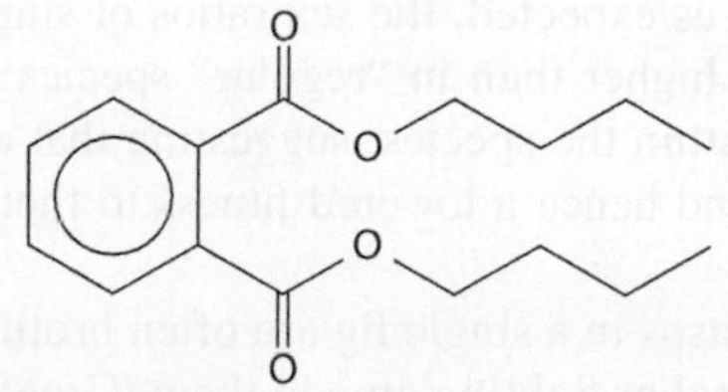

FIGURE 8.5 Dibutyl phthalate. (*NIST Chemistry WebBook*. 1996. [Online Database]. NIST Standard Reference Database Number 69. The National Institute of Standards and Technology (NIST). URL: http://webbook.nist.gov/chemistry/contact.html [accessed August 14, 2009].)

The levoisomer and racemic mixtures of linalool were attractive to the pollinator at high doses, but the dextroisomer was neutral. (+/−)-Limonene and (−)-β-pinene at high doses were even less attractive to the pollinator than clean air and were neutral at low doses, while the racemic mixtures (R)-(+)−, (S)-(−)-limonene were neutral at all doses. In blend tests, all four mixtures of (+/−)-linalool or (S)-(−)-linalool combined with (+/−)-limonene or (−)-β-pinene attracted *C. solmsi marchali* when administered at high doses, and (R)-(+)-linalool and (−)-β-pinene enhanced the attractiveness of (S)-(−)-linalool, while enantiomers of limonene did not. Thus, both the quality and quantity of fig volatiles regulate *C. solmsi marchali* response, and that quality is the main host-finding and floral stage-distinguishing cue for the pollinator. The

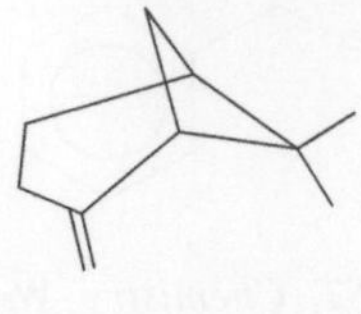

FIGURE 8.6 β-Pinene. (*NIST Chemistry WebBook*. 1996. [Online Database]. NIST Standard Reference Database Number 69. The National Institute of Standards and Technology (NIST). URL: http://webbook.nist.gov/chemistry/contact.html [accessed August 14, 2009].)

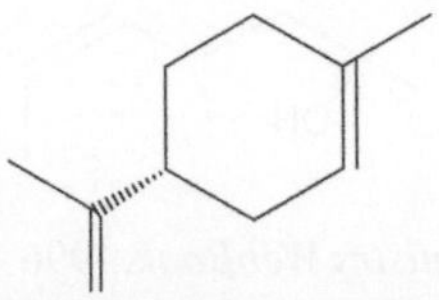

FIGURE 8.7 (+)-Limonene. (*NIST Chemistry WebBook*. 1996. [Online Database]. NIST Standard Reference Database Number 69. The National Institute of Standards and Technology (NIST). URL: http://webbook.nist.gov/chemistry/contact.html [accessed August 14, 2009].)

possibility that some compounds may synergistically enhance effects of some of the attractiveness of the core compounds was also raised (Chen and Song 2008).

Greeff (2002) reminds us that female-biased sex ratios reflect male competition and its attendant inbreeding. Using "a colour dimorphism with a simple Mendelian inheritance," Greeff quantified the mating system of an unusual fig-pollinating wasp in which males disperse to obtain matings on non-natal mating patches (e.g., on nearby leaves). Although, as expected, the sex ratios of single foundresses (ovipositing female wasps) were higher than in "regular" species of fig wasps, there was also profound variance within the species, suggesting that excess males can "avoid local mate competition (and hence a lowered fitness to their mother) by dispersing to other patches."

Male pollinating fig wasps in a single fig are often brothers, but this relatedness does not eliminate dispersal or fighting among them (Greeff et al. 2003). Although the evolution of these behaviors may seem paradoxical and their origins obscure, it does illuminate the latent attraction of these male pollinators to move beyond their ancestral fig, and strike out for greener leaves, taking the thread of evolution with them.

However, the participation of the male wasps in modulating the sex ratio of their kind in a single fig is by no means the only determinant. Their mothers may oviposit in discrete patches within the syconium, able to choose both the sex and the male mating tactics (natal-patch mating or dispersing) of their offspring based only on how many other mothers have used the specific patch before them. These conclusions based on observation of counterfeit wasps of different species of genus *Otitesella* suggest that females respond to population densities at the level of individual figs (Pienaar and Greeff 2003).

Another study (Song et al. 2008) focused on the sex ratio of *Apocryptophagus* sp., a species of nonpollinating counterfeit fig wasps hosted on *Ficus semicordata*.

It was shown that female *Apocryptophagus* sp. started to visit the fig on the third day after pollinator *Ceratosolen gravelyi* oviposited. The female *Apocryptophagus* sp. oviposited on the outside of the fig, their ovipositing lasting 2 days. Male *Apocryptophagus* sp. emerged concurrently with pollinators. The males opened a small hole on the wall of gall where the females developed, and mated with the females. Mated females exited their natal fig, leaving for a new receptive fig. The sex ratio of *Apocryptophagus* sp. was in agreement with local mate competition theory, that is, it was female-biased. Meanwhile, the total number of offspring was proportional to foundress number, while number of offspring per foundress was inversely proportional to foundress number. Finally, the female bias in the sex ratio was correlated negatively with the total number of offspring.

Males may respond to shifting population densities with evolutionary divergence. Polymorphic males may include a type specialized for fighting, and another with wings that are specialized for dispersion. The divergent types recede when conditions change and may even reemerge when needed again. This plasticity and reversibility in the evolution of the fig wasp genus *Philotrypesis* to changing densities is suggestive of complex homeostatic mechanisms within the intrafig ecosystem, affecting polymorphic shifts as a means of adaptation (Jousselin et al. 2004).

The size of the clutch (i.e., the number of wasp eggs within the specific fig) is also a determinant of wasp sex ratio, in that with decreasing clutch size, foundresses produce higher concentrations of sons in their broods. Competition between females correspondingly produces smaller brood size. These regulatory factors suggest that foundresses respond to changes not only in the number of eggs in the clutch, but also to changes in the number of other foundresses (Kjellberg et al. 2005).

Moore et al. (2006) showed that dispersal by male *Platyscapa awekei* pollinating fig wasps is promoted by both low returns in the natal fig and kin competition avoidance, with strategies depending on the interaction between phenotype (body size) and local conditions. They explain the coemergence of male dispersal and fighting behaviors as examples of "why other social interactions are often mixtures of cooperation and conflict," in that they are products of dissimilar selective forces.

In a phylogenetic study of *Otiteselline* fig wasps (Hymenoptera: Chalcidoidea: Agaonidae) associated with *Ficus* in the Afrotropical region using rDNA sequences, the authors (Jousselin et al. 2006) illuminate an area of superimposition of two clades (deriving from common ancestors) of wasps "superimposed on the fig system" and use this to explain why each African fig species usually hosts two *Otiteselline* fig wasp species. Because the phylogenies of the two clades are generally more congruent than expected by chance, they maintain that *Otiteselline* wasp speciation is largely constrained by the diversification of their (*Ficus*) hosts. Further study by this group also revealed difference in ovipositor length between the two *Otiteselline* species coexisting in the same *Ficus* species, corresponding to ecological differences (niches) within a single fig as discussed previously. "The diversification of ecological niches within the fig is probably, with cospeciation, one of the key factors explaining the diversification and maintenance of species of parasites of the fig/pollinator system."

Cospeciation analyses of the figs of selected species within the *Galoglychia* subsection of the *Ficus* genus and their associated fig wasps reveal an intricately complex wasp community of different species (Jousselin et al. 2008). The nonpollinating counterfeit wasps are as constrained by the pollinator wasps as they are by their host. By adapting a randomization test in a supertree context, they reconfirmed the known fact that wasp phylogenies are significantly congruent with each other, and in the process, built a "wasp community" supertree that retrieved *Galoglychia* taxonomic subdivisions.

Thus, wasp host specialization as well as niche saturation within the fig prevent recurrent intrahost speciation and host switching. A comparison of sequence divergence of cospeciating pairs of wasps suggested that the diversification of some pollinating and nonpollinating wasps of *Galoglychia* figs has been synchronous, with the pollinating species being more evolutionarily labile.

In the pollinating fig wasp, *Platyscapa awekei* male dispersal to other loci is less a means to an optimal mating system than a mechanism to reduce competition among brothers. The number of mature offspring a female produces depends on her own heterozygosity and not on that of the offspring, and may be determined by egg and gall quality (Greeff et al. 2009).

Nonetheless, *caprification* is a procedure that, although developed by humans, directly involves the wasps, at least the pollinating types. Female figs that become pollinated by the process of caprification are said by many to be heartier, heavier, and richer than those that developed by the female alone, without the male material.

The term *caprification* contains the prefix for goat (as in Capricorn), probably because these figs containing both male and female elements were thought fit only for goats. In fact, there is evidence that female-only fruits existed from ancient times, perhaps even predating any cultivation.

The process of caprification specifically involves obtaining a branch of a caprifig tree (generally of *Ficus carica*) that is streaming with female pollinating fig wasps. This branch is laid down on the top of an all-female tree, and the wasps are allowed to invade on their own accord. Ongoing arguments for and against caprification in commercial fig growing continue to be heard. On the one hand, caprification is old-fashioned, superfluous, and produces a smaller and inferior fig in every way. On the other hand, caprification is the way of nature, and the resultant figs offer up an indescribably more complex and intricate aroma that confers greater pleasure as well as stimulation of masculine principles and reproductive potency for male and female alike.

If the fig/fig-wasp mutualism has evolved over 60 million years, then it was created even before that. One idea is that *Ficus* emerged in the old southern fused continent, and then drifted northward to the Middle East and throughout the world.

Different pollinating and nonpollinating fig wasps show different degrees of specificity regarding their *Ficus* hosts. In other words, some wasps may show preference for a certain fig type (species), but still demonstrate some ability to pollinate a related type or species. Other fig wasps do not show this flexibility. In one such type of pollinating fig wasp, *Ceratosolen solmsi*, negative deviation from the neutral model of evolution reflects possible selection pressures acting on a particular genetic sequence. The sequence, dubbed Or2, is an orthologous gene in the sense that it appears in multiple species of fig wasps that ostensibly evolved from a common fig-wasp ancestor. It is associated with specialized olfactory functions connected with zeroing in on the specific state of receptivity that it has been allotted by the host,

FIGURE 8.8 *Ficus carica*, young fruit (27.7.2009, Seven Species Garden, Binyamina, Israel, by Helena Paavilainen).

F. hispida. The sequence occurred in the aforementioned pollinating fig wasp and as well in three different nonpollinating fig wasps, specifically *Apocrypta bakeri*, *Philotrypesis pilosa*, and *Philotrypesis* sp., all of which have a high degree of specificity for *Ficus hispida*.

However, the orthologous sequence had a different rate of occurrence among the different wasps. Specifically, expressions of Or2 were restricted to the heads of all wingless male fig wasps, which usually live in the dark cavity of a fig throughout their entire life cycle. However, in the female wasps that must fly from one fig to another for oviposition, and secondarily for pollination, expressions were widely detected in the antennae, legs, and abdomens. Further, compared with the nonpollinating wasps, the Or2 gene in *C. solmsi* showed an elevated rate of substitutions and lower codon usage, the significance of which is still not presently clear (Lu et al. 2009).

For cophylogeny and coevolutionary divergence, that is, codivergence, host specificity of the wasps is a necessary but not sufficient condition. Andrew Jackson (2004) reviewed various mutualistic and antagonistic symbioses between fig trees (*Ficus*:

Moraceae) and chalcid wasps and concluded that these constitute a community in microcosm. His phylogenetic estimates of figs and fig wasps revealed topological correspondence, rendering the fig/fig-wasp microcosm a model system for cophylogeny. Jackson goes on to explain that incongruence between phylogenies from associated organisms can be reconciled through a combination of evolutionary events.

Jackson further delineates cophylogeny mapping to integrate phylogenies by "embedding an associate tree into a host tree, finding the optimal combinations of events capable of explaining incongruence and evaluating the level of codivergence." These "trees" are, of course, mathematical abstractions to describe the events of co-evolution between real *Ficus* trees and their wasps. The review, in short, addresses the results of cophylogeny analysis concerning *Ficus* and discusses the plausibility of different evolutionary events.

Five different associations encompassing figpollinator, figparasite, and pollinator–parasitoid interactions are reconciled in a bioinformatic method that improves on previous comparisons through the use of "jungles" to provide an exhaustive and quantitative analysis of cophylogeny. Such a jungle is a mechanism for "inferring host switches and obtaining all potentially optimal solutions to the reconciliation problem."

Jackson's results support the consensus that figs codiverge significantly with pollinators but not with nonpollinators. However, he notes that pollinators have not been so faithfully codivergent as traditional models have suggested, in that they have also switched host trees or even species and sometimes still do. This adds to the growing realization that evolutionary transitions in the microcosm are more flexible than previously thought.

Jackson's sampling strategy derives from the influence of taxon sets on the fig pollinator and fig parasite jungles. Choice of congruence measure influences significance, supported by spurious significant results for fig parasite and fig-parasitoid jungles. When not possible to realistically weight switches, however, the total number of events required for the reconciliation of these two virtual trees ("total cost") is not suitable as a measure of congruence.

Jackson's approach, in elegant mathematical language, highlights the inherent plasticity of evolution in general and in the fig/fig-wasp mutualism, in particular. It defines the potential zones of interaction between the major players: the fig species, the pollinator wasps, and the nonpollinating (counterfeit) wasps. All possible combinations of interactions occur, and Jackson's trees and jungles offer a preliminary sophisticated understanding of the virtual realms that accrue and control these interactions by strange attractors and complex chaotic confinement within and without the syconium.

Extreme specificity between the pollinating fig wasps (Agaonidae) and their hosts (*Ficus*) results in a situation where each fig tree depends on a single species of tiny wasp. However, this extremis is responsible for the escape of both from the pressure of the specificity. The extremis in mutualism, however, also results in compensatory codivergence and speciation derivative of the mutualism (Lin S. et al. 2008).

Similarly, these species may be particularly sensitive to global climatic changes, and represent perhaps a model for such mutability. In a specific example, the genetic structure of a certain fig-pollinating wasp, dubbed *Ceratosolen* sp. 1, related to

TABLE 8.1
Ficus Species and Their Pollinator and Nonpollinating Wasps

Figwasp Species	True or Counterfeit (Pollinating or Nonpollinating)	*Ficus* Species	References
Alfonsiella fimbriata Waterston	True	*natalensis leprieuri* Miq	Michaloud et al. 2005
Apocrypta bakeri	Counterfeit	*hispida*	Lu et al. 2009
Apocryptophagus sp.	Counterfeit	*semichordata*	*Song et al. 2008*
Blastophaga psenes	True	*carica*	Weiblen 2002
Ceratosolen arabicus	True	*sycomorus*	Compton et al. 1991
Ceratosolen galili	Counterfeit	*sycomorus*	Compton et al. 1991
Ceratosolen gravelyi	True	*semichordata*	Song et al. 2008
Ceratosolen solmsi	True	*hispida*	Chen and Song 2008, Lu et al. 2009, Song et al. 2001
Ceratosolen sp. 1	True	*septica*	Lin R.C. et al. 2008
Ceratosolen fusciceps Mayr	True	*racemosa*	Wang R.W. et al. 2008
Diaziella bizarrea	Counterfeit	*glaberrima*	Zhang et al. 2009
Eupristina sp.	True	*glaberrima*	Zhang et al. 2009
Philotrypesis pilosa	Counterfeit	*hispida*	Lu et al. 2009
Philotrypesis sp.	Counterfeit	*hispida*	Lu et al. 2009
Platyscapa awekei	True	*salicifolia*	Greeff et al. 2009, Moore et al. 2006
Otitesella longicauda	True	*ingens*	Greeff and Ferguson 1999

F. septica in Taiwan, was investigated with attention to partial sequences of the mitochondrial cytochrome c oxidase subunit I (COI) gene (1052 bp) and genotypes at 15 microsatellite loci. The study (Lin R.C. et al. 2008) revealed that the species underwent a 500-fold expansion secondary to climatic oscillations of the last glacial period, that is, the "last glacial maximum" (LGM). These glacial changes, coupled with the dispersal ability of the particular species, resulted in a responsive shaping of the genetic composition of this subtropical fig-pollinating wasp. The example is indicative of how organisms may utilize speciation and codivergence in response to pressures on the macro as well as micro scales, from global parameters to those of individual figs.

As noted earlier, counterfeit wasps may yet provide some poorly understood facilitation of pollination. Of note is the observation that the counterfeit internally ovipositing wasp *Diaziella bizarrea*, which cannot effectively pollinate its host, *F. glaberrima*, or have any specific effect on seed production, nonetheless effected a reduction in the number of progeny of the true pollinator, *Eupristina* sp., emerging from the fig. The results have led to the hypothesis that a relatively limited, but

possibly essential, mutualism exists not only between the *Ficus* species and its pollinator wasp, but also between the pollinator wasp and the counterfeit wasps that
inhabit and oviposit in the same fig (Zhang et al. 2009).

The life spans of female pollinating wasps are as much as doubled when the
wasps are given access to moisture, although their longevity is unaffected by
early exit from the syconium. It may thus be concluded that conflict resolution
in the fig/fig-wasp mutualism is significantly influenced by tropical seasonal-
ity, because wasps may be less able to overexploit ovules in dry periods due to
shorter times available due to their correspondingly shorter life spans (Dunn
et al. 2008a).

The role of the counterfeit wasps in stabilizing the fig/fig-wasp mutualism is further illuminated by the stunning observation that in *Ficus rubiginosa* the offspring
of wasps are subject to attack by counterfeit wasps who oviposit from the outside of
the fig, rather than from the inside as true pollinating fig wasps do. This avoidance
of the outer ovules (leaving them to the fig) is thus crafted largely by the host *Ficus*,
but secondarily by the counterfeit wasps that place an additional pressure on wasps
sniffing around the ovules looking for a place to oviposit. Not only must would-be
outer ovule ovipositors be confronted by long ovule length, but also by attack from
counterfeit wasps ovipositing from outside the syconium. Thus, a kind of mutualism exists between the counterfeit wasps and the *Ficus* host, a mutualism that keeps
pressure on the true pollinating wasps to pollinate in the well-protected center of the
fruit, leaving the outer ovules to the fig, with a small tribute to the nonpollinating
counterfeit facilitator (Dunn et al. 2008b).

Mutualisms are interspecific interactions in which both players benefit. Explaining
their maintenance is problematic, because cheaters should outcompete cooperative
conspecifics, leading to mutualism instability. Monoecious figs (*Ficus*) are pollinated
by host-specific wasps (Agaonidae), whose larvae gall ovules in their "fruits" (syconia). Female pollinating wasps oviposit directly into *Ficus* ovules from inside the
receptive syconium. Across *Ficus* species, there is a widely documented segregation
of pollinator galls in inner ovules and seeds in outer ovules. This pattern suggests that
wasps avoid, or are prevented from ovipositing into, outer ovules, and this results in
mutualism stability. However, the mechanisms preventing wasps from exploiting outer
ovules remain unknown. In *F. rubiginosa*, offspring in outer ovules are vulnerable
to attack by parasitic wasps that oviposit from outside the syconium. Parasitism risk
decreases toward the center of the syconium, where inner ovules provide enemy-free
space for pollinator offspring. The resulting gradient in offspring viability is likely to
contribute to selection on pollinators to avoid outer ovules, and by forcing wasps to
focus on a subset of ovules, reduces their galling rates. This previously unidentified
mechanism may therefore contribute to mutualism persistence independent of additional factors that invoke plant defenses against pollinator oviposition, or physiological constraints on pollinators that prevent oviposition in all available ovules.

Sexual maturation, male reproductive anatomy, and structure of spermatozoa
have been investigated (Fiorillo et al. 2008) in three fig wasp species from genus
Pegoscapus (Hymenoptera, Chalcidoidea). The specificity of the results gives an
idea of the minute measurements and surveying that were achieved, and constitutes
a significant advance in the generative anatomy of male fig wasps.

The three *Pegoscapus* species exhibit identical reproductive tract features, including testes with a single testicular tubule, seminal vesicles, vasa deferentia, accessory glands, and ejaculatory duct. The seminal vesicle consists of two morphologically distinct compartments, though dissimilar from the separate chambers found in other Chalcidoidea. The anterior portion of the seminal vesicle possesses a distinct epithelium and stores the mature spermatozoa, while the posterior region has the thicker muscular sheath that participates in ejaculation. The *Pegoscapus* males are mature at emergence from the ovules. The spermatozoa of *Pegoscapus* are structurally similar to other Chalcidoidea. In *Pegoscapus* sp1. and *Pegoscapus* sp2. the spermatozoa are nearly identical, while *Pegoscapus tonduzi* shows some differentiation. The size of the spermatozoa is approximately 160 μm in *Pegoscapus* sp1. and *Pegoscapus* sp2., and 360 μm in *P. tonduzi*. The extracellular sheath in *Pegoscapus* sp1. and *Pegoscapus* sp2. is thick, whereas that in *P. tonduzi* is very thin, resulting in a large space the extracellular sheath and the nucleus. Nonetheless, the three species allow of definition and of identity of a *Pegoscapus* spermatozoon: seminal vesicle not divided in chambers; absence of acrosomal structure; and central microtubules first to terminate in the microtubular sequence cutoff at the final axonemal portion.

Other work has penetrated more deeply into the mechanisms of mutualism, coevolution, ecology, and divergence. Wang R.W. et al (2008), studying the true pollinating wasp *Ceratosolen fusciceps* Mayr and the host *F. racemosa*, have discussed trade-off between the "reciprocal mutualists," figs, and wasps, in their utilization of the limited resources available in their joint evolutionary venture.

First, they review, and subsequently debunk as inadequate, previous hypotheses aimed at accounting for the mechanisms preventing competition/conflict between the recipient and cooperative actor in the fig/fig-wasp mutualism, namely self-restraint, dispersal, or spatial constraints. Second, they showed the number of viable seeds of figs to be positively correlated with the number of pollinator offspring when the number of vacant female flowers is high and the foundress number is low (two foundresses). They further showed the number of seeds to be negatively correlated with the number of wasp offspring when the number of vacant female flowers is low and the number of foundresses is increased manually (to eight foundresses). The correlation coefficient between viable seeds and wasp offspring (galls) is dependent upon vacant female flower availability. The data indicate the interaction between figs and figwasps to be conditional. Cooperation ensues when local resource availability is plentiful, but conflict occurs when local resource availability is limited. The self-restraint, dispersal, and spatial heterogeneity previously hypothesized to maintain stable cooperation they deem insufficient to prevent the symbionts from utilizing more local resources at the expense of the recipients. Conflict capable of disrupting the cooperation interaction is likely to occur when the local resource is saturated with symbionts. Repression of symbiont increase leading to reduced utilization of local resources in the conflict period enables the homeostatic maintenance and evolution of cooperation.

Going beyond recent studies that examined codivergence between specific fig-wasp genera of pollinating (*Hymenoptera*, Agaonidae, Chalcidoidea) and counterfeit nonpollinating (Chalcid subfamilies with varied natural histories and ecological strategies, e.g., competitors, gallers, and parasitoids), Marussich and Machado (2007) focused on the history of divergence within a fig wasp community comprising of multiple genera of wasps associated with a large number of sympatric fig hosts. Specifically, phylogenetic analyses of mitochondrial DNA sequences (COI) in 411 individuals from 69 pollinating and nonpollinating fig wasp species of five figwasp genera (*Pegoscapus, Idarnes, Heterandrium, Aepocerus, Physothorax*) and figs of 17 *Ficus* species from section Americana were studied. Host-switching and multiple wasp species per host, contrary to traditional understanding, were found to be extremely common among neotropical nonpollinating wasp genera. Nonetheless, cophylogenetic analyses (TREEMAP 1.0, TREEMAP 2.02beta, and parafit) yielded a portrait of codivergence among fig wasps from different ecological guilds, challenging further the classical notion of strict-sense coevolution between figs and their associated wasps. The findings were consistent with molecular studies of other mutualisms that revealed common patterns of diffuse coevolution and asymmetric specialization among the participants.

Chemical mediation by emission of volatile chemicals from the fig trees at different phenological (related to climate) stages of the fig's life and reproductive cycles shapes ovipositing in time by the different wasp species, both the agaonid pollinating and parasitic chalcidoid nonpollinating, involved with a particular *Ficus* species. Such study, here especially of the counterfeit wasps, illustrates a basis for when each species is chemically "called," here either by the monoecious *F. racemosa* or dioecious *F. hispida*, and provides a good model of the functional organization of communities (Proffit et al. 2007). In this manner, chemical mediation facilitates resource partitioning in the counterfeit wasp communities that are linked to the authentic figpollinator mutualisms, and hints at explanations for coexistence in other parasite communities.

Chen et al. (2006) studied the pollination ecology of cultivated *F. pumila* var. *awkeotsang* and noted that it blossomed twice, once in spring and once in autumn. The periods were of longer duration than those of the wild plant, but overall, the figwasps in the area greatly preferred the wild trees.

State-of-the-art molecular phylogenetic trees were employed for analysis of fossil data to generate independent age estimates for fig and pollinator lineages, using both nonparametric rate smoothing and penalized likelihood dating methods. Molecular dating of ten pairs of interacting lineages uniquely established this specific plant–insect codivergence as minimally extending over 60 million years of geological time (Rønsted et al. 2005).

Male true pollinating wasps of *Alfonsiella fimbriata* Waterston bite into the dehiscent anthers of *Ficus natalensis leprieuri* Miq., thus scattering—and effectively wasting—the pollen grains throughout the syconium. Because female pollinators are the only ones to transfer pollen to con-specific trees and collect that pollen from the anthers only, the male wasps' behavior in this regard is antagonistic to pollination! The authors therefore conclude that the behavior is suggestive of an impending host shift for this specific fig/fig-wasp mutualism (Michaloud et al. 2005).

Lopez-Vaamonde et al. (2009) utilize a combination of phylogenetic and biogeographical data to infer the age, the major period of diversification, and the geographic origin of pollinating fig wasps. They carefully consider and accept the possibility of a Gondwanan origin for pollinating figwasps, when the unified southern continent broke up over 60 million years ago and its parts drifted northward. The drift, they argue, transoceanic travel, is a greater factor in the story than previously believed.

The Lopez-Vaamonde view is actually supported by earlier work that also challenged the predominance of the Gondwanaland concept. The entire Moraceae, argue Zerega et al. (2005) backed by their experimental work using phylogenetic inferences from nuclear and chloroplast DNA, molecular dating with multiple fossil calibrations, and independent geological evidence, had diversified by at least the mid-Cretaceous and major clades including the figs may have radiated during the Tertiary after the break-up of Gondwanaland. Molecular evidence together with Eurasian fossils, furthermore, support the idea that early diversification of Moraceae in Eurasia and subsequent migration into the southern hemisphere is at least as plausible as the Gondwanan hypothesis.

An excellent review by Weiblen (2002) provides a look into the entire networking of interlocking fields that make up fig wasp research, or at least did, at the time the article was written. The review has charts to show how much the field had grown in the few years prior to his review, and the years that have followed have seen even a more concentrated outpouring of research activity in the fig wasp arena.

The fig/fig-wasp mutualism has become a cardinal area for interdisciplinary and individual studies in ecology and evolution. It has been and continues to be oft-cited as an appropriate model camp for assessing interactions between species, speciation in response to resource pressure, dispersal, coexistence, coevolution, and co-divergence. From the pharmacological-quest point of view the wasps may not contribute directly to the chemistry of complex drugs, but it is important to keep in mind that this activity is going on constantly in the background, with the fig trees preparing peculiar volatile compounds to usher in certain figwasp species on a revolving basis, and to send others on the way. The figs (i.e., the fruits or the syconia) are for these wasps (especially the wingless males) wombs and tombs, their place of development and where they are interred. The fig itself has ways, barely explored, to dispose and chemically transmute the wasp's remains into its own synergistic milieu, which we, *Homo sapiens*, have attempted to utilize for its own purposes.

Thus, we humans, barely setting out with *Ficus* on a path of coevolution (spanning, what, 12,000 years?), may have quite a lot to learn from our little brothers and sisters, the figwasps, at it with *Ficus* for over 60 million years. In Chapter 9, we will take up the story of our coevolution with *Ficus*, from past successes to present applications and future directions.

REFERENCES

Chen, Y., H. Li, and W. Ma. 2006. [Pollination ecology of cultivated *Ficus pumila* var. *awke-otsang*]. *Ying Yong Sheng Tai Xue Bao* 17: 2403–7.

Chen, C., and Q. Song. 2008. Responses of the pollinating wasp *Ceratosolen solmsi marchali* to odor variation between two floral stages of *Ficus hispida*. *J Chem Ecol*. 34: 1536–44.

Compton, S.G., K.C. Holton, V.K. Rashbrook, S. van Noort, L. Vincent, and A.B. Ware. 1991. Studies of *Ceratosolen galili*, a nonpollinating Agaonid fig wasp. *Biotropica* 23: 188–94.

Dunn, D.W., D.W. Yu, J. Ridley, and J.M. Cook. 2008a. Longevity, early emergence and body size in a pollinating fig wasp—implications for stability in a fig-pollinator mutualism. *J Anim Ecol.* 77: 927–35.

Dunn, D.W., S.T. Segar, J. Ridley, R. Chan, R.H. Crozier, D.W. Yu, and J.M. Cook. 2008b. A role for parasites in stabilising the fig-pollinator mutualism. *PLoS Biol* 6: e59.

Fiorillo, B.S., J. Lino-Neto, and S.N. Báo. 2008. Structural and ultrastructural characterization of male reproductive tracts and spermatozoa in fig wasps of the genus *Pegoscapus* (Hymenoptera, Chalcidoidea). *Micron* 39: 1271–80.

Greeff, J.M. 2002. Mating system and sex ratios of a pollinating fig wasp with dispersing males. *Proc Biol Sci.* 269: 2317–23.

Greeff, J.M., S. van Noort, J.Y. Rasplus, and F. Kjellberg. 2003. Dispersal and fighting in male pollinating fig wasps. *C R Biol.* 326: 121–30.

Greeff, J.M., G.J. Jansen van Vuuren, P. Kryger, and J.C. Moore. 2009. Outbreeding and possibly inbreeding depression in a pollinating fig wasp with a mixed mating system. *Heredity* 102: 349–56.

Greeff, J.M., and J.W. Ferguson. 1999. Mating ecology of the nonpollinating fig wasps of *Ficus ingens*. *Anim Behav.* 57: 215–222.

Grison-Pigé, L., M. Hossaert-McKey, J.M. Greeff, and J.M. Bessière. 2002. Fig volatile compounds—a first comparative study. *Phytochemistry* 61: 61–71.

Jackson, A.P. Cophylogeny of the *Ficus microcosm*. 2004. *Biol Rev Camb Philos Soc* 79: 751–68.

Jousselin, E., S. van Noort, and J.M. Greeff. 2004. Labile male morphology and intraspecific male polymorphism in the *Philotrypesis* fig wasps. *Mol Phylogenet Evol.* 33: 706–18.

Jousselin, E., S. van Noort, J.Y. Rasplus, and J.M. Greeff. 2006. Patterns of diversification of Afrotropical *Otiteselline* fig wasps: Phylogenetic study reveals a double radiation across host figs and conservatism of host association. *J Evol Biol.* 19: 253–66.

Jousselin, E., S. van Noort, V. Berry et al. 2008. One fig to bind them all: Host conservatism in a fig wasp community unraveled by cospeciation analyses among pollinating and nonpollinating fig wasps. *Evolution* 62: 1777–97.

Kjellberg, F., J.L. Bronstein, G. van Ginkel et al. 2005. Clutch size: A major sex ratio determinant in fig pollinating wasps? *C R Biol.* 328: 471–6.

Lin, R.C., C.K. Yeung, and S.H. Li. 2008. Drastic post-LGM expansion and lack of historical genetic structure of a subtropical fig-pollinating wasp (*Ceratosolen* sp. 1) of *Ficus septica* in Taiwan. *Mol Ecol.* 17: 5008–22.

Lin, S., N. Zhao, and Y. Chen. 2008. Co-speciation of figs and fig wasps. *Redai Yaredai Zhiwu Xuebao* 16: 123–7.

Lopez-Vaamonde, C., N. Wikström, K.M. Kjer et al. 2009. Molecular dating and biogeography of fig-pollinating wasps. *Mol Phylogenet Evol.* 52: 715–26.

Lu, B., N. Wang, J. Xiao, Y. Xu, R.W. Murphy, and D. Huang. 2009. Expression and evolutionary divergence of the non-conventional olfactory receptor in four species of fig wasp associated with one species of fig. *BMC Evol Biol.* 9: 43.

Marussich, W.A., and C.A. Machado. 2007. Host-specificity and coevolution among pollinating and nonpollinating New World fig wasps. *Mol Ecol.* 16: 1925–46.

Michaloud, G., N. Bossu-Dupriez, M. Chevolot, and C. Lasbleiz. 2005. Pollen waste and unrelated traits in a fig-fig wasp symbiosis: A new behaviour suggesting a host shift. *C R Biol.* 328: 81–7.

Moore, J.C., A. Loggenberg, and J.M. Greeff. 2006. Kin competition promotes dispersal in a male pollinating fig wasp. *Biol Lett.* 2: 17–9.

NIST Chemistry WebBook. 1996. [Online Database]. NIST Standard Reference Database Number 69. The National Institute of Standards and Technology (NIST). URL: http://webbook.nist.gov/chemistry/contact.html [accessed August 14, 2009].

Pienaar, J., and J.M. Greeff. 2003. Maternal control of offspring sex and male morphology in the *Otitesella* fig wasps. *J Evol Biol.* 16: 244–53.

Proffit, M., B. Schatz, R.M. Borges, and M. Hossaert-McKey. 2007. Chemical mediation and niche partitioning in non-pollinating fig-wasp communities. *J Anim Ecol.* 76: 296–303.

Rønsted, N., G.D. Weiblen, J.M. Cook, N. Salamin, C.A. Machado, and V. Savolainen. 2005. 60 million years of co-divergence in the fig-wasp symbiosis. *Proc Biol Sci.* 272: 2593–9.

Song, B., Y.Q. Peng, J.M. Guan, P. Yang, and D.R. Yang. 2008. [Sex ratio adjustment of a non-pollinating fig wasp species on *Ficus semicordata* in Xishuangbanna]. *Ying Yong Sheng Tai Xue Bao* 19: 588–92.

Song, Q., D. Yang, G. Zhang, and C. Yang. 2001. Volatiles from *Ficus hispida* and their attractiveness to fig wasps. *J Chem Ecol.* 27: 1929–42.

Trivers, R.L., and D.E. Willard. 1973. Natural selection of parental ability to vary the sex ratio of offspring. *Science* 179: 90–2.

Wang, R.W., L. Shi, S.M. Ai, and Q. Zheng. 2008. Trade-off between reciprocal mutualists: Local resource availability-oriented interaction in fig/fig-wasp mutualism. *J Anim Ecol.* 77: 616–23.

Weiblen, G.D. 2002. How to be a fig wasp. *Annu Rev Entomol.* 47: 299–330.

Zerega, N.J.C., W.L. Clement, S.L. Datwyler, and G.D. Weiblen. 2005. Biogeography and divergence times in the mulberry family (Moraceae). *Mol Phylogenetic Evol.* 37: 402–16.

Zhang, F., Y. Peng, S.G. Compton, Y. Zhao, and D. Yang. 2009. Host pollination mode and mutualist pollinator presence: Net effect of internally ovipositing parasite in the fig-wasp mutualism. *Naturwissenschaften* 96: 543–9.

9 Figs and Humans

Our relationship to figs, though ancient by human standards, is relatively new from the fig's point of view. Figs have been good to us, providing food and medicine, shade and shelter, raiment, and even metaphysical inspiration. Figs are reputed to be the most frequently mentioned fruit in the Bible (Flaishman et al. 2008), and an important element of all mythologies and religions of the Mediterranean and Middle

FIGURE 9.1 Young man peddling fresh figs (Haifa, Israel, Pen and ink by Zipora Lansky).

FIGURE 9.2   *Ficus benjamina* in a hothouse (18.7.2009, Loimaa, Finland, by Kaarina Paavilainen).

East. The ease with which cuttings of fig trees can be transported and root when introduced into soil and water has enabled us to propagate figs for our own use thousands of years before we were able to domesticate grains.

For our part, we have also served the fig, helping it to colonize lands far distant from those of its origins. We have succeeded in establishing figs, especially the edible *Ficus carica* and the decorative *F. benjamina*, in countries of many northern climates, from New England to Northern Europe (Condit 1947). We have played an impressive role in figs' evolution, as figs have also figured prominently in our own evolution.

Along the way, there have been fits and starts. For the figs, continents have split off from other continents, and fig wasps have evolved alongside them to create their unique and distinctive environmentally linked system of reproduction. We humans have taken figs to new heights of glory, and have deeply appreciated their virtues and integrated them into our culture. Now, at the threshold of suffering, the consequences of our carelessness with our own instruments of environmental change, we may again turn fig-ward for a leaf or two to cover our shame, to absorb the toxins of our greed and stupidity.

This field is called phytoremediation, and the *Ficus* species are employed in both diagnostic and "therapeutic" functions to "heal" the environmentally ravaged planet. Let us review now briefly some of the work in each of these major areas.

TERRADIAGNOSTICS

Reflectance spectroscopy within the visible-near-IR region of the spectrum has been utilized to measure leaf reflectance spectra of *F. microcarpa* within the highly polluted area of the Guangzhou metropolitan area in China. The method could simultaneously assess pollutants (S, Cd, Cu, Hg, Pb, XCl, XF) in the city's atmosphere in 1985 and 1998. The measurements using this relatively simple and economical

FIGURE 9.3 *Ficus macrocarpa*, decorative bonzai tree (26.7.2009, Hebrew University Botanical Garden, Giv'at Ram, Jerusalem, Israel, by Helena Paavilainen).

method were identical to those obtained by conventional chemical analysis (Wang et al. 2008).

Plants of the *Ficus* genus seem particularly useful as bioindicators for particular minerals, due to their ability to absorb those elements. Specifically, *F. glomerata* Roxb. and *F. retusa* L. were successfully employed in this regard as indicators of Cu, Co, Ni, Pb, Zn, Cr, Fe, Mn, F, and K_2O in a lead-zinc-fluorite mine area (Biswas 2006). Measurements were made with washed plant leaves and young twig samples collected from mineralized and nonmineralized areas.

The essential rationale for employing *Ficus* species as bioindicators derives from their inherent plasticity in response to environmental changes. As an example, *F. religiosa* was studied in regard to reactive oxygen species (ROS), especially hydrogen peroxide, oxidative stress enzymes peroxidase (POX), catalase and glycolate oxidase, and the diurnal variations in stomatal activity by scanning electron microscopy in response to growing in either a normal or adverse habitat, the latter constituting concrete rooftops (Smitha et al. 2009). Of three substrates tested for POX activity (guaiacol, ascorbate, and *o*-dianisidine), *o*-dianisidine was found as the preferred substrate of *F. religiosa* POX with about sevenfold more activity over its counterparts. Cytosolic POX activity showed 11-fold increase over cell-wall-bound POX. Similarly, CAT activity in specimens from the adverse habitat showed about twofold increase during daytime. In short, *F. religiosa* grown on the adverse rooftops showed elevated production of ROS and their scavenging oxidative stress enzymes, direct evidence of drought stress that provides additional insight into the plant's evolutionary ecology.

FIGURE 9.4 *Ficus retusa* (19.7.2009, University of Turku, Botanical Garden, Ruissalo, Turku, Finland, by Kaarina Paavilainen).

FIGURE 9.5 *o*-Dianisidine. (*NIST Chemistry WebBook*. 1996. [Online Database]. NIST Standard Reference Database Number 69. The National Institute of Standards and Technology (NIST). URL: http://webbook.nist.gov/chemistry/contact.html [accessed August 14, 2009].)

Volatile compounds emitted from *F. benjamina* may also react with free radicals, such as nitrate, hydroxyl, and ozone, and in the process help create measurable compounds. These can include ozone itself, as well as distinctive entities such as organic acids (formic and acetic acids), aldehydes (formaldehyde, acetaldehyde, and hexanal) and alcohols (menthol, 1-butanol, 1-pentanol, 2-penten-1-ol, 4-penten-2-ol, and linalool). Although these investigations are still at a very early stage, the measurement of such compounds in the air near such *Ficus* species may provide in the future additional insight into the complex environmental impact of pollutants, specifically containing reactive oxygen species (Souza et al. 2002).

O=O=O

FIGURE 9.6 Ozone. (*NIST Chemistry WebBook*. 1996. [Online Database]. NIST Standard Reference Database Number 69. The National Institute of Standards and Technology (NIST). URL: http://webbook.nist.gov/chemistry/contact.html [accessed August 14, 2009].)

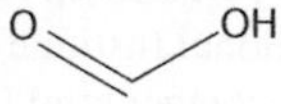

FIGURE 9.7 Formic acid. (*NIST Chemistry WebBook*. 1996. [Online Database]. NIST Standard Reference Database Number 69. The National Institute of Standards and Technology (NIST). URL: http://webbook.nist.gov/chemistry/contact.html [accessed August 14, 2009].)

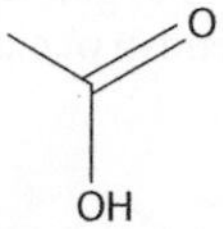

FIGURE 9.8 Acetic acid. (*NIST Chemistry WebBook*. 1996. [Online Database]. NIST Standard Reference Database Number 69. The National Institute of Standards and Technology (NIST). URL: http://webbook.nist.gov/chemistry/contact.html [accessed August 14, 2009].)

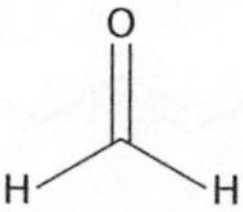

FIGURE 9.9 Formaldehyde. (*NIST Chemistry WebBook*. 1996. [Online Database]. NIST Standard Reference Database Number 69. The National Institute of Standards and Technology (NIST). URL: http://webbook.nist.gov/chemistry/contact.html [accessed August 14, 2009].)

FIGURE 9.10 Acetaldehyde. (*NIST Chemistry WebBook*. 1996. [Online Database]. NIST Standard Reference Database Number 69. The National Institute of Standards and Technology (NIST). URL: http://webbook.nist.gov/chemistry/contact.html [accessed August 14, 2009].)

FIGURE 9.11 Hexanal. (*NIST Chemistry WebBook*. 1996. [Online Database]. NIST Standard Reference Database Number 69. The National Institute of Standards and Technology (NIST). URL: http://webbook.nist.gov/chemistry/contact.html [accessed August 14, 2009].)

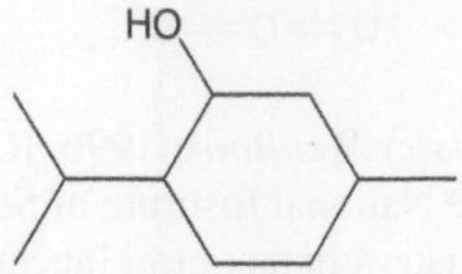

FIGURE 9.12   Menthol. (*NIST Chemistry WebBook*. 1996. [Online Database]. NIST Standard Reference Database Number 69. The National Institute of Standards and Technology (NIST). URL: http://webbook.nist.gov/chemistry/contact.html [accessed August 14, 2009].)

FIGURE 9.13   1-Butanol. (*NIST Chemistry WebBook*. 1996. [Online Database]. NIST Standard Reference Database Number 69. The National Institute of Standards and Technology (NIST). URL: http://webbook.nist.gov/chemistry/contact.html [accessed August 14, 2009].)

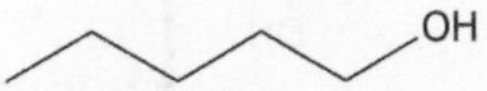

FIGURE 9.14   1-Pentanol. (*NIST Chemistry WebBook*. 1996. [Online Database]. NIST Standard Reference Database Number 69. The National Institute of Standards and Technology (NIST). URL: http://webbook.nist.gov/chemistry/contact.html [accessed August 14, 2009].)

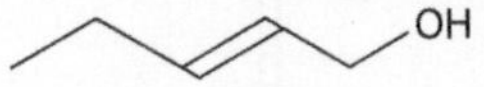

FIGURE 9.15   2-Penten-1-ol. (*NIST Chemistry WebBook*. 1996. [Online Database]. NIST Standard Reference Database Number 69. The National Institute of Standards and Technology (NIST). URL: http://webbook.nist.gov/chemistry/contact.html [accessed August 14, 2009].)

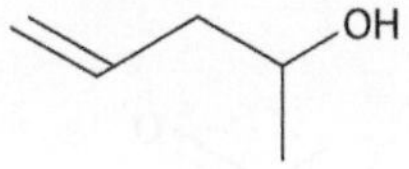

FIGURE 9.16   4-Penten-2-ol. (*NIST Chemistry WebBook*. 1996. [Online Database]. NIST Standard Reference Database Number 69. The National Institute of Standards and Technology (NIST). URL: http://webbook.nist.gov/chemistry/contact.html [accessed August 14, 2009].)

Emission of volatile compounds from *Ficus* spp. is dependent on factors other than pollutants, though, and these must also be factored into such analyses. One of the most important of these factors is water. For example, humidity was shown to be a major factor affecting the efflux of isoprene from the leaves of *F. virgata* (Tambunan et al. 2007). Washing leaf tissue in water prior to measurement will also significantly affect the measurement of minerals, resulting in decreased levels of

FIGURE 9.17 *Ficus carica* (29.7.2009, Hebrew University Botanical Garden, Mount Scopus, Jerusalem, Israel, by Krina Brandt-Doekes).

aluminum (Al), barium (Ba), cadmium (Cd), chromium (Cr), copper (Cu), iron (Fe), manganese (Mn), magnesium (Mg), nickel (Ni), lead (Pb), zinc (Zn), and vanadium (V) as measured from *F. microcarpa* (Oliva and Valdes 2004).

Banyan trees, a broad group of *Ficus* but especially *F. benghalensis*, have been employed as accurate bioindicators of contamination by polychlorinated dibenzo-*p*-dioxins (PCDDs) and dibenzofurans (PCDFs) (Miyata et al. 1997). Another interesting study similarly highlighted *F. benghalensis* as a biomonitor in both highly industrialized and "control" areas in measurements of polycyclic aromatic hydrocarbons (PAHs), with particular attention to seasonal variations. First, as expected, the PAH levels were significantly higher from the urban areas. Second, in both areas within India, the January season was the time of highest levels of the PAHs relative to during the other seasons (summer and monsoon). This latter phenomenon was attributed to medium-molecular-weight PAHs, which increase with respect to both low- and high-molecular-weight PAHs during January, and helps to underscore the sensitivity of *F. benghalensis* leaves as a living biomonitor of air pollution (Prajapati and Tripathi 2008b).

TERRATHERAPEUTICS

In addition to their adaptability as diagnostic tools, *Ficus* species, especially their leaves, also possess an excellent ability to absorb and sequester toxic material in the environment. Although the precise mechanisms of this sequestering ability remain unknown, a good deal of it apparently has to do with a "magnetic" quality of the leaves. The phenomenon, first reported by Pandey et al. (2005) with *F. infectoria* in regard to particulates in the exhaust of internal combustion engines, has also been more recently studied relative to *F. benghalensis* (Prajapati and Tripathi 2008a) as a method to control road-derived respirable particulates. Isothermal remanent magnetization (IRM(300) mT) of three different tree leaves, that is, mango (*Mangifera indica*), sisso (*Dalbergia sisso*), and banyan (*F. benghalensis*), were determined and IRM(300) mT normalized for the leaf area. The normalized 2-D magnetization of leaves was shown by results to be dominantly controlled by leaf morphology and traffic density. The banyan, *F. benghalensis*, leaf showed the highest 2-D magnetization, indicating its superior ability to reduce magnetic particulates. The particle size of the magnetic grains falls in the category of PM2.5, a particle size hazardous to human health due to its capacity to be inhaled deeply into the lungs.

This ability of *Ficus* leaves to remediate the environment through sorption of pollutants was experimentally determined in *F. stipulata* to be enhanced by application of ion beam irradiation to the fig leaves with respect to high adsorption of nitrogen dioxide (Takahashi et al. 2004). A quick overview of the different *Ficus* species tested for sorption of pollutants, and the specific pollutants that they were shown capable of adsorbing, is given in Table 9.1.

Studies on the biosorption of lead from an aqueous solution by powdered *F. religiosa* leaves were carried out in a series of batch experiments to determine biosorption capacity, equilibrium time, optimal pH, and temperature (Qaiser et al. 2009). The maximum biosorption capacity of lead was 37.45 mg/g at an optimal pH of 4. Temperature change in the range of 20°C–40°C affected the biosorption capacity and the maximum removal was observed at 25°C. For continuous experiments, the *Ficus* biomass was immobilized in polysulfone matrix. The release of Ca, Mg, and Na ions during lead biosorption revealed that ion exchange was the major removal mechanism.

TABLE 9.1
Sorption of Pollutants by Leaves of Selected *Ficus* Species

Pollutant	*Ficus* Species	Notes	Reference
Nitrogen dioxide	*stipulata*	Enhanced by application of ion beam radiation	Takahashi et al. 2004
Lead (Pb)	*religiosa*	Mechanism is ion exchange, accompanied by release of Ca, Mg, and Na	Qaiser et al. 2009
Cadmium [Cd (II)], chromium [Cr (III)], and nickel [Ni (II)]	*religiosa*	Removal of target ions according to relative affinity Cd (II)>Cr (III)>Ni (II)	Goyal and Srivastava 2008a
Chromium [Cr(III), Cr(VI)]	*religiosa*	Removal from aqueous media	Goyal and Srivastava 2008b
Cadmium [Cd (II)]	*religiosa*		Goyal et al. 2007
Benzene	*elastica*		Cornejo et al. 1999

The powdered *F. religiosa* leaf has similarly been experimentally employed to measure sorption of other heavy metals besides lead. These have included cadmium (Cd II), chromium (Cr III); (Cr IV), and nickel (II). The most sensitive of these to sorption by the leaf was cadmium. Desorption studies showed increasing ease of metal ion recovery from metal-loaded biomass by the different eluants is of the order nitric acid [0.05 M] > citric acid [0.5 M] > sodium hydroxide [0.05 M] > distilled water (Goyal and Srivastava 2008a). The work suggests the application of such powder in a first-pass method of removing heavy metals from water supplies.

Additional work (Goyal and Srivastava 2008b) employed a scanning electron microscope to study the *F. religiosa* powdered leaf after it had been exposed to the cadmium [Cd (II)]. Morphological changes were indicative of biosorption, and Fourier Transform IR (FTIR) spectrometry showed amino acids–metal interaction to be likely responsible for the metal binding. Kinetics of sorption followed a first-order rate equation.

Cornejo et al. (1999) studied decontamination of air by a battery of plants, including *Pelargonium domesticum*, *Ficus elastica*, *Chlorophytum comosum*, and *Kalanchoe blossfeldiana*. The target was removal of benzene, trichloroethylene (TCE), and toluene from air. Also, in this setting, the *F. elastica* was notable in its ability to remove the organic pollutants, especially benzene.

Phytoremediation of polluted air and water by suitable species of *Ficus* leaves may thus be accomplished both by live trees, in the case of air, and leaf powder, in the case of cleaning water, for example that associated with cooling pools in industrial settings. A specific example involving chromium-polluted water that might have been amenable to *Ficus* remediation was the subject of the popular movie *Erin Brockovitch*. Taking the matter a step further, powdered *Ficus* leaves might also have efficacy as a medicine for facilitating detoxification of heavy metals in humans. This might occur after exposure to toxic elements through chronic or acute poisoning.

ECOLOGY, NUTRITION, AND NOVEL APPLICATIONS

The *Ficus* species, similar to all life-forms, are sensitive to their environments, and evolve to fill particular niches. Traditional use and collection of plant parts is often in tune with temporally linked environmental changes, such as day or night, and through the seasons. An experimental study (Lockett et al. 2000) of native foods consumed in rural northeastern Nigeria included the "cedya," that is, *F. thonningii*, which is valued for its medicinal virtues against bronchitis and urinary tract infections. The parts employed are the fruits and leaves (Iwu 1993). Lockett's study examined both the energy and mineral content of the parts consumed, presumably the fruits. The analyses included protein and fat content and calcium, iron, copper, and zinc. In general, the content of minerals as well as energy in the plant increased during the dry seasons, while the nutritive values of the plants during the rainy season were inferior.

Another innovative use of *Ficus* sp. biomass, especially of the leaves, is as a novel renewable energy source. The leaf and stem of *F. elastica* were so evaluated in comparison to *Artocarpus hirsutus*. *F. elastica* proved to be a superior source of nonpolluting and safe fuel (Palaniraj and Sati 2003), presumably largely due to its rich content of natural latex. Maximum protein (24.5%), polyphenols (4.2%,), oil (6.1%), and hydrocarbon (2.0%) were measured in *F. elastica*, suggesting that its leaf would be an alternative biorenewable source of energy.

Similarly, waste products (leaves, stem) of *F. carica* were shown to be a rich source of the minerals Se, Ge, Zn, Ca, and I, and suitable as a key ingredient for culture medium of medicinal mushrooms of genus *Ganoderma* (*Ganoderma lucidum* and/or *Ganoderma sinense*). The recommended dose of *F. carica* in these settings for improving yield and function of the mushrooms was given as 100–2000 mg/kg media (Liu 2006). These mushrooms are considered valuable for enhancing immunity and for modulating inflammation, with particular efficacy in cancer, and perhaps especially for prostate cancer (Mahajna et al. 2009, Wasser and Weis 1999).

FIGS AND MEDICINE

In the preceding chapters, we recounted numerous medicinal uses for *Ficus* spp. plant parts and their derivatives. However, even in settings where plant parts (e.g., fruits) were combined with other botanical products (e.g., barley meal) or minerals (e.g., vitriol), the combinations usually did not include other *Ficus* parts or their derivatives. That is, we seldom hear of combinations of fig fruits and leaves, fig leaves and roots, etc., and probably never hear of combinations between plant parts of different *Ficus* species. This may or may not be an oversight by physicians over the centuries, or perhaps it is only a deficiency in the faithfulness of their recording. For whatever reason, though, this lack of particular combinations from *anatomically discrete compartments* of single plants may reflect both a mistake and a missed opportunity to establish more powerful synergistic therapeutic products, without, in many cases, much more additional cost. The different compartments within the single plant may be anatomically discrete, but are not necessarily very far physically removed. Concentration on single fractions, such as only latex or only fruits, may be shortsighted. Furthermore, further concentration to highly purified fractions, and especially to single chemical compounds within the fractions, may be particularly shortsighted as the product will miss out on the total synergy within a whole plant.

Whole plant synergy is used in the production of homeopathic remedies (but there are none from any *Ficus* to date) and ancient models of phytotherapeutics. Some research into the synergy existing between compounds from single plants has also recently been completed. Bark, leaves, and roots of the African traditional medical plant *Croton gratissimus* Burch. var. *gratissimus* were studied for antimicrobial properties (van Vuuren and Viljoen 2008). A hydrodistilled essential oil from the leaf was compared in its activity against *Bacillus cereus*, *Candida albicans*, and *Cryptococcus neoformans* to extracts of *C. gratissimus* leaves, stem bark, and roots, each alone, and in various combinations, to establish minimum inhibitory concentrations as well as fractional inhibitory concentrations for combinations of leaf and root (1:1), bark and root (1:1), and leaf and bark (1:1). Combinations were not consistently better than single fractions, but sometimes were, as, for example, in the case of the leaf and root combination against *Cryptococcus neoformans*. Similarly, we demonstrated similar synergies in inhibiting proliferation, invasion, and inflammation in human prostate cancer cells when anatomically discrete fractions of pomegranate seed oil, juice, and peels were combined (Lansky et al. 2005a), and this effect was confirmed when representative pure compounds from those fractions were similarly combined (Lansky et al. 2005b). In the present case, with the entire *Ficus* genus at hand, it could be possible and potentially desirable to combine fractions originating from within the different *Ficus* species, such as the latex of one species, with a fruit extract of another.

Still further complexity may be obtained, of course, by combining complex *Ficus* mixtures with other plant species altogether. As previous discussed, these may include products produced through complex fermentations between *Ficus* extracts and selected medicinal fungi, as well as by mixing extracts of *Ficus* with those of other higher plants, or for that matter, even natural products of mineral, bacterial, or animal origin. Such issues will now hopefully receive some attention from researchers, with the impetus provided by the information contained in this volume.

REFERENCES

Biswas, S.N. 2006. Comparative studies on plants as bioindicator for Cu, Co, Ni, Pb, Zn, Cr, Fe, Mn, F and K2O in lead-zinc-fluorite mine area of Chandidongrh, Distt. Rajnandgaon (C.G.), India. *Asian J Chem.* 18: 991–6.

Condit, I.J. 1947. *The Fig.* Waltham, MA: Chronica Botanica Co.

Cornejo, J.J., F.G. Munoz, C.Y. Ma, and A.J. Stewart. 1999. Studies on the decontamination of air by plants. *Ecotoxicology* 8: 311–20.

Flaishman, M.A., V. Rodov, and E. Stover. 2008. The fig: Botany, horticulture, and breeding. *Hortic Rev.* 34: 113–96.

Goyal, P., M.M. Srivastava, and S. Srivastava. 2007. *Ficus religiosa* leaf powder: A green economical biomaterial for the removal of Cd(II) ions from aqueous solution. *J Nuclear Agr Biol.* 36: 16–27.

Goyal, P., and S. Srivastava. 2008a. Metal biosorption equilibria in a ternary aqueous system using agricultural waste *Ficus religiosa* leaf powder. *Natl Acad Sci Lett (India)* 31: 347–51.

Goyal, P., and S. Srivastava. 2008b. Role of biosorbent *Ficus religiosa* leaf powder (FRLP) in chemical speciation of chromium: A green method for separation. *Archiwum Ochrony Srodowiska* 34: 35–45.

Iwu, M.M. 1993. *Handbook of African Medicinal Plants.* Boca Raton, FL: CRC Press.

Lansky E.P., W. Jiang, H. Mo, L. Bravo, P. Froom, W. Yu, N.M. Harris, I. Neeman, and M.J. Campbell. 2005a. Possible synergistic prostate cancer suppression by anatomically discrete pomegranate fractions. *Invest New Drugs* 23: 11–20.

Lansky, E.P., G. Harrison, P. Froom, and W.G. Jiang. 2005b. Pomegranate (*Punica granatum*) pure chemicals show possible synergistic inhibition of human PC-3 prostate cancer cell invasion across Matrigel. *Invest New Drugs* 23: 121–2, Erratum 23, 379.

Liu, W. 2006. [Culture method and uses of *Ficus carica* or *Gynostemma pentaphyllum* rich in selenium, germanium, zinc, calcium and iodine]. *Faming Zhuanli Shenqing Gongkai Shuomingshu.* Chinese Patent Application CN 2006-10039200 20060323.

Lockett, C.T., C.C. Calvert, and L.E. Grivetti. 2000. Energy and micronutrient composition of dietary and medicinal wild plants consumed during drought: Study of rural Fulani, northeastern Nigeria. *Int J Food Sci Nutr.* 51: 195–208.

Mahajna, J., N. Dotan, B.Z. Zaidman, R.D. Petrova, and S.P. Wasser. 2009. Pharmacological values of medicinal mushrooms for prostate cancer therapy: The case of *Ganoderma lucidum*. *Nutr. Cancer* 61: 16–26.

Miyata, H., O. Aozasa, T. Nakao et al. 1997. Assessment of air pollution by PCDDs and PCDFs in Taiwan using banyan tree leaf as a biomonitoring indicator. *Organohalogen Compounds* 32: 124–9.

NIST Chemistry WebBook. 1996. [Online Database]. NIST Standard Reference Database Number 69. The National Institute of Standards and Technology (NIST). URL: http://webbook.nist.gov/chemistry/contact.html [accessed August 14, 2009].

Oliva, S.R., and B. Valdes. 2004. Influence of washing on metal concentrations in leaf tissue. *Commun Soil Sci Plant Anal.* 35: 1543–52.

Palaniraj, R., and S.C. Sati. 2003. Evaluation of *Artocarpus Hirsute* and *Ficus elastica* as renewable source of energy. *Indian J Agr Chem.* 36: 23–8.

Pandey, S.K., B.D. Tripathi, S.K. Prajapati et al. 2005. Magnetic properties of vehicle-derived particulates and amelioration by *Ficus infectoria*: A keystone species. *Ambio* 34: 645–6.

Prajapati, S.K., and B.D. Tripathi. 2008a. Management of hazardous road derived respirable particulates using magnetic properties of tree leaves. *Environ Monit Assess.* 139: 351–4.

Prajapati, S.K., and B.D. Tripathi. 2008b. Biomonitoring seasonal variation of urban air polycyclic aromatic hydrocarbons (PAHs) using *Ficus benghalensis* leaves. *Environ Pollut.* 151: 543–8.

Qaiser, S., A.R. Saleemi, and M. Umar. 2009. Biosorption of lead from aqueous solution by *Ficus religiosa* leaves: Batch and column study. *J Hazard Mater.* 166: 998–1005.

Smitha, R.B., T. Bennans, C. Mohankumar, and S. Benjamin. 2009. Oxidative stress enzymes in *Ficus religiosa* L.: Biochemical, histochemical and anatomical evidences. *J Photochem Photobiol B.* 95: 17–25.

Souza, S.R., P.C. Vasconcellos, W. Mantovani, and L.R.F. Carvalho. 2002. [Oxygenated volatile organic compounds emitted from leaves of *Ficus benjamina* L. (Moraceae)] *Rev Bras Bot.* 25: 413–18.

Takahashi, M., S. Kohama, and H. Morikawa. 2004. [Induction of plants which repair environment effectively using *Ficus stipulata*]. *Dai 2 Kai Ionbimu Seibutsu Oyo Wakushoppu Ronbunshu* 34–5.

Tambunan, P., S. Baba, and H. Oku. 2007. Effect of humidity level on isoprene emission from leaves of tropical tree *Ficus virgata*. *Nettai Nogyo* 51: 30–4.

van Vuuren, S. F., and A.M. Viljoen. 2008. In vitro evidence of phyto-synergy for plant part combinations of *Croton gratissimus* (Euphorbiaceae) used in African traditional healing. *J Ethnopharmacol.* 119: 700–4.

Wang, J., R. Xu, Y. Ma, L. Miao, R. Cai, and Y. Chen. 2008. The research of air pollution based on spectral features in leaf surface of *Ficus microcarpa* in Guangzhou, China. *Environ Monit Assess.* 142: 73–83.

Wasser, S.P., and A.L. Weis. 1999. Therapeutic effects of substances occurring in higher Basidiomycetes mushrooms: A modern perspective. *Crit Rev Immunol.* 19: 65–96.

10 *Ficus* Postscript

Just as we are about to go to press, we received a very kind communication from Professor Luis Eduardo Luna, who had published the two articles about Peruvian shamanism and their "plant teachers" in the *Journal of Ethnopharmacology* in the early 1980s (Luna 1984a,b). So we asked our publisher, who kindly has allowed us a short appendix to include his comments on the use of the *Ficus* tree(s) known as *Renaco* in Peruvian shamanism. At the same time, we shall very briefly summarize some of the most recent publications on *Ficus* pharmacognosy since we concluded our writing about six months ago.

Professor Luna's comments (6.4.2010):

When I was doing my field work in the '80s there was talk among my informants of two (perhaps three) species of Ficus. *On one hand we have* F. anthelmintica *and* F. insipida, *both called "Dr. Ojé." A bit of the latex was taken with two functions: 1) as a purge (I believe there was in Sweden a doctoral dissertation about* F. anthelmintica*). 2) to "learn" from it, if the traditional Amazonian diet when dealing with the spirit world is observed.*

On the other hand we have the so-called "renacos," trees the grow near lakes, with aerial roots that may go deep into the water. I never identified the plants below genus. Don Emilio, my main informant, told me that this plant was ingested. Most probably either the latex or perhaps the bark was added when preparing ayahuasca. There was always the talk that certain "paleros" (vegetalistas specialized in "palos" or trees) took it in order to visit the underwater realm. Pablo Amaringo, who died recently, did two or three paintings depicting the spiritual aspects of the renaco tree.

Of further interest to the position of the fig in the psychedelic imagination, Harold Greenberg of Eilat, Israel, points out that the popular paisley design of the 1960s is readily appreciated as a stylized fig fruit, as indicated in Figure 10.4.

Advances in *Ficus* pharmacognosy from July 2009 through April 2010 are summarized in the following two supplemental Tables 10.1 and 10.2. Table 10.2 also includes the 2007 paper by Yoon et al., inadvertently omitted from Chapter 4.

FIGURE 10.1   Renaco tree (*Ficus* sp.) by the Yarinacocha Lake, Peru. Note the underwater aerial roots. Photo by Luis Edwardo Luna. Copyright © by Luis Edwardo Luna, 1988.

FIGURE 10.2   (See color insert after page 128.) *Espíritu del Renaco*, by Pablo Amaringo. Gouache on paper. Copyright © by Pablo Amaringo, 1990.

FIGURE 10.3 **(See color insert after page 128.)** *Renacal al Amanecer*, by Anderson Debernardi. Copyright © by Anderson Debernardi.

FIGURE 10.4 Paisley pattern (drawing by Zipora Lansky).

TABLE 10.1

Advances in *Ficus* Chemistry from July 2009 through April 2010

Ficus Species	Plant Part(s)	Compound(s)	Method(s)	Reference
benjamina	Leaves	New triterpenoid: serrat-3-one (Fc-2); and pentacontanyl decanoate (Fc-1), friedelin and β-sitosterol	IR, (1)H-NMR, (13)C-NMR, MS.	Parveen et al. 2009
carica	Leaves	Bergapten, psoralen	High-speed countercurrent chromatography, (1)H-NMR, (13)C-NMR and MS, light petroleum ether extraction, separated with a two-phase solvent system of n-hexane-ethyl acetate-methanol-water (1:1:1:1, v/v).	Chunyan et al. 2009
carica	Leaves, fruits	3-O- and 5-O-caffeoylquinic acids, ferulic acid, quercetin-3-O-glucoside, quercetin-3-O-rutinoside, psoralen and bergapten; 3-O-caffeoylquinic acid, quercetin-3-O-glucoside and oxalic, citric, malic, quinic (leaves only), shikimic and fumaric acids	HPLC/DAD and HPLC/UV.	Oliveira et al. 2009
carica	Latex	Oxalic, citric, malic, quinic, shikimic and fumaric acids; 5 aldehydes, 7 alcohols, 1 ketone, 9 monoterpenes, 9 sesquiterpenes and 3 other compounds, sesquiterpenes most abundant	HS-SPME/GC-IT-MS, HPLC-UV.	Oliveira et al. 2010
hirta	Root	5-methoxyl-4,2′-epoxy-3-(4′,5′-dihydroxyphenyl)-linear pyranocoumarin, and 3-acetyl-3,5,4′-trihydroxy-7-methoxylflavone	1D-, 2D-NMR and HR-ESI-MS.	Ya et al. 2010

TABLE 10.1 (continued)
Advances in *Ficus* Chemistry from July 2009 through April 2010

Ficus Species	Plant Part(s)	Compound(s)	Method(s)	Reference
microcarpa	Leaves	Flavone C-glycosides: orientin, isovitexin-3″-O-glucopyranoside, isovitexin and vitexin	High-speed countercurrent chromatography (HSCCC), medium-pressure liquid chromatography (MPLC), preparative high-performance liquid chromatography (prep-HPLC) on two-phase solvent system: methyl tert butyl ether-ethyl acetate-1-butanol-acetonitrile-0.1% aqueous trifluoroacetic acid; characterized by ESI-MS(n).	Wang et al. 2010
microcarpa	Stem bark	Protocatechuic acid, chlorogenic acid, methyl chlorogenate, catechin, epicatechin, procyanidin B1, and procyanidin B3	ESI-MS, UV, and (1)H- and (13)C-NMR; hexane, ethyl acetate, butanol, and water fractions.	Ao et al. 2010
pandurata	Leaves and stem bark	New compound, 3-O-α-L-arabinopyranosyl-4-hydroxybenzoic acid; α-amyrin acetate; β-amyrone; 3β-acetoxy-20-taraxasten-22-one; α-amyrin; ceryl alcohol; stigmasterol; β-sitosterol; 2α,3α-dihydroxy-lup-20(29)-en-28-oate; ursolic acid; β-sitosterol-3-O-glucoside; protocatechuic acid; betulinic acid; quercetin; quercetin-3-O-β-D-glucoside; kampferol-3-O-β-neohesperidoside; rutin	UV, IR, MS, (1)H- and (13)C-NMR.	Ramadan et al. 2009

(continued on next page)

TABLE 10.1 (continued)
Advances in *Ficus* Chemistry from July 2009 through April 2010

Ficus Species	Plant Part(s)	Compound(s)	Method(s)	Reference
racemosa	Fruits	3-O-(E)-caffeoyl quinate		Jahan et al. 2009
septica	Leaves	New aminocaprophenone alkaloids: ficuseptamines A and B; new pyrrolidine alkaloid, ficuseptamine C; 12 known alkaloids and a known acetophenone derivative	Spectroscopy	Ueda et al. 2009
tsiangii	Stem bark	198 peaks, 57 compounds identified, most prevalent were fatty acids	GC-MS	Duan et al. 2009

TABLE 10.2

Advances in the Functional Pharmacology of *Ficus* spp. from Late 2009 through April 2010

Ficus Species	Plant Part	Pharmacological Action	Details	Reference
benghalensis	Stem bark	Anti-inflammatory, analgesic, lysosomal membrane stabilizing	Methanol extract more active than aqueous in acute (carrageenan induced hind paw edema and acetic acid induced vascular permeability) and subchronic (cotton pellet induced granuloma) models of inflammation.	Thakare et al. 2010
benghalensis	Leaves	Larvicidal against mosquitoes	Methanol extract, after 24 hours.	Govindarajan 2010
benghalensis	Aerial roots	Antihyperglycaemic, improved the diabetic condition at 50 mg/kg(-1)	Oral administration of isolated α-amyrin acetate (α-AA) in normal and diabetic rats and model of type-2 diabetes, i.e., db/db mice.	Singh et al. 2009
benjamina	Leaves	Antinociceptive	Aqueous and alcoholic extracts, analgesiometer test.	Parveen et al. 2009
benjamina	Leaves and fruits	Leaf inhibited all studied viruses, fruit extracts inhibited only VZV—best when extracts added at time of infection, partial inhibition when added post-infection; putative strong interactions with viruses, weak interactions at cell surface	Ethanol extract against Herpes Simplex Virus-1 and -2 (HSV-1 and HSV-2) and Varicella-Zoster Virus (VZV) in vitro	Yarmonlinsky et al. 2009
capensis	Stem bark	Anti-abortifacient: 80 mg/mL significantly (p < 0.05) relaxed uterine smooth muscle, reducing oxytocin, ergometrin and acetylcholine induced contractions	Ethanolic extracts at 40 mg/mL and 80 mg/mL tested ex vivo on a piece of isolated rat uterus pretreated with stilbestrol	Owolabi et al. 2009

(continued on next page)

TABLE 10.2 (continued)
Advances in the Functional Pharmacology of *Ficus* spp. from Late 2009 through April 2010

Ficus Species	Plant Part	Pharmacological Action	Details	Reference
carica	Latex	Antioxidant; weak inhibition of acetylcholinesterase	DPPH (in silico) assessment of effect on nitric oxide and superoxide radicals.	Oliveira et al. 2010
carica	Latex	Antibacterial: methanolic extract active against *Proteus mirabilis* only, ethyl acetate extract inhibited bacteria *Enterococcus fecalis*, *Citobacter freundei*, *Pseudomonas aeruginosa*, *Echerchia coli*, *Proteus mirabilis*; antifungal: ethyl acetate ≈ chloroformic ≈ methanolic fraction against *Candida albicans*, ethyl acetate > methanolic > hexanoic fractions against *Microsporum canis*; methanolic extract paradoxical against *Cryptococcus neoformans*	Methanolic, hexanoïc, chloroformic and ethyl acetate extracts in vitro against five bacteria and seven fungi, using disc-diffusion assay.	Aref et al. 2010
carica	Fruits and leaves	All materials concentration-dependently scavenged DPPH and nitric oxide radicals, only leaves scavenged superoxide radicals, and in all settings, leaves most potent free radical scavengers; no acetylcholinesterase inhibition or antimicrobial action	Antioxidant potential, anticholinesterase activity and antimicrobial action against five bacteria species of the different plant parts checked.	Oliveira et al. 2009

TABLE 10.2 (continued)
Advances in the Functional Pharmacology of *Ficus* spp. from Late 2009 through April 2010

Ficus Species	Plant Part	Pharmacological Action	Details	Reference
carica	Fruit	Crude hot-water soluble polysaccharide (PS) showed higher scavenging activity than water extract (WE) on superoxide radical (EC(50), 0.95 mg/mL) and hydroxyl anion radical (scavenging rate 43.4% at concentration of 4 mg/mL); PS (500 mg/kg) increased carbon particle clearance and serum hemolysin level of normal mice	Antioxidative and immunopotentive activities of WE and PS assays in vitro checked scavenging of DPPH, superoxide and hydroxyl radicals and reducing power. Immune potentiation evaluated using carbon clearance test and serum hemolysin analysis in mice.	Yang et al. 2009
carica	Fruit	Inhibited osteoclastogenesis, p38, NF-κB, expression of NFATc1 and c-Fos, the master regulator of osteoclast differentiation, while activating ERK (putative benefit to osteoporosis, rheumatoid arthritis, periodontal bone absorbtion)	Hexane fraction; RANKL-stimulated RAW264.7 cells and bone marrow-derived macrophages (BMMs).	Park et al. 2009
deltoidea		Reduced HL-60 viability by more than 50% when exposed to $9.6J/cm^2$ of a broad spectrum light when tested at a concentration of 20 µg/mL; photosensitisers likely based on core cyclic tetrapyrrole structure	155 extracts from 93 terrestrial species of plants in Peninsula Malaysia were screened for in vitro photo-cytotoxic activity by means of a cell viability test using a human leukemia cell-line HL60.	Ong et al. 2009

(continued on next page)

TABLE 10.2 (continued)

Advances in the Functional Pharmacology of *Ficus* spp. from Late 2009 through April 2010

Ficus Species	Plant Part	Pharmacological Action	Details	Reference
erecta	Leaf	Decreased mRNA expression of IL-6 and COX-2, and protein levels of COX-2 and PGE2 (putative benefit against osteoporosis)	Hexane and ethyl acetate fractions, MG-63 cells stimulated with IL-1β (10 ng/mL) to induce osteoporotic factors (IL-6, COX-2 and PGE2) and RAW 264.7 cells stimulated with RANKL (100 ng/mL) to induce their differentiation into osteoclasts.	Yoon et al. 2007
exasperata	Leaf	Uterine stimulatory effect equal to oxytocin in presence of atropine, inhibited by diphenhydramine or phentolamine (unlike oxytocin), but significant differences between extract and oxytocin in presence of verapamil and indomethacin all suggest as mechanism activation of histamine H(1)- and/or α-adrenergic receptors, interference with calcium channels and/or stimulation of prostaglandin synthesis in utero	Contractile effect of aqueous extract compared to oxytocin antagonized by atropine, indomethacin, verapamil, phentolamine or diphenhydramine; EC(50) and E(max) determined and analyzed using one-way ANOVA and Dunnett post hoc tests.	Bafor et al. 2010
microcarpa	Stem bark	Antioxidant, inhibition of hyaluronidase	Methanol extract, ethyl acetate fraction strongest of several; in vitro.	Ao et al. 2010
racemosa	Stem bark	Hypoglycemic	Inhibited amylolysis ($P < $ or $= 0.01$) glucose diffusion in vitro and increased ($P < $ or $= 0.01$) glucose transport across yeast cell membrane and isolated rat hemi-diaphragm.	Ahmed and Urooj 2010

TABLE 10.2 (continued)
Advances in the Functional Pharmacology of *Ficus* spp. from Late 2009 through April 2010

Ficus Species	Plant Part	Pharmacological Action	Details	Reference
racemosa	Fruits	Hypoglycemic action in Type 1 diabetes mellitus model rats, antioxidant	DPPH free radical scavenging assay, aqueous/ethanolic extracts.	Jahan et al. 2009
racemosa	Immature fruits	Antioxidant in vitro and in vivo, protection against CCl4 induced lipid peroxidation in vivo	TPC, AOA, RP, DPPH*, O_2*-, *OH scavenging; in vivo evaluation of oxidative stress (lipid peroxidation) and antioxidant defenses (concentration of GSH, as well as CAT and SOD activities) in CCl_4 treated rats.	Verma et al. 2010
religiosa	Bark	Decreased fasting blood glucose; induced catalase (CAT), glutathione peroxidase (GSH-Px), inhibiting superoxide dismutase (SOD), malondialdehyde, restoring glutathione (GSH-reduced form)	Aqueous extract, streptozotocin induced type 2 diabetic rats.	Kirana et al. 2009
religiosa	Bark	Decreased blood sugar, triglycerides, cholesterol, pancreas lipoperoxides; increased body weight and serum insulin and glycogen	Aqueous extract, streptozotocin induced type 2 diabetic rats.	Pandit et al. 2010
septica	Leaves	Cytotoxic action in two cancer cell lines	Methanolic extract, compound (alkaloids) purification.	Ueda et al. 2009
sycomorus	Stem bark	Antiviral: IC50 below 10 µg/mL against HIV-1 (IIIB strain)	60% methanolic extract screened in vitro against Type 1 (HIV-1, IIIB strain) and Type 2 (HIV-2, ROD strain), and against plasmodia	Maregesi et al. 2010
tsiangii	Stem bark (essential oil)	Weak inhibitory effect of ADP-induced platelet aggregation	Born turbidimetric method, rabbit platelets in vitro.	Duan et al. 2009

REFERENCES

Ahmed, F., and A. Urooj. 2010. In vitro studies on the hypoglycemic potential of Ficus racemosa stem bark. *J Sci Food Agric.* 90: 397–401.

Ao, C., T. Higa, H. Ming, Y.T. Ding, and S. Tawata. 2010. Isolation and identification of antioxidant and hyaluronidase inhibitory compounds from Ficus microcarpa L. fil. bark. *J Enzyme Inhib Med Chem.* 25: 406–13.

Aref, H.L., K.B. Salah, J.P. Chaumont, A. Fekih, M. Aouni, and K. Said. 2010. In vitro antimicrobial activity of four Ficus carica latex fractions against resistant human pathogens (antimicrobial activity of Ficus carica latex). *Pak J Pharm Sci.* 23: 53–8.

Bafor, E.E., E.K. Omogbai, and R.I. Ozolua. 2010. In vitro determination of the uterine stimulatory effect of the aqueous leaf extract of Ficus exasperata. *J Ethnopharmacol.* 127: 502–7.

Chunyan, C., S. Bo, L. Ping, L. Jingmei, and Y. Ito. 2009. Isolation and purification of psoralen and bergapten from Ficus Carica L leaves by high-speed countercurrent chromatography. *J Liq Chromatogr Relat Technol.* 32: 136–143.

Duan, S., X. Liu, H. Liang, and Y. Zhao. 2009. [Chemical and biological studies on essential oil from Ficus tsiangii]. *Zhongguo Zhong Yao Za Zhi* 34: 1398–400.

Govindarajan, M. 2010. Larvicidal efficacy of Ficus benghalensis L. plant leaf extracts against Culex quinquefasciatus Say, Aedes aegypti L. and Anopheles stephensi L. (Diptera: Culicidae). *Eur Rev Med Pharmacol Sci.* 14: 107–11.

Jahan, I.A., N. Nahar, M. Mosihuzzaman, et al. 2009. Hypoglycaemic and antioxidant activities of Ficus racemosa Linn. fruits. *Nat Prod Res.* 23: 399–408.

Kirana, H., S.S. Agrawal, and B.P. Srinivasan. 2009. Aqueous extract of Ficus religiosa linn. reduces oxidative stress in experimentally induced type 2 diabetic rats. *Indian J Exp Biol.* 47: 822–6.

Luna, L.E. 1984b. The concept of plants as teachers among four Mestizo shamans of Iquitos, northeastern Peru. *J Ethnopharmacol.* 11: 134–56.

Luna, L.E. 1984a. The healing practices of a Peruvian shaman. *J Ethnopharmacol.* 11: 123–33.

Maregesi, S., S. Van Miert, C. Pannecouque, et al. 2010. Screening of Tanzanian medicinal plants against Plasmodium falciparum and human immunodeficiency virus. *Planta Med.* 76: 195–201.

Oliveira, A.P., L.R. Silva, F. Ferreres, et al. 2010. Chemical assessment and in vitro antioxidant capacity of Ficus carica latex. *J Agric Food Chem.* 58: 3393–8.

Oliveira, A.P., P. Valentão, J.A. Pereira, B.M. Silva, F. Tavares, and P.B. Andrade. 2009. Ficus carica L.: Metabolic and biological screening. *Food Chem Toxicol.* 47: 2841–6.

Ong, C.Y., S.K. Ling, R.M. Ali, et al. 2009. Systematic analysis of in vitro photo-cytotoxic activity in extracts from terrestrial plants in Peninsula Malaysia for photodynamic therapy. *J Photochem Photobiol B.* 96: 216–22.

Owolabi, O.J., Z.A. Nworgu, A. Falodun, B.A. Avinde, and C.N. Nwako. 2009. Evaluation of tocolytic activity of ethanol extract of the stem bark of Ficus capensis Thunb. (Moraceae). *Acta Pol Pharm.* 66: 293–6.

Pandit, R., A. Phadke, and A. Jagtap. 2010. Antidiabetic effect of Ficus religiosa extract in streptozotocin-induced diabetic rats. *J Ethnopharmacol.* 128: 462–6.

Park, Y.R., J.S. Eun, H.J. Choi, et al. 2009. Hexane-soluble fraction of the common fig, Ficus carica, inhibits osteoclast differentiation in murine bone marrow-derived macrophages and RAW 264.7 cells. *Korean J Physiol Pharmacol.* 13: 417–24.

Parveen, M., R.M. Ghalib, S.H. Mehdi, S.Z. Rehman, and M. Ali. 2009. A new triterpenoid from the leaves of Ficus benjamina (var. comosa). *Nat Prod Res.* 23: 729–36.

Ramadan, M.A., A.S. Ahmad, A.M. Nafady, and A.I. Mansour. 2009. Chemical composition of the stem bark and leaves of Ficus pandurata Hance. *Nat Prod Res.* 23: 1218–30.

Singh, A.B., D.K. Yadav, R. Maurya, and A.K. Srivastava. 2009. Antihyperglycaemic activity of alpha-amyrin acetate in rats and db/db mice. *Nat Prod Res.* 23: 876–82.

Thakare, V.N., A.A. Suralkar, A.D. Deshpande, and S.R. Naik. 2010. Stem bark extraction of Ficus bengalensis Linn for anti-inflammatory and analgesic activity in animal models. *Indian J Exp Biol.* 48: 39–45.

Ueda, J.Y., M. Takagi, and K. Shin-ya. 2009. Aminocaprophenone- and pyrrolidine-type alkaloids from the leaves of Ficus septica. *J Nat Prod.* 72: 2181–3.

Verma, A.R., M. Vijayakumar, C.V. Rao, and C.S. Mathela. 2010. In vitro and in vivo antioxidant properties and DNA damage protective activity of green fruit of Ficus glomerata. *Food Chem Toxicol.* 48: 704–9.

Wang, X, Y. Liang, L. Zhu, et al. 2010. Preparative isolation and purification of flavone C-glycosides from the leaves of Ficus microcarpa L. f by medium-pressure liquid chromatography, high-speed countercurrent chromatography, and preparative liquid chromatography. *J Liq Chromatogr Relat Technol.* 33: 462–480.

Ya, J., X.Q. Zhang, Y. Wang, Q.W. Zhang, J.X. Chen, and W.C. Ye. 2010. Two new phenolic compounds from the roots of Ficus hirta. *Nat Prod Res.* 24: 621–5.

Yang, X.M., W. Yu, Z.P. Ou, H.L. Ma, W.M. Liu, and X.L. Ji. 2009. Antioxidant and immunity activity of water extract and crude polysaccharide from Ficus carica L. fruit. *Plant Foods Hum Nutr.* 64: 167–73.

Yarmolinsky, L., M. Zaccai, S. Ben-Shabat, D. Mills, and M. Huleihel. 2009. Antiviral activity of ethanol extracts of Ficus binjamina and Lilium candidum in vitro. *N Biotechnol.* 26: 307–13.

Yoon, W.J., H.J. Lee, G.J. Kang, H.K. Kang, and E.S. Yoo. 2007. Inhibitory effects of Ficus erecta leaves on osteoporotic factors in vitro. *Arch Pharm Res.* 30: 43–9.

Index

Printed and bound by CPI Group (UK) Ltd, Croydon, CR0 4YY

18/10/2024

01776267-0007